Lectures on Global Optimization

Fields Institute Communications

The Fields Institute for Research in Mathematical Sciences

Lectures on Global Optimization

Panos M. Pardalos
Thomas F. Coleman
Editors

American Mathematical Society
Providence, Rhode Island
The Fields Institute
for Research in Mathematical Sciences
Toronto, Ontario

FIELDS

The Fields Institute for Research in Mathematical Sciences

The Fields Institute is a center for mathematical research, located in Toronto, Canada. Our mission is to provide a supportive and stimulating environment for mathematics research, innovation and education. The Institute is supported by the Ontario Ministry of Training, Colleges and Universities, the Natural Sciences and Engineering Research Council of Canada, and seven Ontario universities (Carleton, McMaster, Ottawa, Toronto, Waterloo, Western Ontario, and York). In addition there are several affiliated universities and corporate sponsors in both Canada and the United States.

2000 *Mathematics Subject Classification.* Primary 90C26, 90C27, 90C29, 90C30.

Library of Congress Cataloging-in-Publication Data

Lectures on global optimization / Panos M. Pardalos, Thomas F. Coleman, editors.
p. cm. — (Fields Institute Communications, ISSN 1069-5265 ; v. 55)
Includes bibliographical references.
ISBN 978-0-8218-4485-4 (alk. paper)
1. Mathematical optimization—Industrial applications. I. Pardalos, P. M. (Panos M.), 1954– II. Coleman, Thomas F. (Thomas Frederick), 1950–

QA402.5.L393 2009
519.6—dc22 2009004065

∞ The paper used in this book is acid-free and falls within the guidelines
established to ensure permanence and durability.
This publication was prepared by the Fields Institute.
`http://www.fields.utoronto.ca`
Visit the AMS home page at `http://www.ams.org/`

10 9 8 7 6 5 4 3 2 1 14 13 12 11 10 09

Contents

Preface

A large number of mathematical models in many diverse areas of science and engineering have lead to the formulation of optimization problems where the best solution (globally optimal) is needed. Due to the interdisciplinary nature of global optimization, there has been astonishing progress in this field during the last few decades. Many powerful computational algorithms and new theoretical developments have been introduced to solve a spectrum of hard problems in several disciplines.

In this book, we cover some of the recent important topics in global optimization with emphasis on recent theoretical developments and scientific applications.

During May 11-12, 2007, a workshop on "Global Optimization: Methods and Applications" was held at the Fields Institute. The organizers were Thomas F. Coleman (University of Waterloo), Panos Pardalos (University of Florida), and Stephen Vavasis (University of Waterloo). The purpose of the workshop was to explore recent practical progress made in the area of global optimization, especially in reference to industrial applications. Our goal was to develop connections among researchers and field engineers in this area and raise awareness of current methodologies and approaches among practitioners. This book is based on the presentations at the workshop and reflects the diverse spectrum of theoretical results and applications.

The target audience of the book includes graduate students in mathematics, engineering, and sciences, academic researchers, as well as practitioners who use global optimization for their specific needs and applications.

We would like to take this opportunity to express gratitude to the Fields Institute, the School of Mathematics (University of Waterloo), and the Center for Applied Optimization (University of Florida), for their help in organizing this workshop. We would also like to thank the authors of the papers, the anonymous referees, and Debbie Iscoe of the Fields Institute for helping us with the final typesetting of the book.

Panos M. Pardalos (University of Florida)
Thomas F. Coleman (University of Waterloo)

Fields Institute Communications
Volume **55**, 2009

Extremal Problems for Convex Polygons—An Update

Charles Audet
GERAD and École Polytechnique de Montréal
GERAD and Département de Mathématiques et de Génie Industriel
École Polytechnique de Montréal
Montréal, Québec, H3C 3A7 Canada
Charles.Audet@gerad.ca

Pierre Hansen
GERAD and Département des Méthodes Quantitatives
École des Hautes Études Commerciales
Montréal, Québec, H3C 3A7 Canada
pierre.hansen@gerad.ca

Frédéric Messine
ENSEEIHT-IRIT, UMR-CNRS 5505
2 rue Camichel BP 7122, F-31071
Toulouse Cedex 7, France
Frederic.Messine@n7.fr

Abstract. Consider a convex polygon V_n with n sides, perimeter P_n, diameter D_n, area A_n, sum of distances between vertices S_n and width W_n. Minimizing or maximizing any of these quantities while fixing another defines ten pairs of extremal polygon problems (one of which usually has a trivial solution or no solution at all). In a previous paper, we surveyed research on these problems up to 2005. It appears that geometric reasoning is increasingly complemented by global optimization methods. As several new results have been obtained very recently, we present an update to that survey.

1 Introduction

Extremal problems on convex polygons have attracted attention of geometers and optimizers at least since the work of Zenodorus (circa 200-140 B.C.E.). In a previous paper [6], we surveyed research on that topic up to 2005. As several new results have been obtained very recently we give here an update to that survey.

2000 *Mathematics Subject Classification.* Primary 65K05, 52B60; Secondary 90C20.

The first author was supported in part by NSERC grant 239436-01 and AFOSR grant F49620-01-1-0013.

The second author was done in part in October 2007 during a visit to ENSEEIHT-IRIT, Toulouse, whose support is gratefully acknowledged. It was also supported by NSERC grant 105574-02.

A *polygon* V_n is a closed plane figure with n straight line segments as sides. A *vertex* of V_n is a point at which two sides meet. If the line segment joining two points of V_n is entirely within V_n then V_n is *convex*. If all sides of V_n have equal length, V_n is *equilateral*. If an equilateral polygon has equal inner angles between adjacent sides it is *regular*. The *perimeter* P_n of V_n is the sum of the lengths of its sides. The *diameter* of V_n is the maximum distance between two points of V_n, or, which is equivalent, the length of the longest straight line joining two vertices. The *width* of V_n in direction θ is the distance between two parallel lines perpendicular to θ and supporting V_n from below and above. The *width* of a polygon V_n is the minimum width for all directions θ. The *area* of a triangle is equal to half its width multiplied by the length of the side perpendicular to a minimum width direction. The *area* of a convex polygon with vertices $v_1, v_2, \ldots, v_n$ ranked clockwise is equal to the sum of the areas of the triangles v_1, v_i, v_{i+1}, for $i = 2, \ldots, n-1$.

Two convex polygons V_n and V'_n are *isoperimetric* if they have the same perimeter. They are *isodiametric* if they have the same diameter. Isoperimetric problems for convex polygons consist in finding the extremal polygons for some quantity such as area, diameter, sum of distances or width while keeping the perimeter fixed (say at 1). Isodiametric problems are defined similarly.

Consider a convex polygon V_n with n sides. Let A_n denote its area, P_n its perimeter, D_n its diameter, S_n the sum of distances between all pairs of its vertices and W_n its width. Maximizing and minimizing any of these quantities while fixing another one defines ten pairs of extremal problems. Usually, one problem from each pair has a trivial solution or no solution at all.

Simple formulations of extremal problems for convex polygons are easily obtained by denoting the consecutive vertices of the $n-$gon V_n by $v_i = (x_i, y_i)$. Then

. $A_n = \left|\frac{1}{2}\sum_{i=1}^{n}(y_{i+1} - y_i)(x_{i+1} + x_i)\right|$,
. $P_n = \sum_{i=1}^{n} \|v_{i+1} - v_i\|$,
. $D_n = \max_{i<j} \|v_i - v_j\|$,
. $S_n = \sum_{i<j} \|v_i - v_j\|$,
. $W_n = \min_i \max_{j \neq i, i+1} \frac{|(y_{j+1}-y_j)x_i + (x_j - x_{j+1})y_i + x_{j+1}y_j - x_j y_{j+1}|}{\|v_{j+1} - v_j\|}$,

where $i+1$ and $j+1$ are taken modulo n, $|.|$ denotes absolute value and $\|.\|$ the Euclidean norm. Expressions of the objective function and constraints easily follow. Note that several other formulations are also used in the literature.

Three main geometrical concepts, which we now define, are used in several of the results reported on below. A *Reuleaux polygon* [34] is not a usual polygon but has a convex polygonal basis on an odd number n of vertices, and is such that each pair of consecutive vertices has an opposite (not necessary distinct) vertex at equal distance of both of them. Then, the Reuleaux polygon is obtained by replacing each edge by a circular arc centered at the corresponding opposite vertex. Reuleaux polygons have the properties of being of constant width, and hence, due to Barbier's theorem, of perimeter $\pi \times D_n$. A *clipped regular Reuleaux polygon* on $n = m2^s$ vertices, with m an odd integer greater than 1 and not necessarily prime, and s is a positive integer is obtained as follows: (i) build a regular convex polygon with m' vertices, where m' is an odd factor of m; (ii) replace each edge of this polygon by a circular arc as explained above; (iii) introduce into each successive arc of the Reuleaux polygon so obtained $\frac{m2^s}{m'} - 1$ additional vertices, at equal distances each one from the next; (iii) replace each sub-arc bounded by two successive vertices by the corresponding chord (or in other words take the convex hull of the set of vertices). This construction is illustrated in Figure 1 for $n = 6 = 3 \times 2^1$.

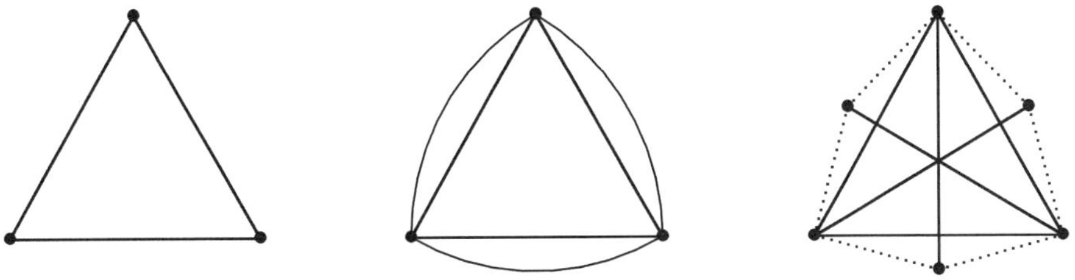

Equilateral triangle Reuleaux triangle Clipped Reuleaux hexagon

Figure 1 Construction of a clipped regular Reuleaux hexagon

The *diameter graph* $\mathcal{G} = (V, E)$ of a polygon V_n, has the same vertex set V as V_n and as edge set E all pairs of vertices at maximum mutual distance of V_n. Any diameter graph can be represented by a linear thrackleation [37]; i.e., in such a way that all edges have line segments of length one and any two edges intersect at a common extremity or at a point inside both of them.

In the survey [6], we present a Table specifying for each extremal convex polygon problem for which values of n it has been solved or remains open. An updated Table is given in the conclusion section of this paper. The other sections follow the same outline as in the survey [6]: in Section 2, devoted to isoperimetric problems, we first discuss the proof given by Larcher and Pillichshammer [18] of a conjecture on S_n given in [6]. We then consider recent results on the width W_n [5]: an upper bound, attained for all n which contain an odd factor is given, the maximal width quadrilateral is determined and solutions very close to the optimal ones are provided for $n = 8, 16$ and 32. Section 3 deals with isodiametric problems. Here, the main new result is a proof, due to Foster and Szabo [16], of a conjecture of Graham [17]: for even n the diameter graph of the polygon with maximum area and unit diameter is composed of a cycle with $n - 1$ vertices and a pending edge. This result eliminates the vast majority of cases considered in previous optimality proofs. Foster and Szabo [16] also ask two questions: (i) is the optimal diameter graph symmetric with respect to the pending edge, and (ii) is the maximal area polygon unique for each even $n \geq 6$. We use the Foster and Szabo result we determine upper bounds on the area of polygons with unit diameter, assuming symmetry of the optimal solutions, for $n = 10$ and 12. We thus confirm the near-optimality of the solutions proposed by Mossinghoff [25, 26] in these cases. A new result, also due to Mossinghoff [27], concerns the equilateral polygons with unit diameter and maximal perimeter when $n = 2^s$ with $s \geq 3$ and integer: a family of polygons is obtained for which it is shown that the perimeter is very close to an upper bound and that the error decreases very rapidly when s and n increase. Another problem for which new results have been obtained is the maximization of S_n [2]: the cases $n = 6$ and $n = 7$ are solved and correspond arbitrarily closely to configurations based on a triangle and on a Tamvakis [35] qualidrateral. A lower bound of the maximal value of S_n for all n follows. On the basis of further computational experiments, we conjecture that it is best possible for all n up to 20, but not always.

2 Isoperimetric problems

The oldest isoperimetric problem for convex polygons: *Find for all n which convex polygon V_n with unit perimeter has maximum area* was answered by Zenodorus (circa 200-140 b.c.). The optimal polygons are the regular ones.

A second isoperimeteric problem, first addressed in 1922 by Reinhardt [33] is: *Find for all n which polygon with unit perimeter has the smallest diameter.* As this problem is equivalent to finding a small polygon with longest perimeter, it will be discussed in Section 3.

2.1 Minimizing and maximizing the sum of distances between vertices. Based on computer experimentations, lower and upper bounds on S_n were conjectured by the present authors in [6]. This conjecture was recently proved in two theorems by Larcher and Pillichshammer [18].

Theorem 2.1 (Larcher, Pillichshammer) *For any $n \geq 3$ and any convex polygon with n vertices and with unit perimeter in the Euclidean plane we have*

$$S_n \leq \frac{1}{2}\left\lfloor\frac{n}{2}\right\rfloor\left\lceil\frac{n}{2}\right\rceil,$$

where for $n \geq 4$ the inequality is strict.

The proof relies on partitioning the set of edges joining pairs of vertices of the polygon into convex qualidraterals, triangles and isolated edges, then bounding the length of these geometric figures termwise and summing. Assume vertices $v_1, v_2, \ldots, v_n$ of the polygon V_n to be arranged clockwise. For n even, the partition is into all convex quadrangles on vertices v_i, v_j $v_{i+\frac{n}{2}}$ and $v_{j+\frac{n}{2}}$ for $1 \leq i \leq j \leq \frac{n}{2}$ as well as the edges of the perfect matching on v_i, $v_{i+\frac{n}{2}}$ for $1 \leq i \leq \frac{n}{2}$. For n odd, the partitioning is into all convex quadrangles on vertices v_i, v_j $v_{i+\frac{n+1}{2}}$ and $v_{j+\frac{n+1}{2}}$ for $1 \leq i \leq j \leq \frac{n-1}{2}$ and all convex triangles on vertices $v_i, v_{\frac{n+1}{2}}$ and $v_{i+\frac{n+1}{2}}$. Then, as these figures are enclosed in the polygon V_n their perimeter is not more than 1 and individual edges are not longer than $\frac{1}{2}$. Summing gives the upper bound which can be almost attained by a polygon arbitrarily close to a straight line of length $\frac{1}{2}$, with vertices $v_1, v_2, \ldots, v_{\lfloor\frac{n}{2}\rfloor}$ at one hand and $v_{\lfloor\frac{n}{2}\rfloor+1}, \ldots, v_n$ at the other hand.

Theorem 2.2 (Larcher, Pillichshammer) *For any $n \geq 3$ and any convex polygon with n vertices and with unit perimeter in the Euclidean plane we have*

$$S_n \geq \frac{n-1}{2},$$

where for $n \geq 4$ the inequality is strict.

The proof involves a more complicated decomposition of edges than the previous one, into subsets which may contain parts of edges and not complete ones [18].

Remark: The partitioning technique used in the proof of Theorem 2.1 was applied previously by Pillichshammer [30] to obtain upper bounds for S_n for small polygons V_n.

Remark: Several generalizations are possible. Larcher and Pillichshammer [19] consider the general case of functions $f : [0, \frac{1}{2}] \to \mathbb{R}_0^+$ such that $f(x)/x$ attains its maximum in $x = \frac{1}{2}$. Then the upper bound

$$S_n(f) \le f\left(\frac{1}{2}\right) \left\lfloor \frac{n}{2} \right\rfloor \left\lceil \frac{n}{2} \right\rceil.$$

holds and the optimal solutions are similar to those in Theorem 2.1

2.2 Maximizing the width. This problem is expressed as: *Find for all n which polygon with unit perimeter has the largest width.* We recently obtained the following result [5]:

Theorem 2.3 *Let V_n be a polygon with $n \ge 3$ vertices, unit perimeter and width w. Then*

$$w \le \frac{1}{2n} \cot \frac{\pi}{2n}.$$

The proof is obtained by majorizing w in the following mathematical program where the h_i are the width of V_n in the direction perpendicular to the edges. The α_i are the angles at vertices of the star polygon built on V_n. For each edge $v_i v_{i+1}$ let c_i denotes its length, $v_j(i)$ the opposite vertex, h_i the width of the polygon in the direction perpendicular to $v_i v_{i+1}$ and α_i the angle of $v_i v_j(i) v_{i+1}$, see Figure 2.

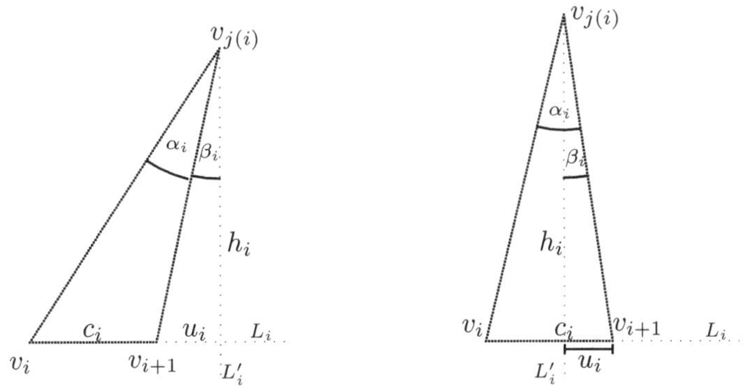

Figure 2 Upper bound for h_i

The width is then determined from the following mathematical program, where symbols are represented in Figure 2:

$$\begin{aligned} w = \max_{h,c,\alpha} \quad & h \\ \text{s.t.} \quad & h \le h_i \qquad i = 1, 2, \dots, n \\ & \textstyle\sum_{i=1}^n \alpha_i = \pi \\ & \textstyle\sum_{i=1}^n c_i = 1 \\ & c \ge 0, \ \alpha \ge 0. \end{aligned}$$

by a series of majorizations.

The bound is attained for regular polygons V_n^r with an odd number of vertices and by clipped regular Reuleaux polygons for $n = m2^s$ where m is an odd, not necessarily prime, integer and s is a positive integer.

Examples of clipped regular Reuleaux polygons with maximal width and $n = 6$, 10 and 12 sides are given in Figure 3. If m is a composite number, there are several equivalent solutions. An example with three optimal solutions for a pentadecagon ($n = 15$) is given in Figure 4.

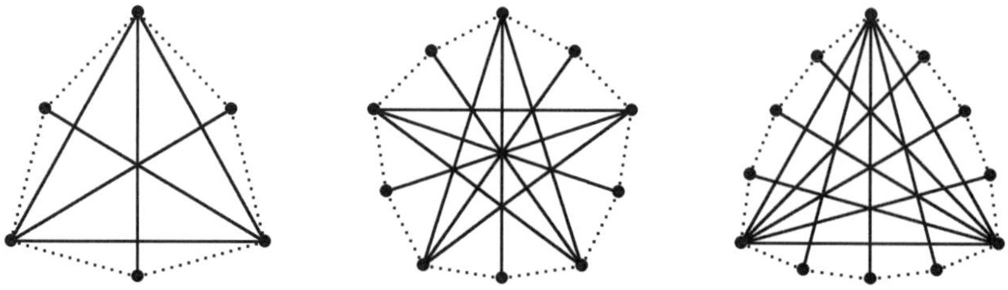

Optimal hexagon $w_6 = \frac{1}{12}\cot(\frac{\pi}{12}) \approx 0.3110\ldots$ Optimal decagon $w_{10} = \frac{1}{20}\cot(\frac{\pi}{20}) \approx 0.3157\ldots$ Optimal dodecagon $w_{12} = \frac{1}{24}\cot(\frac{\pi}{24}) \approx 0.3165\ldots$

Figure 3 Clipped Reuleaux polygons with maximal width, even n and $n \neq 2^s$

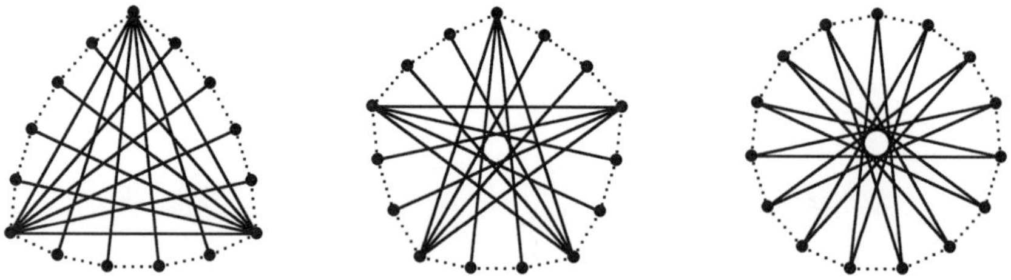

With an equilateral Reuleaux triangle — With a equilateral Reuleaux pentagon — Regular pentadecagon

Figure 4 Three pentadecagons of maximal width $w_{15} = \frac{1}{30}\cot(\frac{\pi}{30}) \approx 0.3171\ldots$

The optimal polygon for $n = 4$ is not the square since its width $\frac{1}{4}$ is inferior to the width of the equilateral triangle $\frac{\sqrt{3}}{6}$. The optimal quadrilateral has a width of $\frac{1}{4}\sqrt{(6\sqrt{3} - 9)}$. The proof consist in determining analytically the optimal symmetrical quadrilateral, thus obtaining a lower bound, and then, solving the following mathematical program in the general case, see Figure 5 for the definitions of the variables.

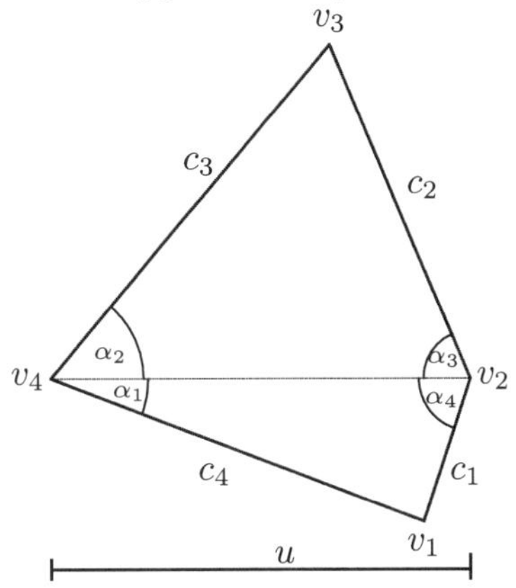

Figure 5 Variables defining any quadrilateral

$$
\begin{array}{ll}
\max\limits_{w,c,\alpha,\alpha_3,\alpha_4} & w \\
\text{s.t.} & w \le u\sin(\alpha_4) \ge c_2\sin(\alpha_3+\alpha_4) \\
& w \le u\sin(\alpha_3) \ge c_1\sin(\alpha_3+\alpha_4) \\
& w \le c_4\sin(\alpha_1+\alpha_2) \ge u\sin(\alpha_2) \\
& w \le c_3\sin(\alpha_1+\alpha_2) \ge u\sin(\alpha_1) \\
& c_1+c_2+c_3+c_4 = 1 \\
& c_4\sin(\alpha_1) = c_1\sin(\alpha_4) \\
& c_3\sin(\alpha_2) = c_2\sin(\alpha_3) \\
& u = c_4\cos(\alpha_1)+c_1\cos(\alpha_4) \\
& u = c_3\cos(\alpha_2)+c_2\cos(\alpha_3) \\
& 0 \le \alpha_1 \le \alpha_2 \le \pi \\
& 0 \le u \le \frac{1}{2} \\
& 0 \le c_i \le \frac{1}{2} \qquad i = 1,2,3,4.
\end{array}
$$

Using the global optimization code IBBA [20, 21, 22, 23] based on Branch and Bound in continuous variables and interval analysis [24, 32], we proved in about 36 hours of computing time on a PC with 1.86GHz cycle time and 1GB of memory, that there is no solution with a value larger than $\frac{1}{4}\sqrt{(6\sqrt{3}-9)}$ plus an error of 10^{-4}.

For $n = 8, 16$ and 32 , we obtained local optima of the following mathematical program using the procedure *fmincon* of MatLab^{TM}:

$$
\begin{array}{lll}
\max\limits_{z,\lambda,x,y} & z & \\
\text{s.t.} & z \le \sum\limits_{(i,k)\in L} \lambda_{i,k} \frac{((y_{i+1}-y_i)x_k+(x_i-x_{i+1})y_k+(x_{i+1}y_i-x_iy_{i+1}))^2}{(x_{i+1}-x_i)^2+(y_{i+1}-y_i)^2} & \forall i \in N \\
& \sum\limits_{(i,k)\in L} \lambda_{i,k} = 1 & \forall i \in N \\
& \sum\limits_{i\in N} \sqrt{(x_{i+1}-x_i)^2+(y_{i+1}-y_i)^2} = 1 & \\
& \lambda_{i,k} \ge 0 & \forall (i,k) \in L.
\end{array}
\qquad (2.1)
$$

This program has only n^2+1 variables and $2n+1$ constraints. These polygons obtained are denoted by U_n^*. Computing times on a 1.6GHz PC with 512MB were of about 1 seconds, 20 seconds and 30 minutes respectively. Best values found for the widths are given in Table 1; for the corresponding vertex coordinates, see [5]. As the best solutions found were symmetric with respect to a vertical axis only points with $x_i \geq 0$ are described. Observe the small values of the errors and their rapid decrease with increase of n. Note that starting points were the n-gons denoted by U_n in [26]. Other starting points, e.g. randomly generated ones, did not lead to improved solutions. Values of approximate solutions for $n = 8, 16$ and 32 are presented in Table 1, best values being in bold.

Table 1 Values of the widths of some approximate solutions

	$n = 8$	$n = 16$	$n = 32$
$w(V_n^r)$	0.3017767	0.3142087	0.3172866
$w(U_n^*)$	**0.3141144**	**0.3172849**	**0.3180542**
Upper Bound	0.3142087	0.3172866	0.3180542
Error	$< 9.43 \times 10^{-5}$	$< 1.66 \times 10^{-6}$	$< 1.43 \times 10^{-8}$

3 Isodiametric problems

3.1 Maximizing the area. This problem was studied by many authors since the seminal work of Reinhardt [33] in 1922. Optimal solutions are known for all odd n to be the regular polygons, as already noted by Reinhardt; this is also the case for the square and an infinity of other quadrilateral polygons. Graham [17] enumerated diameter configurations, which are caterpillars or odd cycles with (possibly) pendent edges [13, 37] and determined the optimal hexagon. The optimal octagon was found by the present authors in [9] using a similar approach and the QP program for non-convex quadratic programming [1, 3]. In these last two cases hexagons and octagons with areas about 3.92% and 2.795% above those of the corresponding regular polygons were obtained.

Mossinghoff provided [25, 26] heuristic solutions for polygons with n even, with a guaranteed bound on the error which decreases very rapidly when n increases.

Recent progress was not numerical but theoretical: an important conjecture of Graham [17], open for 25 years, was proved by Foster and Szabo [16].

Theorem 3.1 (Foster and Szabo) *If, for $m \geq 3$, a polygon with $2m$ vertices, has unit diameter and maximum area, then its diameter graph must be a cycle of length $2m-1$ with an additional edge attached to a vertex of the cycle.*

The proof relies upon several intermediate results:

(i) *if $n \geq 5$ the diameter graph of a polygon with n vertices, unit diameter and maximal area cannot be a path.* This step is proved by exhibiting a convex transformation of the polygon. Hence if the diameter graph is a path the area of the corresponding polygon can be augmented by this convex transformation until it contains a cycle.

(ii) *if $n \geq 5$ the diameter graph of a polygon with n vertices, unit diameter and maximal area cannot be a tree, and hence must contain a cycle.* Again, a convex transformation is exploited.

(iii) *If a polygon V_n has n vertices, diameter d and a k-cycle in its diameter graph, then its area A satisfies*

$$A \leq \frac{d^2}{2}\left(k\cos\frac{\pi}{k}\tan\frac{\pi}{2k} - k\sin\frac{\pi}{k} + n\sin\frac{\pi}{n}\right).$$

Equality holds if and only if k divides n, the polygon is equilateral and every vertex is the end-point of at least one diameter. This new bound on the area of a clipped Reuleaux polygon is obtained from a bound on the area of a Reuleaux polygon, due to Firey [15]:

$$A \leq \frac{d^2}{2}\left(k\cos\frac{\pi}{k}\tan\frac{\pi}{2k} + \pi - k\sin\frac{\pi}{k}\right).$$

Equality holds if and only if the Reuleaux polygon V_k is regular. Then, Foster and Szabo observe that, as the star polygon obtained by joining vertices to the opposite ones has a sum of angles equal to π, all triangles corresponding to pairs of successives vertices and the opposite one can be juxtaposed around a semi-circle, see Figure 6.

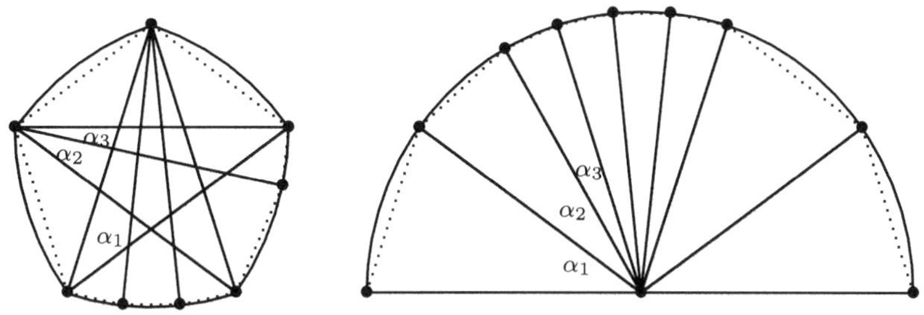

Figure 6 Placing the n segments around a semicircle

The sum of the clipped areas is then at least $\frac{d^2}{2}\left(\pi - n\sin\frac{\pi}{n}\right)$ and the result follows by subtraction in Firey's formula.

(iv) A key corollary is: *for $m \geq 3$, if a polygon with $2m$ sides of unit diameter, has a cycle in its diameter graph of length at most $2m-3$, then its area A satisfies:*

$$A \leq \frac{2m-3}{2}\cos\frac{\pi}{2m-3}\tan\frac{\pi}{2(2m-3)} - \frac{2m-3}{2}\sin\frac{\pi}{2m-3} + m\sin\frac{\pi}{2m}.$$

Then to prove the theorem it suffices to show that if the cycle in the diameter graph has length less than $2m-3$ (note this cycle cannot have the even length $2m-2$), one can construct a polygon V_n with an area exceeding the bound given above. A family of polygons V_n for all even n having this property is then constructed by a sophisticated specific procedure.

An immediate consequence of the Foster and Szabo theorem is that most cases for the diameter graphs examinated in the optimality proofs for n even do not need to be considered anymore; i.e., 9 out of 10 cases for $n = 6$ and 30 out of 31 for $n = 8$. This opens up new possibilities of numerical resolution. To illustrate, taking into account Foster and Szabo's theorem, and assuming symmetry of the optimal polygon obtained, we have determined the symmetric polygons with unit diameter and maximum area with $n = 10$ and $n = 12$ vertices. The non-convex quadratic program (where the variables are illustrated in Figure 7) for $n = 10$ is the following:

$$
\begin{array}{lll}
\max\limits_{x,y} & x_4 + x_1y_2 - y_1x_2 + x_2y_3 - 0.5y_2 + 0.5y_4 - y_3x_4 & \\
\text{s.t.} & x_4 + x_1y_2 - y_1x_2 + x_2y_3 - 0.5y_2 + 0.5y_4 - y_3x_4 \geq 0.749 & \\
& x_1^2 + y_1^2 \leq 1 & x_1^2 + y_1^2 - 2y_1 = 0 \\
& x_2^2 + y_2^2 \leq 1 & x_2^2 + y_2^2 - 2y_2 \leq 0 \\
& y_3^2 \leq 0.75 & y_3^2 - 2y_3 \leq -0.25 \\
& x_4^2 + y_4^2 \leq 1 & x_4^2 + y_4^2 - 2y_4 \leq 0 \\
& x_1^2 + y_1^2 + x_2^2 + y_2^2 - 2x_1x_2 - 2y_1y_2 \leq 1 & \\
& x_1^2 + y_1^2 + y_3^2 - x_1 - 2y_1y_3 \leq 0.75 & \\
& x_1^2 + y_1^2 + x_4^2 + y_4^2 - 2x_1x_4 - 2y_1y_4 \leq 1 & \\
& x_1^2 + y_1^2 + x_2^2 + y_2^2 + 2x_1x_2 - 2y_1y_2 \leq 1 & \\
& x_1^2 + y_1^2 + y_3^2 + x_1 - 2y_1y_3 \leq 0.75 & \\
& x_1^2 + y_1^2 + x_4^2 + y_4^2 + 2x_1x_4 - 2y_1y_4 = 1 & \\
& x_2^2 + y_2^2 + y_3^2 - x_2 - 2y_2y_3 \leq 0.75 & \\
& x_2^2 + y_2^2 + x_4^2 + y_4^2 - 2x_2x_4 - 2y_2y_4 \leq 1 & \\
& x_2^2 + y_2^2 + x_1^2 + y_1^2 + 2x_1x_2 - 2y_1y_2 \leq 1 & \\
& x_2^2 + y_2^2 + y_3^2 + x_2 - 2y_2y_3 = 0.75 & \\
& x_2^2 + y_2^2 + x_4^2 + y_4^2 + 2x_2x_4 - 2y_2y_4 = 1 & \\
& y_3^2 + x_4^2 + y_4^2 - x_4 - 2y_3y_4 \leq 0.75 & \\
& y_3^2 + x_1^2 + y_1^2 + x_1 - 2y_1y_3 \leq 0.75 & \\
& y_3^2 + x_2^2 + y_2^2 + x_2 - 2y_2y_3 = 0.75 & \\
& y_3^2 + x_4^2 + y_4^2 + x_4 - 2y_3y_4 \leq 0.75 & \\
& x_4^2 + y_4^2 + x_1^2 + y_1^2 + 2x_1x_4 - 2y_1y_4 = 1 & \\
& x_4^2 + y_4^2 + x_2^2 + y_2^2 + 2x_2x_4 - 2y_2y_4 = 1 & \\
& x_4^2 + y_4^2 + y_3^2 + x_4 - 2y_4y_3 \leq 0.75 & \\
& x_4^2 + y_4^2 + x_4^2 + y_4^2 + 2x_4^2 - 2y_4^2 \leq 1 & \\
& 0 \leq y_1 \leq y_2 \leq y_3 \leq y_4 \leq 1 & \\
& 0 \leq x_1 \leq x_2 \leq 0.5 & 0 \leq x_4 \leq 0.5.
\end{array}
$$

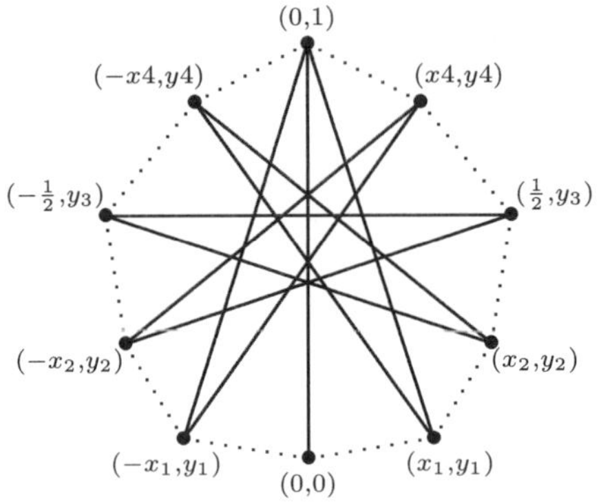

Figure 7 Symmetric dodecagon with cycle of length 9 and pending edge

These programs were solved with the NewQP algorithm of Sylvain Perron [28] which corresponds to an efficient implementation of the QP algorithm of Audet, Hansen, Jaumard and Savard [1, 3] when the density of quadratic terms is small.

The algorithm builds an outer-approximation of the feasible region, and thus provides an upper bound on the optimal value.

With computing times of 103 minutes and 27.5 hours (on a 3.06 GHz computer with 2 GB of RAM under Centos OS), the algorithm showed that $A_{10}^* \leq 0.749144$ and $A_{12}^* \leq 0.76102$. These values are between those of the upper bound $A_n \leq \frac{n}{2}\cos\left(\frac{\pi}{n}\right)\tan\left(\frac{\pi}{2n}\right)$ [33, 26] (0.7532 and 0.763), and the values of heuristic local solutions of Mossighoff [26] ($A_{10}^* \geq 0.7491373459$, $A_{12}^* \geq 0.7607298734$).

Foster and Szabo [16] also mention two open questions: (i) *Is there, for all $m \geq 3$, a unique polygon with $2m$ sides of unit diameter and maximal area?*; (ii) *If so, is it symmetric about the diameter that is not in the cycle of the diameter graph.* They note that this is true for $m = 3$ and 4 ([17, 9]).

3.2 Maximizing the perimeter. This problem, already mentioned in its equivalent form in the previous section, may be expressed as follows: *Find for all n which unit diameter polygon has the largest perimeter.*

Clipped regular Reuleaux polygons are optimal when n is odd, as already proved when $n = m2^s$ with m odd. The remaining cases are $n = 2^s$ for $s \geq 2$, and s integer. The case $n = 4$ was solved by Tamvakis [35] and the case $n = 8$ by the present authors [4].

Again for this problem, Mossinghoff [25, 26] provided heuristic solutions for polygons with n a power of two, with a guaranteed bound on the error which decreases very rapidly when s and n increase.

For *equilateral polygons* when n is a power of two, the octagon with unit diameter and maximum perimeter was determined by the present authors together with Sylvain Perron [8], improving a result of Vincze [36]. Mossinghoff [27] presents several attempts followed by successful constructions for near-optimal perimeter equilateral polygons with unit diameter and $n = 16$ and 32. These results culminate in a general construction of such polygons for all $n = 2^s$ with $s \geq 4$ and the proof of the following result:

Theorem 3.2 (Mossinghoff) *Suppose $n = 2^s$ with $s \geq 4$, let $\overline{P}_n$ denote the upper bound $2n\sin\frac{\pi}{2n}$ [33] on the perimeter of a convex n−gon with unit diameter, and let V_n^r denote the regular n−gon with unit diameter. Then there exists a convex, equilateral polygon, $V_n^{\prime r}$ whose perimeter $P(V_n^{\prime r})$ satisfies*

$$P(V_n^{\prime r}) - P(V_n^r) = \frac{\pi^3}{8n^2} + O\left(\frac{1}{n^4}\right)$$

and

$$\overline{P}_n - P(V_n^{\prime r}) = \frac{3\pi^4}{n^4} - 9\frac{\pi^5}{n^5} + O\left(\frac{1}{n^6}\right).$$

(Some symbols have been changed for unification of their use in this paper).

Moreover, Mossinghoff shows that his construction leads to an improved solution, in the case $n = 28$, to a problem of Bateman and Erdös [10]: *find the configuration of n points in the plane, mutually distanced by at least one, with minimum diameter.* To this effect, 12 points are added in the near-optimal polygon with 16 sides.

3.3 Maximizing the width. A third isodiametric problem has attracted some attention. It can be described as follows: *Find for all n which convex polygon with unit diameter has the largest width.*

Bezdek and Fodor [11] provided an upper bound for width for all n, and proved that when n has an odd prime divisor this bound is attained by clipped regular Reuleaux polygons. Moreover they showed that the Tamvakis quadrilateral [35] is the quadrilateral with unit diameter and maximal width. Complete results are still not available for larger powers of 2 than 4.

3.4 Maximizing the sum of distances. The fourth isodiametric problem that we consider was studied by a few authors. It can be expressed as follows: *Find for all n which convex polygon with unit diameter has the largest sum of distances between pairs of vertices.*

This problem appears to have been first posed by L. Fejes Tóth [14]. It can be stated mathematically as

$$\begin{array}{ll} \max\limits_{v_1,v_2,\ldots,v_n} & S_n = \sum_{i,j=1}^{n} \|v_i - v_j\| \\ \text{s.t.} & \|v_i - v_j\| \le 1, i,j = 1,2,\ldots,n. \end{array} \tag{3.1}$$

Bounds were obtained by Pillichshammer [30, 31] and are attained for the cases $n = 3,4$ and 5. The optimal polygons are the regular ones for $n = 3$ and 5 and the Tamvakis quadrilateral for $n = 4$. (Note that Pillichshammer also explored maximization of the sum of squared Euclidean distances between vertices [29].)

Very recently, the cases $n = 6$ and $n = 7$ were solved by the present authors together with Anthony Guillou and Sylvain Perron [2].

It is shown that, as the objective function is convex, the optimal solutions have diameters belonging to the families of diameter configurations for given n, described above. Then, the problem can be expressed as a non-convex quadratic program with non-convex quadratic contraints as follows:

$$\begin{array}{ll} \max\limits_{d,v} & \sum\limits_{i<j} d_{ij} \\ \text{s.t.} & d_{ij}^2 = \|v_i - v_j\|, \forall (i,j) = \{1,2,\ldots,n\}^2, \\ & 0 \le d_{ij} \le 1, \text{ if edge } (i,j) \notin \mathcal{G} \text{ and else } d_{ij} = 1 \end{array}$$

where $\mathcal{G}$ is the corresponding diameter graph.

The optimal polygons are very different than for the cases $n = 3,4,5$: in that the diameter graphs fold to an approximate simple polygon with a small number of vertices. For $n = 6$, it is arbitrary close to an equilateral triangle in which each vertex corresponds to a cluster of two vertices of the hexagon. For $n = 7$, it corresponds to a Tamvakis quadrilateral in which each vertex of the triangle in the diameter graph corresponds to a cluster of two vertices of the heptagon and the vertex of the diameter graph to one vertex. These are illustrated in Figure 8.

The NewQP algorithm [1, 3, 28] was applied for each diameter configurations, and in some cases several times as auxiliary problems had to be solved in order to obtain the global solution in a reasonable time. These results are shown to imply the following lower bound on the maximum sum of distances of pairs vertices of a

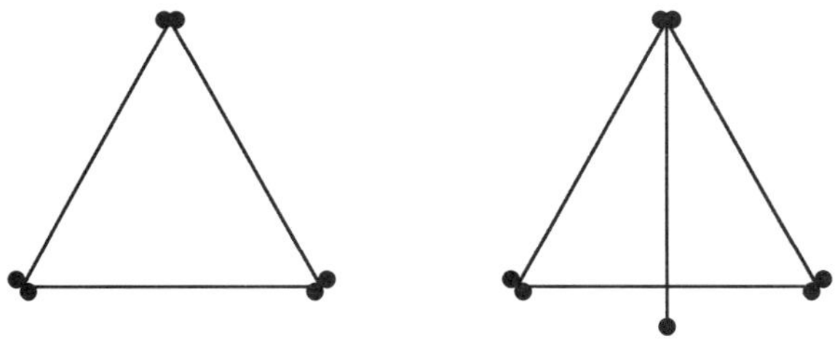

Figure 8 Optimal hexagon and heptagon for S_n

polygon with unit diameter:

$$\begin{cases} 3q^2 & \text{if } n \equiv 0[3] \text{ and } n \leq 30, \\ 3\left((q-1)q + 4(q-1)\sin\frac{\pi}{12} + \sqrt{3} - 1\right) & \text{if } n \equiv 0[3] \text{ and } n \geq 30, \\ 3q^2 + q + 4q\sin\frac{\pi}{12} & \text{if } n \equiv 1[3], \\ 3q^2 + 3q + 4q\sin\frac{\pi}{12} + 1 & \text{if } n \equiv 2[3], \end{cases}.$$

where $q := \lfloor \frac{n}{3} \rfloor$. In view of further numerical experiments with the *fmincon* of MatLab^{TM}, leading to local optima, these bounds appear best possible for $n = 3q \leq 51$, or $n = 3q + 1 \leq 16$ or $n = 8$ or $n = 11$, but not for some larger values of n.

4 Conclusion and open problems

Work on extremal problems for polygons is substantial and rapidly increasing. Many problems remain open or unexplored.

Several results have been obtained since the survey [6] was published. A main theoretical result is Foster and Szabo's [16] proof of a conjecture of Graham [17]. Another conjecture, about minimum and maximum sum of distances between vertices for polygons with unit perimeter, was proved also, [18]. Polygons with minimum width and unit perimeter were determined for all n which are not a power of two greater than 4 [6]. Approximate solutions were obtained in some of the remaining cases, i.e., $n = 8, 16$ and 32, [6]. Heuristic solutions with a guaranteed bound on the error were found for equilateral polygons with maximum perimeter, unit diameter and $n = 2^s$, with $s \geq 2$ and integer, [27]. Hexagons and heptagons with maximum sum of distances between vertices and unit diameter, were determined numerically,[2].

The current state of the art is summarized in the updated Table 2.

As in the survey [6], results in Table 2 above the main diagonal corresponds to general polygons and those below to the equilateral ones. New results are in bold. The mention *Approx.* means that approximate solutions with a guaranteed bound are provided. *Conj.* means that a conjecture on the optimal polygons and the corresponding objective values has been stated.

Table 2 Values of n for which extremal polygons have been determined

	P_n	D_n	A_n	S_n	W_n
$P_n = 1$	—	min n with odd factor and $n = 4, 8$ Approx. $n = 2^s$	max all n	max **and** min **for all n** [18]	max **n with odd factor and n = 4** **Approx. n = 8, 16, 32** [6]
$D_n = 1$	max n with odd factor and $n = 4, 8$ **Approx. n = 2^s**[27]	—	max odd n and $n = 4, 6, 8$* Approx. $n = 2 \times s$	max $n =$ $3, 4, 5,$ **6, 7** **Conj. n ≤ 20**	max n with odd factor and $n = 4$
$A_n = 1$	min all n	min odd n $n = 4$	—	min open	max open
$S_n = 1$	open	min $n = 3, 5$	open	—	open
$W_n = 1$	min **n with odd factor** [6]	min n with odd factor	open	open	—

* characterization of diameter graph obtained, [16].

References

[1] C. Audet, *Optimisation globale structurée: propriétés, équivalences et résolution*, Thèse de doctorat, École Polytechnique de Montréal, Canada, 1997, http://www.gerad.ca/~charlesa/.

[2] C. Audet, A. Guillou, P. Hansen, F. Messine, S. Perron *The Small Hexagon and Heptagon with Maximum Sum of Distances between Vertices*, Les Cahiers du GERAD G-2008-20, 2008.

[3] C. Audet, P. Hansen, B. Jaumard, G. Savard, *A branch and cut algorithm for nonconvex quadratically constrained quadratic programming*, Mathematical Programming, Vol. 87, No. 1, Series A, pp. 131–152, 2000.

[4] C. Audet, P. Hansen, F. Messine, *The small octagon with longest perimeter*, Journal of Combinatorial Theory, Series A, Vol. 114, pp. 135–150, 2007.

[5] C. Audet, P. Hansen, F. Messine, *Isoperimetric Polygons of Maximal Width*, to appear in Discrete & Computational Geometry, DOI 10.1007/s00454-008-9103-9, 2008.

[6] C. Audet, P. Hansen, F. Messine, *Extremal problems for convex polygons*, Journal of Global Optimization, Vol. 38, pp. 163–179, 2007.

[7] C. Audet, P. Hansen, F. Messine, *Quatre petits octogones*, MATAPLI, N. 1, pp. 63–75, 2006.

[8] C. Audet, P. Hansen, F. Messine, S. Perron, *The minimum diameter octagon with unit-length sides: Vincze's wife's octagon is suboptimal*, Journal of Combinatorial Theory, Series A, Vol. 108, pp. 63–75, 2004.

[9] C. Audet, P. Hansen, F. Messine, and J. Xiong, *The largest small octagon*, Journal of Combinatorial Theory, Series A, Vol. 98, No. 1, pp. 46–59, 2002.

[10] P. Bateman, P. Erdős, *Geometrical extrema suggested by a lemma of Besicovitch*, American Mathematical Monthly, Vol. 58, pp. 306–314, 1950.

[11] A. Bezdek and F. Fodor, *On convex polygons of maximal width*, Archiv der Mathematik, Vol. 74, No. 1, pp. 75–80, 2000.

[12] B. Datta, *A discrete isoperimetric problem*, Geometriae Dedicata, Vol. 64, pp. 55–68, 1997.

[13] P. Erdős, *On sets of distances on n points*, American Mathemtical Monthly, Vol. 53, pp. 248–250, 1946.

[14] L. Fejes Tóth, *Uber eine Punktverteilung auf der Kugel*, Acta Mathematica Hungarica, Vol. 10, pp. 13–19, 1959.

[15] W.J. Firey, *Isoperimetric ratios of Reuleaux polygons*, Pacific Journal of Mathematics, Vol. 10, pp. 823–830, 1960.

[16] J. Foster, T. Szabo, *Diameter graphs of polygons and the proof of a conjecture of Graham*, Journal of Combinatorial Theory, Series A, Vol. 114, No. 8, pp. 1515–1525, 2002.

[17] R. L. Graham, *The largest small hexagon*, Journal of Combinatorial Theory, Series A, Vol. 18, pp. 165–170, 1975.

[18] G. Larcher, F. Pillichshammer, *The sum of distances between vertices of a convex polygon with unit perimeter*, American Mathematical Monthly, Vol. 115, pp. 350–355, 2008.

[19] G. Larcher, F. Pillichshammer, *An extremum problem for convex polygons*, Anz. Österreich. Akad. Wiss. Math.-Natur. Kl., Vol. 215, pp. 173–176, 2006.

[20] F. Messine, *Méthodes d'optimisation globale basées sur l'analyse d'intervalles pour la résolution de problèmes avec contraintes*, Thèse de doctorat, INPT-ENSEEIHT, Toulouse, 1997, www.enseeiht.fr/~messine.

[21] F. Messine *Deterministic Global Optimization using Interval Contraint Propagation Techniques*, RAIRO Oper. Res., Vol. 38, No. 4, pp. 277–294, 2004.

[22] F. Messine, *A Deterministic Global Optimization Algorithm for Design Problems*, in C. Audet, P. Hansen, G. Savard (editors), Essays and Surveys in Global Optimization, Kluwer, pp. 267-294, 2005.

[23] F. Messine, J.L. Lagouanelle, *Enclosure Methods for Multivariate Differentiable Functions and Application to Global Optimization*, Journal of Universal Computer Science, Vol. 4, No. 6, Springer-Verlag, pp. 589-603, 1998.

[24] R.E. Moore, *Interval Analysis*, Prentice Hall, Englewood Cliffs, NJ, 1966.

[25] M.J. Mossinghoff, *A $1 problem*, American Mathematical Monthly, Vol. 113, pp. 385–402, 2006.

[26] M.J. Mossinghoff, *Isodiametric Problems for Polygons*, Discrete and Computational Geometry, Vol. 36, pp. 363–379, 2006.

[27] M.J. Mossinghoff, *An Isodiametric Problem for Equilateral Polygons*, Tapas in Experimental Mathematics, T. Amdeberhan and V. Moll, eds., Contemp. Math. 457, Amer. Math. Soc., pp. 237–252, 2008.

[28] S. Perron, *Applications jointes de l'optimisation combinatoire et globale*, Thèse de doctorat, École Polytechnique de Montréal, Canada, 2004.

[29] F. Pillichshammer, *On the sum of the squared distances in the Euclidean plane*, Archiv der Mathematik, Vol. 74, pp. 472–480, 2000.

[30] F. Pillichshammer, *A note on the sum of distances under a diameter constraint*, Archiv der Mathematik, Vol. 77, pp. 195–199, 2001.

[31] F. Pillichshammer, *On extremal point distributions in the Euclidean plane*, Acta Mathematica Hungarica, Vol. 98, pp. 311–321, 2003.

[32] H. Ratschek, J. Rokne, *New Computer Methods for Global Optimization*, Ellis Horwood, Chichester, 1988.

[33] K. Reinhardt, *Extremale polygone gegebenen durchmessers*, Jahresber. Deutsch. Math. Verein, Vol. 31, pp. 251–270, 1922.

[34] F. Reuleaux, *The Kinematics of Machinery*, translation of german original, New York: Dover, 1963.

[35] N.K. Tamvakis, *On the perimeter and the area of the convex polygon of a given diameter*, Bull. Greek Math. Soc., Vol. 28, pp. 115–132, 1987.

[36] S. Vincze, *On a geometrical extremum problem*, Acta Sci. Math. Szeged, Vol. 12, pp. 136–142, 1950.

[37] D.R. Woodall *Thrackles and Deadlock*, In *Combinatorial Mathematics and Its Applications* (D. J. A. Welsh, Ed.), Academic Press, New York, 1971.

Fields Institute Communications
Volume **55**, 2009

Finite Algorithms for Global Minimization of Separable Concave Programs

Xiaowei Bao
Department of Chemical and Biomolecular Engineering
University of Illinois at Urbana-Champaign
Urbana, Illinois 61801 USA
xbao2@uiuc.edu

Nikolaos V. Sahinidis
Department of Chemical Engineering
Carnegie Mellon University
Pittsburgh, Pennsylvania 15213 USA
sahinidis@cmu.edu

Abstract. When applied to continuous global optimization problems, branch-and-bound algorithms are infinite processes, in the sense that the lower and upper bounds they provide converge but not necessarily become equal in a finite number of iterations. In this chapter, we review elements of four finite algorithms for global minimization of separable concave functions over a polytope. Even though a naive finite algorithm exists for this class of global optimization problems, the algorithms addressed here are not exhaustive and illustrate how certain algorithmic constructs serve to overcome some of the challenges in continuous global optimization.

1 Introduction

Branch-and-bound is one of the most popular methods for global optimization. In this algorithm, the feasible region is partitioned into successively refined elements through branching, while partition elements are eliminated if it is determined that they are suboptimal or infeasible. This procedure is typically depicted as a search tree, where each node represents a partition element (or subproblem), which is created by subdividing its parent node. Branch-and-bound algorithms terminate when all subproblems are fathomed, thus ensuring that a globally optimal solution is obtained. A major question about branch-and-bound methods is whether they can terminate in a finite number of iterations. In binary integer programming (BIP), branching on any binary variable creates two subproblems. As a result,

---

2000 *Mathematics Subject Classification.* Primary 54C40, 14E20; Secondary 46E25, 20C20.

This work was supported in part by the Joint NSF/NIGMS Initiative to Support Research in the Area of Mathematical Biology under NIH award GM072023.

for a BIP problem with n binary variables, the number of nodes in a branch-and-bound tree is at most $2^{n+1} - 1$, which implies that the global optimum can be found in finitely many iterations. However, for continuous nonlinear programming, branching on a continuous variable may generate infinite partitions and branch-and-bound methods can be potentially infinite. Results pertaining to finite termination of branch-and-bound algorithms for continuous global optimization problems have been rather sparse in the literature. Finiteness for these problems was obtained based on carefully chosen operations, and explicit consideration of special problem structure.

In this chapter, we review some finite algorithms for global optimization. We restrict attention to finite rectangular branch-and-bound algorithms for separable concave programming over a polytope (SCP). SCP has unique properties that make it an interesting problem for branch-and-bound algorithms. Even though a naive extreme point enumeration algorithm for SCP is finite, branch-and-bound algorithms for this problem are not finite unless the branching and bounding techniques are carefully designed. In the next section, we introduce SCP and provide a brief overview of its applications and algorithms for solving it. In Section 3, a prototypical branch-and-bound algorithm is presented, followed by a discussion of convergence and finiteness properties in Section 4. Four concrete implementations of the prototypical branch-and-bound algorithm are addressed in Sections 5, 6, and 7. The first two of these sections, respectively, present finite algorithms due to Soland [32] and Shectman and Sahinidis [31]. Section 7 presents two algorithms that are due to Al-Khayyal and Sherali [2]. Conclusions are drawn in Section 8.

2 Separable concave programming over a polytope

SCP has some attractive mathematical properties that make it seemingly more tractable than general nonlinear global optimization and is one of the most studied problems in global optimization. The general concave minimization problem over a polytope can be formulated as follows:

$$\begin{aligned} \text{(SCP)} \quad \min \quad & f(x) \\ \text{s.t.} \quad & x \in X \end{aligned}$$

where $X = D \cap C \subset R^n$; $C = \prod_{j=1}^{n} C_j$ is a rectangular set, $C_j = [l_j, u_j] \subset R$ with $-\infty < l_j,\ u_j < \infty$; $D = \{x : Ax \leq b\}$, A is a real $m \times n$ matrix with rank $m < n$; $f(x) = \sum_{j=1}^{n} f_j(x_j)$, where, for each j, the function $f_j : C_j \to R$ is a lower semicontinuous and concave; and the bounded feasible domain X is assumed to be nonempty.

The following properties of SCP are of interest for the purposes of this chapter:

1. The convex envelope of the objective function $f(x) = \sum_{j=1}^{n} f_j(x_j)$ over any rectangular set $C^k = \prod_{j=1}^{n} C_j^k \subset C$ is the sum of the convex envelopes of the univariate functions $f_j(x_j)$, each taken over its respective interval $C_j^k = [l_j^k, u_j^k]$. Hence, the convex envelope of $f(x)$ over C^k is $g^k(x) = \sum_{j=1}^{n} g_j^k(x_j)$, where $g_j^k(x_j)$ is the convex envelope of concave function $f_j(x_j)$ over C_j^k. It is obvious that $g_j^k(x_j)$ is a linear function as follows:

$$g_j^k(x_j) = f_j(l_j^k) + \frac{f_j(u_j^k) - f_j(l_j^k)}{u_j^k - l_j^k}(x_j - l_j^k)$$

2. The feasible set X is convex since it is the intersection of two convex sets C and D.
3. There exists an extreme point of X that solves the problem.

Concave minimization problems arise naturally in various applications settings, including product and process planning, plan location (or site selection), cost network flow problems, production and location problems, inventory management, transportation planning, assignment problems, bid evaluation, economies of scale, and fixed charge problems. Direct applications of concave programming can be found in [16, 26, 3, 5, 12, 18, 21, 28, 40].

In addition to direct applications, some other types of optimization problems can be transformed into concave minimization problems. Examples include bilinear programming [13], 0-1 integer programming [11], complementary problems [23, 33], and max-min problems [8].

While fast solution of SCP is desired, SCP is a well-known NP-hard problem from the computational complexity point of view even for some special cases such as minimizing a quadratic concave function over the unit hypercube [27]. The major difficulty in solving concave minimization problems in practice is caused, in general, by the large number of local optima and the lack of simple criteria to distinguish between local and global optima. Even the simple problem of minimizing a concave function over a rectangle in R^n could have an exponential number of extreme points that are local minima [19].

The first landmark concave minimization paper was introduced by Tuy in 1964 [36]. Although it was shown later that the algorithm proposed in this paper does not converge [41], the Tuy cuts and the conical subdivision ideas that were introduced in Tuy's 1964 paper continue to influence today's development of algorithms in global optimization. Murty in 1968 [25] proposed the first extreme point ranking method that can find an exact global optimum for concave minimization problems in a finite number of enumerations, which can be used entirely on its own or combined with other methods to form hybrid approaches. Falk and Soland in 1969 [10] presented the first successive partition algorithm for separable concave minimization using rectangular partition elements. The rectangular subdivision rules proposed in this paper are used frequently in more general global optimization problems since then. Other influential algorithms for this problem were developed by Falk and Hoffman in 1976 [9], Tuy and Thai in 1983 [39], and Thieu in 1988 [34]. Surveys of algorithms for concave minimization can be found in [18, 4]. The most important solution strategies for SCP are based on enumeration, successive approximation, and successive partitioning.

Enumerative methods work by ranking the extreme points of the polytope and represent an obvious way to solve the concave minimization problem when the feasible region is polyhedral. Murty [25] developed such an extreme point ranking approach. Surveys of enumerative methods are given in [24, 7]. Due to difficulties caused by degeneracy and the sheer number of extreme points to be enumerated, application of purely enumerative methods in practice has been confined to problems that are either small or have some special structure. In the past decade, enumerative methods have been combined with cutting plane methods that eliminate feasible portions that could not contain better solutions than a given solution. The commonly used cuts are Tuy's concavity cuts [36], that are deep and simple to construct [18].

In successive approximation methods, a sequence of problems that are used to approximate the original one are solved. There are several ways to approximate: outer approximation, inner approximation, and successive underestimation [18]. Outer approximation generates a sequence of nested relaxation sets that contain the feasible region [14, 34]. When applied to large scale problems, these methods require a considerable amount of effort to solve the sequence of relaxed concave minimization problems they generate [17]. Inner approximation, also called polyhedral annexation, generates a sequence of expanding polyhedral approximations of the feasible region [37]. Inner approximation seems to outperform pure outer approximation methods [17]. An interesting polyhedral annexation scheme that incorporates dimension reduction is developed in [38]. Successive underestimation uses underestimators of the objective function over a bounded polyhedral feasible region [9, 29].

The successive partition methods and their branch-and-bound variants are currently the most popular and most successful methods for solving concave minimization problems. These methods can be classified in three groups according to the partitioning techniques used. The conical subdivision for concave minimization was first introduced by Tuy in 1964 [36] and followed by the first convergent algorithm by Thoai and Tuy in 1980 [35]. The simplicial subdivision was introduced by Hosrt in 1976 [15]. Both conical and simplicial subdivisions are natural for concave minimization problems because the convex envelope for concave functions taken over a simplex is readily computable and actually linear. As mentioned above, the rectangular subdivision for separable concave minimization was developed by Falk and Soland in 1969 [10]. When the concave objective function is separable, such as that in SCP, the convex envelope is easy to compute over a rectangular set. From a computational standpoint, the rectangular branch-and-bound algorithm of Shectman and Sahinidis [31] represents the current state-of-the-art for SCP and has led to the solution of large-scale problems.

3 General branch-and-bound algorithm

Branch-and-bound methods are widely used to solve various global optimization problems [18]. In this section, we introduce the general branch-and-bound algorithm, which is useful for understanding the special algorithms described in Sections 5, 6, and 7. These finite algorithms are based upon branch-and-bound techniques for solving concave minimization problems and can be viewed as concrete implementations of the prototype branch-and-bound algorithm.

The basic ideas of branch-and-bound methods applied to a global minimization problem are as follows:

- A possibly relaxed feasible set that contains the original problem's feasible region is subdivided (partitioned) into finitely many subsets.
- For each subset, lower and, if possible, upper bounds for the objective function are determined. The "overall" lower (upper) bound is defined to be the minimum of lower (upper) bounds over all partition elements.
- Subsets that cannot possibly contain a globally optimal solution for the problem at hand are eliminated (fathomed).
- If the overall lower and upper bounds satisfy a termination criterion, the algorithm terminates and gives a (near-)global optimum.

- In every iteration of the algorithm, some the subsets are selected, and partitioned to yield a refined partition.

Branch-and-bound methods provide a complete search over the feasible region, and therefore are guaranteed to terminate at a global optimum or "close" to a global optimum. However, the partitioning and bounding processes involved in branch-and-bounds methods will not stop until some termination criteria are satisfied. In general, branch-and-bound algorithms are infinite in the sense that the lower and upper bounds they provide converge but not necessarily become equal in a finite number of iterations.

In the remainder of this section, a formal description of a general branch-and-bound method for SCP is presented. Before presenting the prototypical procedure, we give a formal definition for partitioning:

Definition 3.1 Let M_0 be a subset of R^n and $I = \{1, \ldots, n\}$ be a finite set of indices. A set $\{M_i : i \in I\}$ of subsets of M_0 is said to be a *partition* of M_0 if

$$M_0 = \cup_{i \in I} M_i \text{ and } M_i \cap M_j = \partial M_i \cap \partial M_j \quad \forall i, j \in I,\ i \neq j$$

where ∂M_k denotes the boundary of set $M_k,\ k \in I$.

In practice, partitioning techniques generate simple polyhedral sets, such as simplices, rectangles, and cones. The algorithms considered in this chapter are all based on rectangular partitions. In a rectangular partition, the initial (relaxed) feasible set M_0 and all partition elements are n-dimensional rectangles. Partitioning of M_0 is usually carried out by using $(n-1)$-dimensional hyperplanes that pass through a point and parallel to the $(n-1)$-dimensional faces.

The following algorithmic description is primarily based upon [18, 31]. By convention, we take the infima and minima over an empty set as $+\infty$.

Initialization. Choose a relaxed set M_0 such that $M_0 \supseteq X$. Let $\mathcal{P}_0 = \{M_0\}$ be the initial partition of M_0. Set initial overall upper bound $\alpha_0 = \infty$ and lower bound $\beta_0 = -\infty$.

Step k. At the beginning of Step k, we have the current partition $\mathcal{P}_{k-1}$ of subsets of M_0 that are still of interest, the current overall upper and lower bounds α_{k-1}, β_{k-1} satisfying $\beta_{k-1} \leq \min f(X) \leq \alpha_{k-1}$. The algorithm then proceeds as follows:

1. Select an element $M_k \in \mathcal{P}_{k-1}$ from $\mathcal{P}_{k-1}$.
2. For each element M_k, bound the objective function $f(x)$ of the problem over M_k. Determine a lower bound $\beta(M_k)$ and an upper bound $\alpha(M_k)$ such that
$$\beta(M_k) \leq \inf_{x \in M_k \cap X} f(x) \leq \alpha(M_k).$$
If $\alpha(M_k) < \infty$, choose $\tilde{x} \in X$ such that $\alpha(M_k) = f(\tilde{x})$. Set the overall lower bound
$$\beta_k = \min_{M \in \mathcal{P}_{k-1}} \beta(M).$$
If $\alpha(M_k) < \alpha_{k-1}$, set the upper bound $\alpha_k = \alpha(M_k)$, and the current best solution $\bar{x} = \tilde{x}$; otherwise, $\alpha_k = \alpha_{k-1}$.
3. Fathom all $M \in \mathcal{P}_{k-1}$ with $\beta(M) \geq \alpha_k$ (or $\beta(M) \geq \alpha_k - \epsilon,\ \epsilon > 0$).
4. Terminate if $\mathcal{P}_{k-1} = \emptyset$.

5. Otherwise, select an element $M_i \in \mathcal{P}_{k-1}$, construct a rectangular partition $\bar{\mathcal{P}}_i$ of M_i through a point in M_i. Delete the new partition elements $M \in \bar{\mathcal{P}}_i$ for which $M \cap X = \emptyset$ or where $\min f(X)$ cannot occur. Add the remaining new partition elements $\bar{\mathcal{P}}_i'$ to $\mathcal{P}_{k-1}$ to form $\mathcal{P}_k = \mathcal{P}_{k-1} \cup \bar{\mathcal{P}}_i'$ and repeat Step k.

Based on the algorithm described above, $\{\alpha_k\}$ and $\{\beta_k\}$ are monotonically nonincreasing and nondecreasing sequences, and the limits $\alpha = \lim_{k\to\infty} \alpha_k$ and $\beta = \lim_{k\to\infty} \beta_k$ exist. Additionally, we have $\alpha \geq \min f(X) \geq \beta$. The algorithm is said to be *convergent* if $\alpha_k - \beta_k \to 0$.

When the algorithm utilizes the fathoming criterion $\beta(M) \geq \alpha_k - \epsilon$, $\epsilon > 0$, the solution thus obtained is an ϵ-global optimum. Many concave minimization algorithms in the literature terminate in finitely many iterations for $\epsilon > 0$. However, most algorithms in the literature may require infinite iterations to terminate if $\epsilon = 0$. In recent years, convergence to an exact global optimum in finitely many iterations using branch-and-bound has attracted considerable attention. Recent finite algorithms for $\epsilon = 0$ for SCP include the rectangular algorithm of Shectman and Sahinidis in 1998 [31], the simplicial algorithm of Locatelli and Thoai in 2000 [22], and the purification approach of Al-Khayyal and Sherali in 2000 [2]. Finiteness issues of branch-and-bound algorithms are becoming of interest in other contexts as well, including stochastic programming [1], linear multiplicative programming [20], and nonconvex quadratic programming [30, 6].

Various branch-and-bound algorithms differ from each other in the three basic operations: bounding (computing lower and upper bounds), selection (choosing subsets M_k), and branching (partitioning M_i to construct $\bar{\mathcal{P}}_i$). Specific choices of those operations are of crucial importance as far as convergence and computational efficiency of the overall algorithm are concerned.

4 Finiteness and convergence conditions

In this section, we review some related preliminary material on finiteness and convergence conditions for branch-and-bound methods.

First, we introduce the concept of consistent and finitely consistent bounding operation, and a related finiteness theorem for branch-and-bound procedures ([18, pp. 126-127]):

Definition 4.1 A bounding operation is called *finitely consistent* if, at every step, any unfathomed partition element can be further refined, and if any decreasing sequence $\{M_k\}$ of successively refined partition elements is finite.

Theorem 4.1 *In a branch-and-bound procedure, suppose that the bounding operation is finitely consistent. Then the procedure terminates after finitely many steps.*

On the other hand, if a branch-and-bound procedure is infinite, it must generate at least one infinitely nested decreasing sequence $\{M_k\}$ of successively refined partition elements:

Definition 4.2 A bounding operation is called *consistent* if at every step any unfathomed partition element can be further refined, and if any infinitely decreasing sequence $\{M_k\}$ of successively refined partition elements satisfies

$$\lim_{q\to\infty} (\alpha_{k_q} - \beta(M_{k_q})) = 0$$

In order to establish a consistent bounding operation, a convenient choice for the partition (or subdivision) is to make sure the generated sequence $\{M_k\}$ converges to a singleton, which results in the concept of exhaustive subdivision introduced by Thoai and Tuy in 1980 [35]:

Definition 4.3 A subdivision is called *exhaustive* if $\delta(M_q)_{q\to\infty} \to 0$ for all decreasing subsequences $\{M_q\}$ of partition elements generated by the subdivision, where $\delta(M)$ is the diameter of M measured by the Euclidean distance.

In rectangular subdivision, an example of exhaustive subdivision is bisection of the longest edge, whereby the n-dimensional rectangle M is divided by an $(n-1)$-dimensional hyperplane passing through the midpoint of one of the longest edges of M.

We also introduce a bound improving node selection rule because the consistent bounding operation alone is not sufficient for convergence of the branch-and-bound procedure. An additional requirement must be imposed on the selection operation, in order to guarantee that a partition element containing the actual lower bound is selected for further partitioning every finitely many steps [18, pp. 130-131]:

Definition 4.4 A selection operation is said to be *bound improving* if, at least each time after a finite number of steps, M_k satisfies the relation

$$\beta(M_k) = \beta_k$$

It can be shown that bound improvingness and consistency suffice to guarantee convergence of branch-and-bound [18]:

Theorem 4.2 *Consider the infinite branch-and-bound procedure and suppose that the bounding operation is consistent and the selection operation is bound improving. Then, the procedure is convergent:*

$$\lim_{k\to\infty} \alpha_k = \min f(X) = \lim_{k\to\infty} \beta_k.$$

In the following sections, we address four specific finite rectangular branch-and-bound algorithms for SCP and outline their proofs of finiteness. Whenever a solution of a linear programming problem is required over a polytope, we assume that an extreme point optimal solution is obtained.

5 Soland algorithm

Soland's algorithm [32] is a simplified version of the algorithm in [10]. The difference is that the linear underestimators for lower bounding are solved over the whole feasible region of the original SCP problem instead of the local partition elements. This specific bounding procedure is essential for this finite algorithm.

5.1 Algorithm. The algorithm determines the bounds on each partition element ($M_k = C^k \cap D$) $C^k = \prod_{j=1}^n C_j^k$ with $C_j^k = [l_j^k, u_j^k]$ using the linear convex envelope $g^k(x)$ defined in Section 2. The lower bound $\beta(M_k)$ is a relaxation of the typical lower bound $\min_{x\in D\cap C^k} g^k(x)$:

$$\beta(M_k) = \min_{x\in D\cap C} g^k(x) \le \min_{x\in D\cap C^k} g^k(x) \le \min_{x\in D\cap C^k} f(x).$$

The constraints for the LPs are the same at all stages. The upper bound is $\alpha(M_k) = f(x^{k*})$, where x^{k*} is the optimal solution of the LP and may not be in C^k.

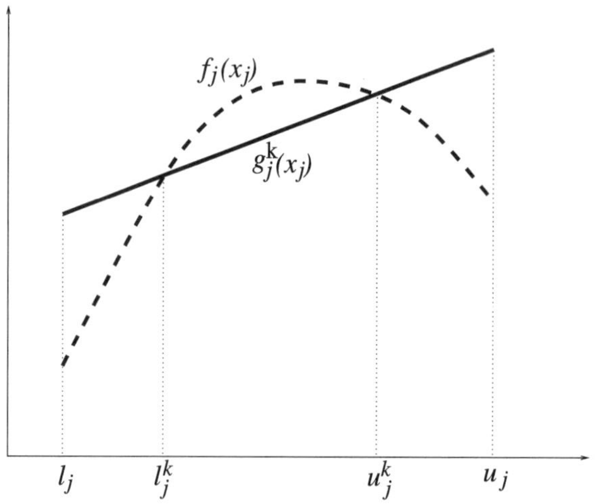

Figure 1 Soland's relaxed lower bounding

In a rectangular branching scheme, one must decide on the branching variable as well as the branching point. In this algorithm, the variable x_j that maximizes $f_j(x_j^{k*}) - g_j^k(x_j^{k*})$ at the optimal solution x^{k*} is selected as branching variable while x_j^{k*} is selected as the branching point.

The fathoming and termination criteria used in this algorithm are exact ($\epsilon = 0$). Any selection operation can be used.

5.2 Proof of finiteness of Soland's algorithm. It is clear that, if the algorithm terminates at any finite stage k with $\alpha_k = \beta_k$, the solution is an exact global optimal solution to SCP. The finiteness is shown by two steps.

First, we show that any partition element is either fathomed or can be further refined by the branching operation. As shown in Figure 1, for any concave function $f_j(x_j)$ and the corresponding convex envelop $g_j^k(x_j)$ over $[l_j^k, u_j^k]$, $f_j(x_j) - g_j^k(x_j) > 0$ only if $x_j \in]l_j^k, u_j^k[$. Furthermore, for the lower bound $\beta(M_k) = \sum_j g_j^k(x_j^{k*})$ and upper bound $\alpha(M_k) = \sum_j f_j(x_j^{k*})$, either $\beta(M_k) \geq \alpha(M_k)$ or $f_j(x_j^{k*}) - g_j^k(x_j^{k*}) > 0$ for at least one j. Therefore, at each stage, either the partition element is fathomed or the coordinate x_j^{k*} that maximizes $f_j(x_j^{k*}) - g_j^k(x_j^{k*})$ is in the interior of its corresponding edge.

The next step in the proof is to show that no infinite sequence of partition elements C^k can be generated. To illustrate the special structure of the partition elements C^k, a two-dimensional example is shown in Figure 2, where the rectangle represents C^0, and the line V_1V_5 represents the linear constraint in D. The five solid points are vertices of $D \cap C$. Starting with $C^0 = C$, the box is subdivided to two subsets by a line parallel to the axis and passing through a vertex. The chosen vertex must have at least one coordinate V_j in the interior of C_j. Hence, there are totally eight points (five solid points and three open circle points) that could be vertices of the partition elements. At stage 0, if x^{0*} is at V_2, V_3 or V_4, the algorithm terminates with $\alpha_0 = \beta_0$. Assuming $x^{0*} = V_1$, two subsets C^{11} and C^{12} are obtained. The partition element C^{11} is fathomed unless $x^{11*} = V_5$ so the only possible partitions for C^{11} is C^{21} and C^{22}, which cannot be subdivided by any line

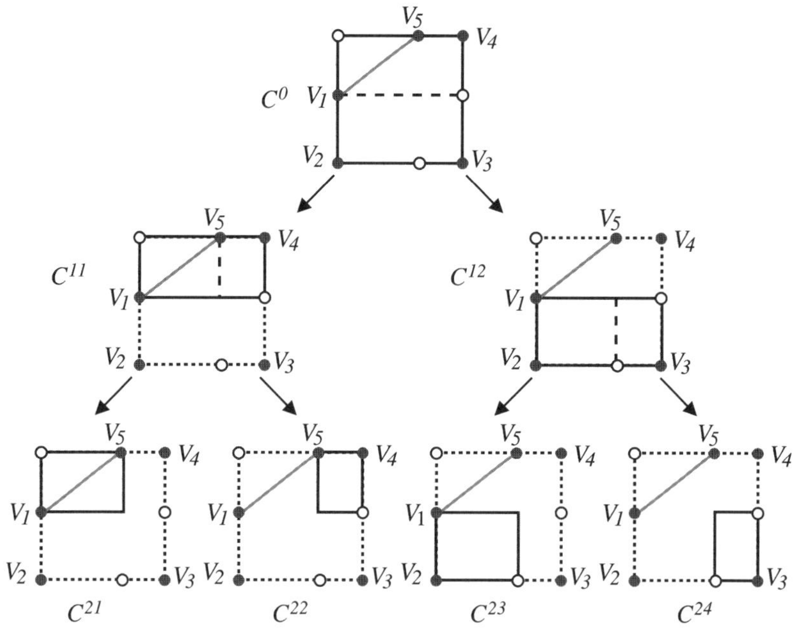

Figure 2 The partition tree for a two dimensional example

passing through a vertex. Partitioning of C^{12} is similar. Therefore, the algorithm always terminates in finitely many iterations. We next provide a formal proof of this step.

In general, let the vertex set of $D \cap C$ be $\{v^1, \ldots, v^m\}$ where m is a finite number. The set $E = \prod_{j=1}^n E_j = \prod_{j=1}^n \{l_j, u_j, v_j^1, \ldots, v_j^m\}$, is a finite set of points. In the algorithm, the initial partition is the rectangular set C, the vertices of which are $\prod_{j=1}^n \{l_j, u_j\}$. The feasible set is always subdivided at a point x_j^{k*} on edge x_j, where $x_j^{k*} \in E_j$ because x^{k*} is the solution of a linear program minimizing $g^k(x)$ over the polyhedral set $D \cap C$ and, therefore, x^{k*} must be a vertex of $D \cap C$. Hence, the vertices of all partition elements C^k are elements of E and the number of possible partition elements C^k that may be generated is finite. Therefore, the branch-and-bound algorithm terminates after finitely many steps. □

Although the lower bounds used are not as tight as those obtained by minimizing $g^k(x)$ over $D \cap C^k$ in each iteration k, this algorithm generates a sequence of LP subproblems that share the same constraints, which can be advantageous in a computational implementation.

6 Shectman and Sahinidis' algorithm

The crucial part, responsible for finiteness, in the algorithm proposed by Shectman and Sahinidis [31] is the branching scheme.

6.1 Algorithm. The lower bounding used in this algorithm is tighter than that in Soland's algorithm. On each partition element ($M_k = C^k \cap D$), with $C^k = \prod_{j=1}^n C_j^k$ and $C_j^k = [l_j^k, u_j^k]$, the lower bound $\beta(M_k)$ is the optimal value of the linear programming problem:

$$\beta(M_k) = \min_{x \in D \cap C^k} g^k(x) \leq \min_{x \in D \cap C^k} f(x).$$

Both the linear convex envelope and the constraints for the LP change for each partition element. The upper bound is also $\alpha(M_k) = f(x^{k*})$ where x^{k*} is the optimal solution of LP in $D \cap C^k$.

The selection operation used is bound improving (Definition 4.4): select an unfathomed node having the least lower bound $\beta(M_k) = \beta_k$.

For the branching operation, two branching methods are combined together in order to ensure finiteness. Let N be a predetermined finite number. At any Step k, if k is a multiple of N, an exhaustive branching operation is used: select a concave variable x_j corresponding to a longest edge of C^k as branching variable, and select the midpoint of this edge as the branching point. If k is not a multiple of N, the branching variable is chosen to be the variable x_j that maximizes $f_j(x_j^{k*}) - g_j^k(x_j^{k*})$. Branching is then performed at the midpoint of the corresponding edge unless the current best solution $\bar{x}_j$ is in the interior of the current box. In particular, if $\bar{x}_j \in]l_j^k, u_j^k[$, then branching is done at $\bar{x}_j$.

6.2 Proof of finiteness of the Shectman–Sahinidis algorithm. The proof of finiteness for this algorithm contains three main parts.

First, it is shown that the branching operation is exhaustive and the bounding operation is consistent. Since an exhaustive branching operation is used at least every finitely many, it is easy to show that the generated sequence $\{C^k\}$ converges to a singleton for nonlinear variables. Furthermore, if x_j is nonlinear in C^{k_q}, $u_j^{k_q} - l_j^{k_q} < \epsilon$ can be satisfied after a finite number of iterations for any small positive number $\epsilon > 0$. Consistency (Definition 4.2) is ensured by the bounding process and the exhaustiveness of the branching operation. From Theorem 4.2, it follows that the algorithm is convergent and a global optimum is contained in the sequence $\{C_{k_q}\}$ as $q \to \infty$.

The second part of the proof is to show that a global optimum of SCP can be identified in finitely many iterations. Let C^s be a box containing a global optimum x^*, d be any feasible direction in C^s, and x^A be the point that the ray $x^* + \lambda d$, $\lambda > 0$ intersects $D \cap C$. Exhaustiveness guarantees that a small box C^s can be generated in finitely many iterations such that, for all x_j nonlinear in C^s, $x_j^A > u_j^s$, or $x_j^A < l_j^s$, or $x_j^A = x_j^*$. Under this condition, Shectman and Sahinidis use concavity of $f(x)$ and the construction of $g(x)$ to show that, in such a small box C^s, the directions that increase the objective function $f(x)$ of SCP also increase the linear relaxation objective function $g(x)$. As a result, a solution of the LP relaxation is also a solution of the SCP. Hence, the global optimum can be identified by this LP relaxation in finitely many iterations.

Finally, branching on the incumbent whenever possible ensures that the algorithm goes no further than a finite number of iterations once a global optimum is identified. It is shown in [31] that, after at most a finite number of iteration, $\bar{x}$ will be a vertex of a partition element and the relaxation gap between the LP and SCP will disappear and the algorithm terminates. □

Based on the above proof, Shectman and Sahinidis [31] extend a guarantee of finiteness to all branch-and-bound algorithms for SCP that satisfy the following conditions:

1. For all nested sequences C^{k_q}, any coordinate x_j that is nonlinear converges to a singleton. This can be ensured by bound improving selection and exhaustive branching operation.
2. If a subproblem contains a global optimum, the relaxation gap can be reduced to zero in finitely many iterations. This can be guaranteed by proper underestimators that help identify the global solution point, and branching operations through that point in order to close the gap.

According to these conditions, several existing convergent branch-and-bound algorithms can be turned to finite algorithms via small modifications.

7 Al-Khayyal and Sherali's algorithms

Many branch-and-bound algorithms are convergent and can terminate with an ϵ-global optimum after a finite number of iterations for $\epsilon > 0$. Al-Khayyal and Sherali [2] propose two methods to augment these algorithms to guarantee finite termination with an exact global optimum for pseudoconcave polynomial and jointly bilinear programs. In this chapter, we consider the application of these algorithms in the context of SCP.

7.1 First Al-Khayyal–Sherali algorithm. The key ideas of the first method are the purification scheme and ϵ-fathoming.

The bounding operation used in this algorithm is any consistent bounding operation with an additional purification step for the upper bound so that the best solutions are always vertices of X:

Purification step

Given a point $x^k \in X$ found at step k as a candidate for the upper bound, such as the x^{k*} in the Shectman–Sahinidis algorithm, do the following:

1. If there are n linearly independent constraints active at x^k, x^k is a vertex of X, let $\tilde{x} = x^k$ and stop.
2. Otherwise, find a direction d in the null space of the active constraints at x^k.
3. If $\nabla f(x^k)^T d > 0$, $d = -d$, else $d = d$. Move along d until blocked by another constraint of X. Repeat this step.

Since $f(x)$ is concave, $f(x^k + \lambda d) \le f(x^k) + \lambda \nabla f(x^k)^T d \le f(x^k)$, $\forall \lambda > 0$, this step finds an upper bound which is a vertex of X and at least as good as x^k. The purification step would repeat at most n times.

The fathoming criterion is $\beta(M) \ge \alpha_k - \epsilon$, where ϵ is a predetermined positive real number depending on the problem. Although this criterion involves the use of a positive ϵ as opposed to an $\epsilon = 0$ that was used in the Soland and Shectman–Sahinidis algorithms, Al-Khayyal and Sherali proved that their algorithm can still obtain an exact global optimum. The proof is outlined below.

Any bound improving selection operation and exhaustive subdivision can be used as long as the algorithm terminates finitely with an ϵ-global optimum for $\epsilon > 0$.

7.2 Proof of finiteness of the first Al-Khayyal–Sherali algorithm. By exhaustiveness and consistency, the algorithm terminates after finite iterations when $\epsilon > 0$ is used. Then, it suffices to show that the solution obtained by this algorithm is an exact global optimum with a proper selection of ϵ.

It is equivalent to show that nodes with $\beta(M) \geq f(\bar{x}) - \epsilon$ do not admit a feasible solution $\hat{x} \neq \bar{x}$ such that $\hat{x}$ is a vertex of X and $f(\hat{x}) < f(\bar{x})$, where $\bar{x}$ is the current best solution and a vertex of X by the purification step. This is proved by showing that there exists $\epsilon > 0$, such that for any two vertices x, y in X with $f(x) \neq f(y)$, we have $|f(x) - f(y)| > \epsilon$. The existence and value of ϵ is based on the discreteness of the set of vertices of X and the polynomiality of the objective function $f(x)$. This algorithm works only for concave minimization problems with polynomial objective functions. The value of ϵ depends on the degree of the polynomial objective function as given in [2]. □

7.3 Second Al-Khayyal–Sherali algorithm. The second method is based on the fact that any variable x_j is either at the boundary of C_j or basic for D in the interval C_j. This approach utilizes a convergent algorithm with a subdivision operation that bisects the longest edge every finitely many iterations. For convenience, it is assumed that all coefficients in $D \cap C$ are integral.

In the bounding step, an augmentation routine is executed before computing the lower bound, where L is a predetermined positive number.

Augmentation routine

Given is the current partition element $M_k = C^k \cap D$ with $C_j^k = [l_j^k, u_j^k]$. Let $S_l = \{j : l_j^k = l_j\}$, $S_u = \{j : u_j^k = u_j\}$, and $S_0 = \{j : l_j^k = u_j^k\}$. Then, do the following:

1. If $|S_l \cup S_u| < n - m$, fathom the node.
2. If $|S_l \cup S_u| = n - m$, and $S_l \cap S_u = \emptyset$, obtain a solution at the boundary of D by letting x_j be nonbasic at its lower (upper) bounds for $j \in S_l$ $(j \in S_j)$. If the solution is feasible, let it be $\tilde{x}$. Otherwise, fathom the node.
3. For any $j \in S_l \cup S_u$, if $u_j^k - l_j^k \leq 2^{-L}$, fix $x_j^k = l_j^k, u_j^k = l_j^k$ for $j \in S_l$ and $x_j^k = u_j^k, l_j^k = u_j^k$ for $j \in S_u$. If $|S_0| \geq n - m$, fix x_j for $j \in S_0$. If this yields a feasible solution, update $\tilde{x}$, if necessary, and fathom the node. If $S_0 = S_l \cup S_u$, fathom the node.

The existence of a positive value of L is guaranteed by the discreteness of the polyhedron D. For example, for $D = \{x : Ax = b\}$, $L \geq 1 + \log_2 \Delta$, where Δ is the largest absolute determinant of all bases of A.

7.4 Proof of finiteness of the second Al-Khayyal–Sherali algorithm. The algorithm will terminate after finitely many iterations because if a node is not fathomed by Steps 1 or 2 in the augmentation routine, it must satisfy $u_j^k - l_j^k \leq 2^{-L}$ and $|S_l \cup S_u| \geq n - m$ after finitely many iterations because of the bisection of the longest edge every finitely many iterations. Then, Step 3 in the augmentation routine will fathom this node.

To show that the algorithm yields a global optimum, one only needs to show that the operations used to fathom the node in the augmentation routine do not eliminate possible optimal solutions. The proof is based on the property that the optimal solutions of SCP can only be at vertices of $D \cap C$. The fathoming criteria in Steps 1 and 2 are obviously valid because at least $n - m$ boundary constraints from C are needed to form a vertex of $D \cap C$. For Step 3, it is shown in Al-Khayyal and Sherali [2] that the variable x_j for any optimal solution can only be at the boundary if $u_j^k - l_j^k \leq 2^{-L}$ while L is determined by the polyhedron D. □

Both algorithms utilize the properties that SCP admits a vertex optimal solution and that the set of vertices of $D \cap C$ is discrete. The values of ϵ or L in these algorithms could be extremely small, thus preventing their practical application to high-dimensional problems.

8 Conclusions

Four finite rectangular branch-and-bound algorithms for global minimization of SCP problems are discussed in this chapter. This review covers essential algorithmic constructs that lead to finiteness and the theoretical aspects of these tools. The extension of these techniques to obtain finite termination of global optimization algorithms for more general continuous nonlinear programs remains a major open question.

References

[1] Ahmed, S., Tawarmalani, M. and Sahinidis, N. V. *A finite branch-and-bound algorithm for two-stage stochastic integer programs*, Math. Program. **100** (2004), 355–377.

[2] Al-Khayyal, F. A. and Sherali, H. D. *On finitely terminating branch-and-bound algorithms for some global optimization problems*, SIAMJ. Optim. **10** (2000), 1049–1057.

[3] Benson, H. P. *Concave minimization: Theory, applications and algorithms, In R. Horst and P. M. Pardalos (eds.)*, Handbook of Global Optimization, Kluwer Academic Publishers, Boston, MA (1995), 43–148.

[4] ———, *Deterministic algorithms for constrained concave minimization: A unified critical survey*, Naval Res. Logist. **43** (1996), no. 6, 765–795.

[5] Bretthauer, K. M. *Capacity planning in networks of queues with manufacturing applications*, Math. Comput. Modelling **21** (1995), no. 12, 35–46.

[6] Burer, S. and Vandenbussche, D. *A finite branch-and-bound algorithm for nonconvex quadratic programming via semidefinite relaxations*, Math. Program. (2006), DOI 10.1007/s10107-006-0080-6.

[7] Dyer, M. E. *The complexity of vertex enumeration methods*, Math. Oper. Res. **8** (1983), 381–402.

[8] Falk, J. E. *A linear max-min problem*, Math. Program. **5** (1973), no. 169-188.

[9] Falk, J. E. and Hoffman, K. R. *A successive underestimation method for concave minimization problems*, Math. Oper. Res. **1** (1976), no. 3, 251–259.

[10] Falk, J. E. and Soland, R. M. *An algorithm for separable nonconvex programming problems*, Management Sci. **15** (1969), 550–569.

[11] Giannessi, F. and Niccolucci, F *Connection between nonlinear and integer programming problems*, Sympos. Math. **19** (1976), 161–176.

[12] Guisewite, G. M. *Network Problems, In R. Horst and P. M. Pardalos (eds.)*, Handbook of Global Optimization, Kluwer Academic Publishers, Boston, MA (1995), 609–678.

[13] Hansen, P. , Juamard, B. and Lu, S-H. *Global optimization of univariate Lipschitz functions: I. Survey and properties*, Math. Program. **55** (1992), 251–272.

[14] Hoffman, K. *A method for globally minimizing concave functions over convex sets*, Math. Program. **20** (1981), 22–32.

[15] Horst, R. *An algorithm for nonconvex programming problems*, Math. Program. **10** (1976), 312–321.

[16] ———, *On the global minimization of concave functions–introduction and survey*, OR Spectrum **6** (1984), 195–205.

[17] Horst, R. and Thoai, N. V. *Modification, implemenation and comparison of three algorithms for globally solving linearly constrained concave minimization problems*, Computing **42** (1989), 271–289.

[18] Horst, R. and Tuy, H. *Global Optimization: Deterministic Approaches*, Third ed., Springer Verlag, Berlin, 1996.

[19] Kalantari, B. *Quadratic functions with exponential number of local maxima*, Oper. Res. Lett. **5** (1986), 47–49.

[20] Kuno, T. *A finite branch-and-bound algorithm for linear multiplicative programming*, Comput. Optim. Appl. **20** (2001), 119–135.

[21] Liu, M. L., Sahinidis, N. V. and Shectman, J. P. *Planning of chemical process networks via global concave minimization, In I. E. Grossmann (ed.)*, Global Optimization in Engineering Design, Kluwer Academic Publishers, Boston, MA (1996), 195–230.

[22] Locatelli, M. and Thoai, N. V. *Finite exact branch-and-bound algorithms for concave minimization over polytopes*, J. Global Optim. **18** (2000), no. 2, 107–128.

[23] Mangasarian, O. L. *Characterization of linear complementarity problems as linear programs*, Math. Program. Study **7** (1978), 74–87.

[24] Matheiss, T. H. and Rubin, D. S. *A survey and comparison of methods for finding all vertices of convex polyhedral sets*, Math. Oper. Res. **5** (1980), 167–185.

[25] Murty, K. G. *Solving the fixed charge problem by ranking the extreme points*, Oper. Res. **16** (1968), no. 2, 268–279.

[26] Pardalos, P. M. and Rosen, J. B. *Methods for global concave minimization: A bibliographic survey*, SIAM Review **28** (1986), 367–379.

[27] Pardalos, P. M. and Schnitger, G. *Checking local optimality in constrained quadratic programming is np-hard*, Oper. Res. Lett. **7** (1988), no. 1, 33–35.

[28] Park, S. S., Kim, T. and Hong, Y. *Production allocation and shipment policies in a multiple-manufacturer-single-retailer supply chain*, International Journal of Systems Science **37** (2006), no. 3, 163–171.

[29] Rosen, J. B. *Global minimization of a linearly constrained concave function by partition of feasible domain*, Math. Oper. Res. **8** (1983), no. 2, 215–230.

[30] Shectman, J. P. *Finite Algorithms for Global Optimization of Concave Programs and General Quadratic Programs*, Ph.D. thesis, Department of Mechanical & Industrial Engineering, University of Illinois, Urbana, IL, October 1999.

[31] Shectman, J. P. and Sahinidis, N. V. *A finite algorithm for global minimization of separable concave programs*, J. Global Optim. **12** (1998), 1–36.

[32] Soland, R. M. *Optimal facility location with concave costs*, Oper. Res. **22** (1974), 373–382.

[33] Thieu, T. V. *Relationship between bilinear programming and concave porgamming*, Acta Math. Vietnam. **2** (1980), 106–113.

[34] ———, *A finite method for global minimizing a concave function over an unbounded polyhedral convex set and its applications*, Acta Math. Hungar. **52** (1988), no. 1-2, 21–36.

[35] Thoai, N. V. and Tuy, H. *Convergent algorithms for minimizing a concave function*, Math. Oper. Res. **5** (1980), 556–566.

[36] Tuy, H. *Concave programming under linear constraints*, Dokl. Akad Nauk **159** (1964), 32–35.

[37] ———, *On polyhedral annexation method for concave minimization, In L. J. Leifman (ed.)*, Functional analysis, Optimization and Mathematical Economics, Oxford University Press (1990), 248–260.

[38] ———, *Polyhedral annexaton, dualization and dimension reduction technique in global optimization*, J. Global Optim. **1** (1991), no. 3, 229–244.

[39] Tuy, H. and Thai, N. Q. *Minimizing a concave function over a compact convex set*, Acta Math. Vietnam. **8** (1983), 13–20.

[40] Yi, G. , Kim, Y. H. and Lee, E. S. *Optimal design of multisite batch-storage network under scenario-based demand uncertainty*, Chem. Eng. Comm. **194** (2007), no. 10-12, 1297–1327.

[41] Zwart, P. B. *Nonlinear programming: Counterexamples to two global optimization algorithms*, Oper. Res. **21** (1973), 1260–1266.

Fields Institute Communications
Volume **55**, 2009

Solving Min-Max Multi-Depot Vehicle Routing Problem

John Carlsson
Institute of Computational and Mathematical Engineering
Stanford University, Stanford, CA 94305 USA
johnnyc@stanford.edu

Dongdong Ge
Department of Management Science and Engineering
Stanford University, Stanford, CA 94305 USA
dongdong@stanford.edu

Arjun Subramaniam
Department of Management Science and Engineering
Stanford University, Stanford, CA 94305 USA
arjun.subramaniam@gmail.com

Yinyu Ye
Department of Management Science and Engineering
and, by courtesy, Electrical Engineering
Stanford University, Stanford, CA 94305, USA
yinyu-ye@stanford.edu

Abstract. The Multi-Depot Vehicle Routing Problem (MDVRP) is a generalization of the Single-Depot Vehicle Routing Problem (SDVRP) in which vehicle(s) start from multiple depots and return to their depots of origin at the end of their assigned tours. The traditional objective in MDVRP is to minimize the sum of all tour lengths, and existing literature handles this problem with a variety of assumptions and constraints. In this paper, we explore the notion of minimizing the maximal length of a tour in MDVRP ("the min-max MDVRP"). We present two heuristics in this paper. The first heuristic is a linear programming-based approach with global improvement. The second one, the region partition heuristic, is proved to be asymptotically optimal and is potentially useful for general network applications. A comparison of the computational implementations for different heuristics is presented.

1 Introduction

The Vehicle Routing Problem (VRP) has been one of the central topics in optimization since Dantzig and Ramser proposed the problem in 1959 [8]. A simple

2000 *Mathematics Subject Classification.* Primary 54C40, 14E20; Secondary 46E25, 20C20.
This research is supported in part by the Boeing Company.

general model of VRP can be described as follows: a set of service vehicles need to visit all customers in a geographical region with the minimum cost.

In the Single-Depot Vehicle Routing Problem (SDVRP), multiple vehicles leave from a single location (a "depot") and must return to that location after completing their assigned tours. The Multi-Depot Vehicle Routing Problem (MDVRP) is a generalization of SDVRP in which multiple vehicles start from multiple depots and return to their original depots at the end of their assigned tours. The traditional objective in MDVRP is to minimize the sum of all tour lengths, and existing literature handles this problem with a variety of assumptions and constraints.

In this paper, we explore the notion of minimizing the maximal length of a tour in a MDVRP ("the min-max MDVRP") by both theoretical analysis and heuristic design. To the best of our knowledge, no prior exploration of the min-max MDVRP has been published, but this formulation is advantageous for a number of applications. Solving the traditional MDVRP gives a collection of routes that minimizes the total distance travelled by vehicles, but we have no bound on the time to completion (the "makespan") of all routes.

Although many heuristics have been designed to solve the SDVRP, few have focused on solving the min-max MDVRP. We are unaware of any existing specific heuristic or theoretical result for the min-max MDVRP, especially in the large-sized case. In general literature, researchers have made extensive efforts to solve various versions of VRPs. A typical idea is a two-stage heuristic(for example: Wren and Holliday [19]): construct an initial solution, then apply a number of local improvements. The initial solutions in common heuristics are usually constructed by naive approaches, for example, the nearest neighbor assignment. Gillet and Johnson [9] proposed a clustering heuristic which used a sweep heuristic at each depot. Golden *et al.* [11] presented two heuristics for the MDVRP; the second one specifically solved large-sized problems by a two-stage algorithm which assigns nodes to depots first and then built TSP tours for each vehicle. Chao *et al.* [6] proposed a simple initialization heuristic followed by a refinement, which gained the best performance in many benchmark problems. In 2002, Giosa *et al.* [10] demonstrated a series of 2-stage heuristics to tackle the MDVRPTW. Recently, Lim and Wang [14] gave a one-stage approach to the MDVRP with the constraint each depot only has a fixed number of vehicles. In theory Baltz *et al.* [2, 3] presented a probabilistic analysis of the optimal solution for the MDVRP. Tansini [18] provided a polynomial-time approximation scheme (PTAS) for the MDVRP similar to Arora's PTAS algorithm for the TSP.

Define $\mathcal{N}$ to be the set of all customers; $S_i (\subseteq \mathcal{N})$ to be the subset of customers assigned to vehicle i; and $TSP(S)$ to be the minimal TSP tour length that visits all customers in set S. The formulation of the min-max MDVRP is given as follows.

$$\begin{array}{ll} \text{minimize} & \lambda \\ \text{subject to} & TSP(S_i) \leq \lambda, \ \forall i, \\ & \cup S_i = \mathcal{N}. \end{array}$$

In this paper we first give a theoretical analysis for the optimal solution of the min-max MDVRP by developing lower and upper bounds. We show that the optimal solution to the min-max MDVRP is asymptotically close to the optimal TSP tour of all customers divided by the number of depots when the ratio of customers to depots is high.

We then develop two heuristics to the min-max MDVRP. The first heuristic is a linear programming-based heuristic with a global load balancing technique. By starting from a simple linear programming-based algorithm, we rapidly assign customers to depots and generate TSP routes for each vehicle. The solution is further improved in a post refinement period in which we keep changing certain parameters globally. Next, noticing that a convex equitable region partition yields an even division of points (i.e., if we divide the service region into a set of subregions with equal area, then each subregion will contain – asymptotically – the same number of points), we propose a fast approximation algorithm to generate good initial solutions for the min-max MDVRP. The solution can then be improved with common local improvement procedures. Moreover, this region partition method, for which we demonstrate both theory and implementation, is potentially useful in many network design problems with regular topology structures, such as the minimum spanning forest problem or the Chinese postman problem with multiple agents.

The simulation results of these two methods and a routine heuristic are presented in the performance analysis section. The implementation results demonstrate the efficiency of our heuristic methods. Moreover, these methods handle instances of tens of thousands of customers with ease, a scalability property which is becoming increasingly important as networks expand in size. Finally, we summarize our results and address future research issues.

2 Asymptotic bounds on the min-max MDVRP

2.1 Analysis of optimal solutions. Given its practical importance, researchers have made compelling efforts to develop heuristics for the MDVRP. In addition, exploring the theoretical bounds of the MDVRP has also become an intriguing topic. For the MDVRP whose objective is to minimize the total tour lengths, Baltz *et al.* [2, 3] established the asymptotic convergence of the optimal solution for the uniformly distributed data. In this section, we provide lower and upper bounds for the min-max MDVRP in a planar graph and conclude the asymptotic convergence for a broad class as the problem size expands. To provide context, we will first review known results for the probabilistic TSP. The most celebrated discovery is the *BHH* theorem (1959) by Beardwood, Halton and Hammersley.

Theorem 2.1 *Suppose that X_i's, $1 \leq i \leq n$, are independent and identically distributed random points uniformly distributed in a unit cube $[0,1]^d, d \geq 2$. Then with probability one, the length $TSP(\{X_1, X_2, \cdots X_n\})$ of an optimal TSP tour traversing all points satisfies:*

$$\lim_{n\to\infty} TSP(\{X_1, X_2, \cdots, X_n\})/n^{\frac{d-1}{d}} = \alpha(d),$$

where $\alpha(d)$ is a positive constant depending on d.

Assume that all customers and depots are uniformly and randomly distributed in a unit square in this paper. So define $\alpha \equiv \alpha(2)$ in our discussion. Denote the set of depots by $\mathcal{D} = \{D_1, D_2, \cdots, D_m\}$, $|\mathcal{D}| = m$. Denote the set of vehicles by $\mathcal{V} = \{V_1, V_2, \cdots, V_k\}$, $|\mathcal{V}| = k \geq m$. Denote the set of nodes(or customers) by $\mathcal{N} = \{N_1, N_2, \cdots, N_n\}$, $|\mathcal{N}| = n$. We will use node and customer interchangeably in this paper.

Denote the value of an optimal solution to the min-max MDVRP by L, i.e., the length of the longest tour. Denote the largest distance between an arbitrary pair of points in two different sets A and B by $d(A, B)$, i.e., $d(A, B) = \max_{x \in A, y \in B} ||x - y||$.

First we have the following observation:

Lemma 2.2 *For a general planar graph (points do not necessarily follow any distribution),*

$$\frac{TSP(\mathcal{D} \cup \mathcal{N}) - TSP(\mathcal{D})}{k} \leq L \leq \frac{TSP(\mathcal{N})}{k} + 2 * d(\mathcal{D}, \mathcal{N}).$$

Proof Consider the lower bound first. In an optimal pattern of the min-max MDVRP, assume that the sum of all tour lengths is S_{opt}. Noticing that L is the longest tour, we have

$$k \cdot L \geq S_{opt}.$$

Add an optimal TSP tour of all the *depots* to an optimal solution. Now each node or depot in the graph has an even degree, which implies an Euler tour. This Euler tour can be reduced to a feasible TSP tour for the set of all the depots and nodes. So we have

$$S_{opt} + TSP(\mathcal{D}) \geq TSP(\mathcal{D} \cup \mathcal{N}).$$

Therefore,

$$L \geq \frac{TSP(\mathcal{D} \cup \mathcal{N}) - TSP(\mathcal{D})}{k}.$$

For the upper bound, consider a tour partition heuristic.

Heuristic 1 Tour Partition Heuristic

1. Generate an optimal TSP tour for all the nodes.
2. Partition this tour into k subtours $t_1, t_2, \cdots, t_k$ of approximately equal length.
3. Connect both the starting and ending nodes of subtour t_i to the depot where vehicle i stays.

This heuristic generates a feasible solution for the MDVRP. The maximal tour length, which is an upper bound of L, is at most

$$\frac{TSP(\mathcal{N})}{k} + 2 * d(\mathcal{D}, \mathcal{N}).$$

□

The proof implies that this lower bound holds even when L is the average length of all the tours, so it is not essentially tight. However, when the ratio of nodes to depots is high, we can derive the asymptotic bounds for L.

Corollary 2.3 *(a) When $k = o(\sqrt{n})$,*

$$\lim_{n \to \infty} \frac{L}{\sqrt{n/k}} = \alpha.$$

(b) For the non-fixed MDVRP, i.e., a vehicle can start from and return to arbitrary different depots, if $n = \Omega(k \log^2 k)$ and $k = m$,

$$\lim_{n \to \infty} \frac{L}{\sqrt{n/k}} \leq \alpha.$$

Proof (a) According to the BHH theorem, for arbitrarily small ϵ, when number n is sufficiently large,

$$\frac{\alpha((1-\epsilon)\sqrt{m+n}-(1+\epsilon)\sqrt{m})}{k} \le L \le \frac{\alpha(1+\epsilon)\sqrt{n}}{k} + 2d(\mathcal{D},\mathcal{N}) + \epsilon. \tag{2.1}$$

Noticing that $m \le k$ and $k = o(\sqrt{n})$, we know that ratio $\sqrt{m}/k$ is less than 1, while ratio $\frac{\sqrt{m+n}}{k}$ goes to infinity as parameter n goes to infinity. On the other side, $d(\mathcal{D},\mathcal{N})$ is bounded by a constant while $\sqrt{n}/k$ goes to infinity as n goes to infinity.

(b) Consider the tour partition heuristic. Then at step 3 in Heuristic 1, for each subtour t_i, instead of connecting its starting and ending nodes to the depot where vehicle i stays, we respectively build a perfect matching between depots and starting points of subtours, and another one between depots and ending points of subtours.

The min-max perfect matching for uniformly distributed points has a bound provided by Leighton and Shor in 1989 [13]: if X_i's and Y_i's are independent and uniformly distributed points on a unit square for $1 \le i \le r$, then there exists a constant C, such that

$$\min_{\sigma\in\mathcal{P}} \max_{1\le i\le r} ||X_i - Y_{\sigma(i)}|| \le Cr^{-\frac{1}{2}}(\log r)^{\frac{3}{4}} \tag{2.2}$$

with probability one as $r \to \infty$, where $\mathcal{P}$ is the set of all permutations of $\{1, 2, \cdots, r\}$.

Apply this bound to the two perfect matchings separately. We get $L \le \frac{\alpha(1+\epsilon)\sqrt{n}}{k} + 2Ck^{-\frac{1}{2}}(\log k)^{\frac{3}{4}}$. With the assumption $n = \Omega(k\log^2 k)$, we complete the proof.

□

2.2 Bounds by region partition heuristics. In this section, we will present a region partition heuristic with theoretically good performances, and estimate upper bounds for the case $k = \Omega(\sqrt{n})$.

Before presenting the heuristic, we need several lemmas to review some facts:

Theorem 2.4 *(Convex Region Partition Theorem)* [4] *Given k points in a convex bounded planar polygon, it is always possible to find a partition of the domain into k equal-area convex polygons, with exactly one point in each face.*

Bespamyatnikh *et al.* [4] proved this convex region partition theorem for both continuous and discrete versions, and presented an algorithm for its discrete version. An algorithm for the continuous version, i.e. the convex region partition described in the theorem was recently revealed by Carlsson and Armbruster [1].

For the time being, we assume that an equitable partition exists and develop a theoretical rationale for the desirability of such a partition. The next two lemmas will provide us important backgrounds to analyze the region partition approach.

Lemma 2.5 *(Occupancy Lemma)* [15] *Randomly put n balls into m bins with equal probability. If parameter $n = \Omega(m\log^2 m)$, then the number of balls in the bin holding the least balls has an asymptotic performance as ratio $\frac{n}{m}$. So does the number of balls in the bin holding the most balls.*

Proof This fact can be proved by Chernoff bounds. We will give a sketch of the proof here.

We only consider the case $n = m\log^2 m$ here. For an arbitrary bin, define x_i is 1 if the ith ball falls into this bin, and 0 otherwise. Let $S_i = x_1 + x_2 + \cdots + x_i$. Note that the expected number of balls falling into a bin is $\log^2 m$. From the Chernoff bounds [12],

$$P = Prob(S_n \leq \frac{n}{m}(1 - 2\sqrt{\frac{2}{\log m}})) \leq e^{-nD(p-px||p)},$$

where $p = \frac{1}{m}$, $x = 2\sqrt{\frac{2}{\log m}}$, $D(a||b) = a\log\frac{a}{b} + (1-a)\log\frac{1-a}{1-b}$.
When m is sufficiently large, we have

$$e^{-nD(p-px||p)} \leq e^{-n(px^2/4)} = \frac{1}{m^2}.$$

Therefore, the probability that one bin has less than $\frac{n}{m}(1 - 2\sqrt{\frac{2}{\log m}})$ balls is at most $mP \leq \frac{1}{m}$.

Similarly, by applying another Chernoff bound, we can also prove that the probability that one bin has more than $\frac{n}{m}(1 + 2\sqrt{\frac{2}{\log m}})$ balls is at most $\frac{1}{m}$. □

Simple scaling arguments in [20] show that the BHH theorem still holds even if the unit cube is replaced by an arbitrary compact subset K. This fact suggests that the limit is independent of the shape of compact set K.

Lemma 2.6 *(generalized BHH theorem [20]) In particular, if X_i's are i.i.d with uniform distribution on a compact set K of Lebesgue measure one, the BHH theorem still holds.*

We consider a region partition heuristic for the min-max MDVRP in the case $k = m$.

Heuristic 2 Region Partition Heuristic

1. Divide the region into k equitable convex polygons such that each polygon has exactly one vehicle inside.
2. Find an optimal TSP tour for each vehicle and the nodes in the same subregion.

Lemma 2.7 *If parameter k is a constant, with high probability, the solution L generated by Heuristic 2 satisfies:*

$$\lim_{n\to\infty} \frac{L}{\sqrt{n/k}} = \alpha.$$

Proof Assume that the region with the most nodes has $N_{max} = n/k + t$ nodes. From the Occupancy lemma, we know with high probability, $\lim_{n\to\infty} \frac{t}{n/k} = 0$. Therefore, by the generalized BHH theorem, we know:

$$L \leq \alpha(1+o(1))\sqrt{N_{max}}\sqrt{\frac{1}{k}} \leq \alpha(1+o(1))\sqrt{(1+o(1))\frac{n}{k}}\sqrt{\frac{1}{k}} \leq \alpha(1+o(1))\frac{\sqrt{n}}{k}.$$

On the other hand, since the region with the most nodes has at least n/k nodes, the claim is true:

$$\lim_{n\to\infty} \frac{L}{\sqrt{n/k}} = \alpha.$$

□

If vehicles are more than depots available ($k > m$), we use the perturbing idea to generate their routes. In this procedure, if there are two or more vehicles on the same depot, we only need to slightly relocate vehicles evenly distributed on a small circle encircling that depot (Figure 1). For a sufficiently small perturbation, the algorithm remains feasible, and the optimal value only changes marginally.

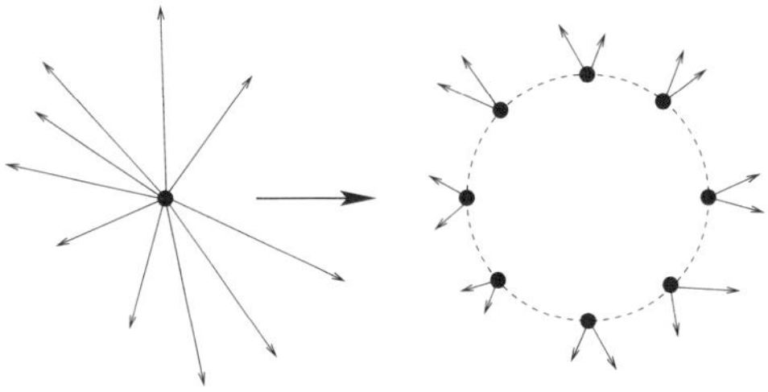

Figure 1 The perturbation method, and its subsequent effects on assignment and routes generation. Here the routes are suggested by arrows emanating from each depot node.

If ratio n/k is smaller, for example, $n = O(k \log k)$, it is harder to predict the optimal solution. Motivated by the Occupancy lemma, we can still derive an asymptotic upper bound. The following theorem illustrates such a case:

Theorem 2.8 *Assume that $k = m$, i.e., all vehicles are i.i.d. in the cube. If the number of vehicles is proportional to the number of nodes, i.e., $k = \lambda n$, $0 < \lambda < 1$, then with high probability, as $k \to \infty$,*

$$L \leq (\frac{\alpha}{\lambda} + 2\sqrt{2}\log n)\frac{1}{\sqrt{n}}.$$

Proof We divide the unit square into $n/\log^2 n$ smaller squares each of which has the side length $\log n/\sqrt{n}$. Then, the Occupancy lemma implies that any small square has at most $\log^2 n + O(\log n)$ nodes and at least $\lambda(\log^2 n - O(\log n))$ vehicles with probability one as $n \to \infty$. Apply the tour partition heuristic to the square with the most customers. According to the generalized BHH theorem and Lemma 2.2, we have

$$L \leq \frac{\alpha\sqrt{\log^2 n + O(\log n)}}{\lambda(\log^2 n + O(\log n))}\frac{\log n}{\sqrt{n}} + 2\sqrt{2}\frac{\log n}{\sqrt{n}}.$$

□

Applying the similar argument, we obtain the upper bound $(\frac{\alpha}{\lambda} + 2\sqrt{2}\log n)\frac{1}{\sqrt{n}}$ for the more general case $k = \Omega(\sqrt{n})$:

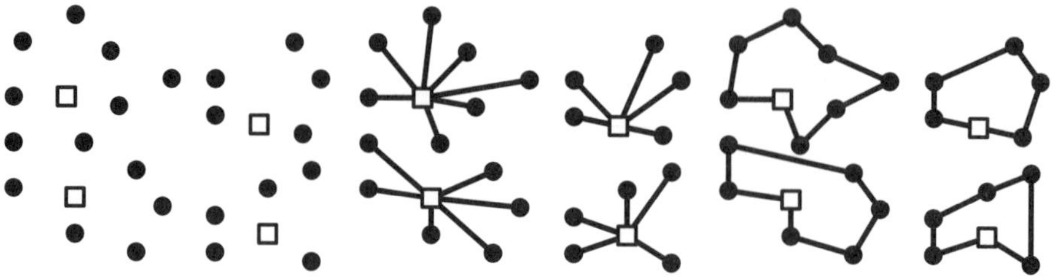

Figure 2 An illustration of the initial routes. The initial input (a), the assignment (b), and the construction of TSP tours (c).

Corollary 2.9 *Assume that all vehicles are uniformly distributed in the cube and* $k = \Omega(\sqrt{n})$, *then with high probability, as* $k \to \infty$,

$$L \leq \frac{\alpha\sqrt{n}}{k} + 2\sqrt{2}\frac{\log k}{\sqrt{k}}.$$

3 An LP-based load balancing heuristic

We present a linear programming-based heuristic in this section, which uses a *global* load balancing technique during post-refinement periods. Our heuristic is based on a *load balancing* idea: the load for each vehicle is usually well balanced in an optimal plan, so each vehicle is assigned the similar working load in terms of the number of nodes to serve. This intuition is reasonable considering that the distribution of nodes in our model is uniform. We first balance the numbers of nodes assigned to depots by linear programming and generate the initial TSP tours, then keep improving the solution by global adjustment of these numbers instead of traditional local search procedures.

Assume that all customers are uniformly distributed on a square and the locations of depots are randomly distributed. A depot is allowed to have more than one vehicle. Each vehicle must return to the depot from which it originates. First we assume that each depot has exactly one vehicle, and no two depots share a location (this assumption will be generalized later). Suppose that each node $X_j, 1 \leq j \leq n$, is located at (x_j, y_j), and each vehicle $V_i, 1 \leq i \leq k$, is located at (v_i, w_i). The distance c_{ij} between points X_j and V_i is $||(x_j, y_j) - (v_i, w_i)||$.

Define variable x_{ij} to be a binary variable indicating if node X_j is assigned to vehicle V_i or not. We build an integer program for clustering as follows.

$$\begin{aligned} (LP): \text{minimize} \quad & \sum_{1\leq i\leq k, 1\leq j\leq n} c_{ij}x_{ij} \\ \text{subject to} \quad & \sum_{1\leq j\leq n} x_{ij} = \frac{n}{k}, \ \forall i, \\ & \sum_{1\leq i\leq k} x_{ij} = 1, \ \forall j. \\ & x_{ij} \in \{0,1\}, \ \forall i,j. \end{aligned}$$

We assume that ratio n/k is an integer(the fractional case will be discussed later). Noticing that its linear program relaxation (LPR) is a typical network flow model, a vertex solution to LPR will be integral. Therefore, we can directly solve LPR for the initial assignment.

A sketch of the heuristic is as follows:

Heuristic 3 LP-based Load Balancing Heuristic

1. Build LPR and solve it.
2. Build a TSP tour for each vehicle by the Concorde solver. Assume that their lengths are sorted in a descending order $\{L_1, \cdots, L_k\}$.
3. If $(L_1 - L_k)/L_k \leq r$, with r a small constant, or if the linear program has been run many times, stop and output a current best solution. Otherwise, decrease the number of nodes assigned to vehicles having longer tours and increase the number of nodes assigned to vehicles having shorter tours (we will explain details later). This way, a new revised LP is generated. Return to step 1 and repeat the procedure.

An illustration of the heuristic in one loop is shown in figure 2.

There are several details we need to explain in our algorithm.

Note 1: If ratio n/k is fractional, assume that $n = pk + q$ where parameter p and q are integers and parameter q is the residue. Then we assign the first q vehicles to $p+1$ nodes and assign each of the remaining vehicles p nodes. Thus the number of nodes assigned to a vehicle is always an integer.

Note 2: Constant r in step 3 is a ratio threshold to measure the output, which can be adjusted according to the need for accuracy. In most cases, setting $r = 25\%$ is ideal to achieve a satisfactory result.

Note 3: At step 3, if $(L_1 - L_K)/L_K \leq r$, we adjust the number of nodes assigned to each depot by the following procedure:

- Build the sets $L^+ = \{i : L_i - \bar{L} > 0\}$, $L^- = \{i : L_i - \bar{L} < 0\}$, where parameter $\bar{L}$ is the average of tour length.
- Without loss of generality, assume that $s = |L^+| \geq |L^-|$. For any $i \in L^+$, define the *relative drift* $d_i^+ = \frac{L_i - \bar{L}}{\bar{L}}$. Decrease the number of nodes assigned to depot i by $[c * \frac{d_i}{r}]$(c is a constant empirically decided. We set $c = 2$ in implementation). Sort elements in set L^- in the descending order of their tour lengths. Then define $d_i^- = \frac{\bar{L} - L_i}{\bar{L}}$ for the first s elements in set L^-. Increase the number of nodes in the same way.
- If the number of nodes removed from set L^+ is more than the number of nodes added to set L^-, add one node to each element in set L^- sequentially starting from the depot which has the fewest nodes, keep doing that until nodes are balanced. Repeat the similar procedure for the contrary case.

3.1 Multiple vehicles on a depot. We have assumed that each depot includes exactly one vehicle in all the heuristics aforementioned. However, we do not complicate matters by allowing multiple vehicles to lie on the same depot. We need to extend our heuristic to this general case. If we still follow the original procedures and assign the same number of nodes to each vehicle instead of each depot, nodes will be assigned to each such vehicle randomly and the solution quality will be poor.

In order to work around this difficulty for vehicles in a same depot, we perturb the locations of vehicles located at the same depot. Each vehicle is relocated to a new location lying in a small circle around that depot. Keep them evenly distributed

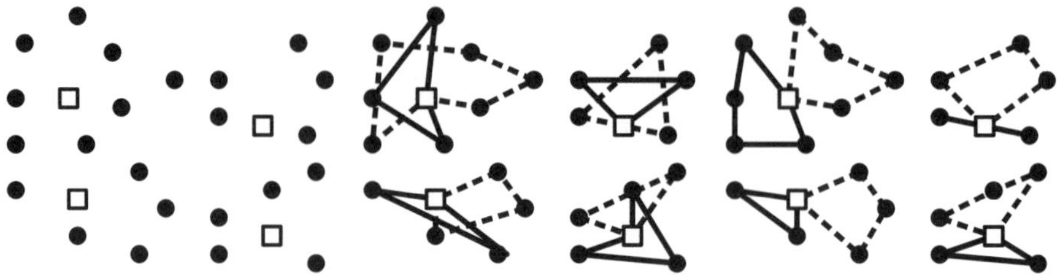

Figure 3 The case in which multiple vehicles reside in a single depot. The left is the initial node set given to the problem (we have 8 vehicles and 4 depots) and the middle represents a solution without the perturbation applied. The right represents a solution with the perturbation applied; it is clearly preferable, as vehicle routes do not intersect.

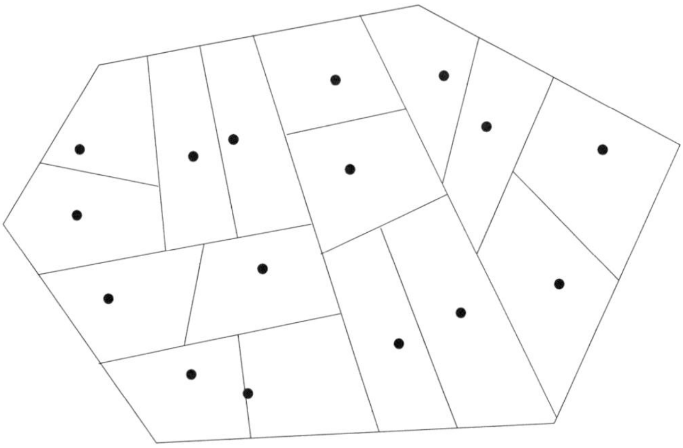

Figure 4 An equitable partition of a convex polygon.

on the circle. This small change will dramatically improve the final solution (Figure 3).

4 A gegion partition heuristic

The second heuristic we present is the region partition heuristic (Heuristic 2) based on the algorithm in [1] that takes as input a convex planar region C and a set of m depots $\mathcal{D} = \{D_1, \ldots, D_m\}$, and outputs a partition of C into m convex subregions satisfying the following properties:

- Each subregion is convex.
- Each subregion contains one depot D_i.
- Each subregion has equal area.

Note that a Voronoi diagram satisfies the first two properties, but not the third. An example of such a partition is shown in figure 4.

As mentioned previously, the requirement that each subregion have equal area ensures that the lengths of the TSP tours of each region be equal asymptotically [20]. This result applies to any compact subset of $\mathbb{R}^2$ with Lebesgue measure one, but

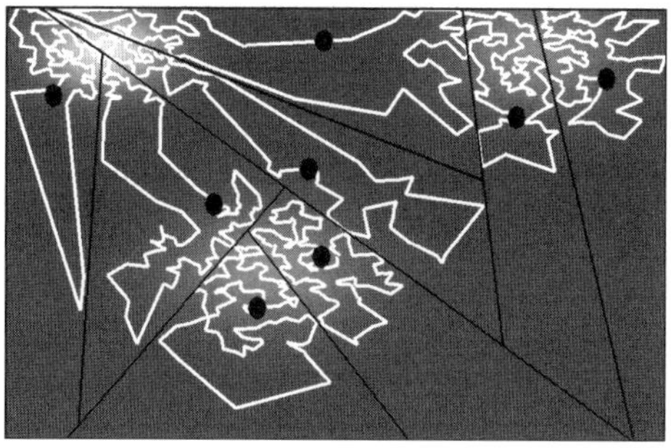

Figure 5 An equitable partition with respect to a probability density and a depot set. Here we construct sub-regions C_i satisfying the property that $\int_{C_i} \sqrt{f(x)}\,dx$ is equal across all sub-regions. The BHH theorem guarantees that all sub-regions will have the same induced TSP tours, up to $o\left(\sqrt{n_i}\right)$, the number of points contained in each region.

we require that these regions be convex for practical purposes. The shortest path between any two cities in a subregion R_i is a straight line, and we want to ensure that the route taken from one city to the next lie in the same service region and therefore be feasible.

This algorithm has a further advantage in that it can be readily applied to situations with i.i.d. non-uniformly distributed points, when a density function is known. In this situation, we replace the equal-area constraint with a constraint involving the probability density function given, as shown in figure 5. See [1] for a full description.

Many network structures, like minimum spanning trees and Voronoi graphs, have conclusions similar to that of the travelling salesman tour, i.e., their lengths converge asymptotically in a manner similar to the BHH theorem. For example, Steele [17] described a similar result for Minimum Spanning Tree (MST). Region partition is therefore a potentially powerful tool not only in vehicle routing but also in many network applications, such as finding the minimum spanning forest of a node set with fixed roots. One can potentially solve the Chinese postman problem with multiple agents by equitably partitioning a city map with respect to all road mass contained (figure 6). This ignores the underlying combinatorics of the problem but can likely serve as a useful initial guess in a local-search algorithm when connectivity in the graph is high.

It is worth extra effort to investigate how to implement partition quickly. Bespamyatnikh *et al.* [4] gave the proof for the existence of a convex region partition and an algorithm with runtime $O(N^{\frac{4}{3}} \log N)$ for the discrete version in which the dividing subjects are red and blue points. Parameter N is the sum of the number of nodes and sides of the initial polygon. Given that an area can be approximated within any ϵ factor by taking enough sampling points, we can implement their algorithm to obtain a near-optimal convex subdivision. We have also been searching for a quick method to realize the algorithm. Our best result is the realization of an equitable partition of a polygon and a set of n points in $O(Nn \log N)$ operations.

Figure 6 Equitably partitioning a city map so that each convex sub-region contains a depot and the same total amount of road.

5 Performance and analysis

No specific benchmark problems for the min-max MDVRP are available in the literature. Most available MDVRP benchmarks have time window constraints, and their sizes are generally small, varying between $50 \sim 1000$(see [22]). Thus, we generated 13 MDVRP benchmark sets ourselves and the results are reported in this section. These sets vary in size with respect to the number of depots (2 to 20) and number of nodes (100 to 2048), and in the location of depots and nodes. Each set contains 4 randomly generated instances of the same size. The data presented represent the average values obtained over the 4 instances for each benchmark set. Tests were conducted on a Pentium 4 1.8GHz/512MB notebook. The implementation was done in Matlab with the COPL LP Solver [21] and the Concorde TSP solver [7].

In table 1 and 2, we compare the initial solutions provided by three heuristics: the LP-based heuristic, the region partition heuristic and a traditional nearest-neighbor approach that can be stated as follows:

Heuristic 4 Nearest Neighbor Heuristic

1. Assign each node to its nearest depot.
2. Build a TSP route for each vehicle accordingly.
3. Run the local improvement method to improve the solution.

We have also implemented a modified version of a local improvement procedure proposed by Chao *et al.* [6] called "1-point movement". In table 3 and 4, we present

Table 1 Summary of first iteration solutions for three heuristics

			LP-TSP				Region Partition			
Dataset	Depots	Nodes	Max	Min	Mean	Time	Max	Min	Mean	Time
d2n100	2	100	472.61	391.12	431.87	1.07	429.22	399.90	414.56	1.10
d2n500	2	500	870.76	847.10	858.93	41.70	875.66	843.35	861.00	28.34
d5n100	5	100	242.12	159.50	197.15	1.13	209.24	162.26	190.24	2.20
d5n500	5	500	378.95	332.04	356.64	6.43	377.97	334.34	356.03	9.06
d5n1000	5	1000	542.37	467.60	503.22	35.65	522.34	473.58	497.20	36.73
d10n1000	10	1000	283.32	225.20	252.86	11.03	277.63	233.97	255.43	41.11
d10n2000	10	2000	380.10	331.38	355.08	115.34	378.11	317.47	353.58	180.76
d16n256	16	256	161.22	75.21	106.87	12.16	140.74	70.26	105.51	21.47
d16n512	16	512	182.11	106.46	132.74	10.95	172.55	106.58	131.59	22.32
d16n1024	16	1024	206.95	149.55	173.64	17.31	196.16	145.83	169.19	57.66
d16n2048	16	2048	263.61	207.90	231.87	91.03	253.85	206.56	229.62	230.51
d20n1000	20	1000	172.29	117.07	139.24	24.95	170.34	118.31	138.16	84.88
d20n2000	20	2000	207.23	167.49	184.83	83.02	222.11	160.57	185.62	177.91

Table 2 *Continued:* Summary of first iteration solutions for three heuristics

			Nearest Neighbor			
Dataset	Depots	Nodes	Max	Min	Mean	Time
d2n100	2	100	574.85	260.88	417.85	0.35
d2n500	2	500	1097.33	715.18	906.25	48.33
d5n100	5	100	268.58	95.23	186.30	0.20
d5n500	5	500	519.58	223.48	369.35	5.00
d5n1000	5	1000	1101.95	163.28	511.98	295.65
d10n1000	10	1000	451.18	83.50	257.03	11.68
d10n2000	10	2000	688.58	178.15	361.55	351.40
d16n256	16	256	186.23	23.03	91.40	1.38
d16n512	16	512	265.48	32.13	122.93	2.33
d16n1024	16	1024	419.10	50.58	167.35	12.55
d16n2048	16	2048	496.25	72.25	230.63	243.13
d20n1000	20	1000	278.38	36.50	134.63	8.75
d20n2000	20	2000	368.88	57.25	183.35	49.10

the final outputs of four heuristics. The first heuristic uses the LP-based load balancing heuristic (Heuristic 3) to distribute the load evenly among vehicles. The other three heuristics all use the modified 1-point movement improvement method in their post optimization periods, but they start with different initial assignments. The second heuristic uses the same initial assignment as the first one. The third heuristic starts with the region partition method, and the fourth heuristic starts with the nearest neighbor assignment.

The running time of the LP-based load balancing heuristic is primarily comprised of the execution time of the LP and TSP solvers. The running time of the region partition with local improvement method (Heuristic 2) mainly consists of the running time of the region partition heuristic and TSP solvers. For small to medium-sized benchmarks, the main bottleneck for all heuristics is finding a near optimal TSP tour for each depot using the Concorde solver.

Table 3 Summary for improved heuristics

			LP-TSP + Load Balancing				LP-TSP + Local Improvement			
Dataset	Depots	Nodes	Max	Min	Mean	Time	Max	Min	Mean	Time
d2n100	2	100	458.35	403.23	430.80	9.18	448.93	443.28	446.10	6.65
d2n500	2	500	868.60	846.50	859.53	192.35	865.00	863.63	864.33	187.33
d5n100	5	100	207.08	172.08	190.58	8.85	229.45	174.68	215.50	7.83
d5n500	5	500	377.28	332.20	357.08	57.90	374.65	357.33	368.03	60.85
d5n1000	5	1000	531.85	467.73	501.05	192.58	534.73	488.08	520.13	202.33
d10n1000	10	1000	277.83	229.28	252.93	84.88	279.90	224.80	262.23	86.10
d10n2000	10	2000	378.65	331.60	355.43	875.45	379.70	335.68	365.88	1098.43
d16n256	16	256	114.63	88.08	101.98	68.60	142.63	75.55	112.75	61.23
d16n512	16	512	142.15	118.70	131.55	78.75	171.95	108.08	144.38	73.60
d16n1024	16	1024	187.98	155.00	171.98	153.88	201.05	150.58	178.10	134.05
d16n2048	16	2048	254.75	209.05	231.55	870.18	261.83	207.18	235.10	860.70
d20n1000	20	1000	151.05	124.20	138.20	192.15	166.33	117.25	145.65	166.93
d20n2000	20	2000	201.43	168.25	184.80	601.60	205.03	168.75	189.13	612.70

Table 4 *Continued:* Summary for improved heuristics

			Region Partition+Local Imp				Nearest Neighbor+Local Imp			
Dataset	Depots	Nodes	Max	Min	Mean	Time	Max	Min	Mean	Time
d2n100	2	100	416.70	414.03	415.38	5.73	528.75	313.88	421.30	6.60
d2n500	2	500	873.15	869.18	871.15	152.38	913.35	851.85	882.60	135.03
d5n100	5	100	204.80	187.00	199.40	8.98	235.20	141.35	204.73	7.58
d5n500	5	500	371.88	339.00	359.00	62.18	447.48	261.73	381.85	72.90
d5n1000	5	1000	518.33	499.83	511.43	247.50	858.75	253.08	548.78	321.43
d10n1000	10	1000	270.50	238.30	260.30	165.03	391.08	124.85	282.80	150.03
d10n2000	10	2000	375.50	335.80	363.25	923.58	574.48	173.93	362.43	1206.25
d16n256	16	256	132.33	80.08	111.85	74.18	156.93	23.68	102.60	58.80
d16n512	16	512	168.15	106.93	143.68	75.03	204.70	35.38	136.63	64.08
d16n1024	16	1024	191.68	151.65	176.75	182.20	321.70	57.45	184.38	188.48
d16n2048	16	2048	250.95	207.40	233.63	1062.70	414.53	90.43	238.30	1011.68
d20n1000	20	1000	163.70	118.13	143.43	229.15	244.88	41.20	144.58	242.35
d20n2000	20	2000	219.38	162.65	191.60	543.58	335.48	59.93	209.18	617.83

We have made the following observations based on the simulation results presented in the tables.

First, the region partition heuristic with local improvement produced the best solutions for 7 out of 13 instances, and the LP-based load balancing heuristic generated the best solutions for 5 out of 13 instances. They both maintained a competitive running time, as seen in table 3 and 4.

Second, according to table 1 and 2, both LP-based and region partition heuristics produce notably better initial routes than the traditional nearest neighbor approach. They are solid candidates for the initial assignment.

Third, when the number of depots is small, the region partition heuristic tends to outperform the other three heuristics, and its running time remains low. The LP-based load balancing heuristic performs better as the number of depots is increased.

Fourth, the region partition heuristic with local improvement and the LP-based load balancing heuristic generate quality solutions for the min-max MDVRP at the minor expense of the sum of the tour lengths.

Finally the region partition heuristic with local improvement generates a higher total cost than the LP-based load balancing heuristic in most instances, though it obtains the best results for the min-max MDVRP in more instances than the latter. The difference of objective functions between usual MDVRP and the min-max MDVRP motivates different types of algorithms.

6 Conclusion

This paper has presented the first study to the min-max MDVRP. We explored the theoretical asymptotic property of the optimal solution, which gave rise to the region partition heuristic. The region partition idea has potential applications beyond the vehicle routing problem. An LP-based heuristic with global improvement was also presented to solve very large-sized problems effectively. We concluded by reporting our simulation results.

Many interesting topics remain after this first attempt on the min-max MD-VRP. For example, from a theoretical point of view, an accurate estimation of the optimal solution in large-scale problems for the general case is still open. Our work can serve as starting points for further exploration as well.

The techniques used in our algorithms provide advantages over traditional local search methods. For example, since the LP-based load balancing heuristic does not inherently require that the points be scattered in a Euclidean space, it can be adapted as a backend to existing map software to solve VRPs with actual geographic locations. In addition, the region partition algorithm can be applied to problems outside the scope of VRP. An exact convex region partition algorithm will be helpful not only for high-quality initial solutions to MDVRPs, but also for solving a class of network problems with similar structures and a min-max objective function.

References

[1] B. Armbruster, J. Carlsson. *Finding equitable convex partitions of points in a polygon efficiently.* ACM Trans. Algorithms. To appear.

[2] A. Baltz, D. P. Dubhashi, A. Srivastav, L. Tansini, S. Werth. *Probabilistic Analysis for a Multiple Depot Vehicle Routing Problem.* Foundations of Software Technology and Theoretical Computer Science, 25th International Conference(FSTTCS) (2005), 360–371.

[3] A. Baltz, D. P. Dubhashi, A. Srivastav, L. Tansini, S. Werth. *Probabilistic Analysis for a Multiple Depot Vehicle Routing Problem.* Random Structures Algorithms (2007), 206–225.

[4] S. N. Bespamyatnikh, D. G. Kirkpatrick, J. Snoeyink. *Generalizing Ham Sandwich Cuts to Equitable Subdivisions.* Symposium on Computational Geometry (SoCG) (1999), 49–58.

[5] J. Carlsson, D. W. Kim, Y. Ye. *Solving the Min-max Multi-depot Vehicle Routing Problem.* Presented at INFORMS (2007).

[6] I.-M. Chao, B. Golden, E. Wasil. *A new heuristic for the multidepot vehicle routing problem that improves upon best-known solutions.* Amer. J. Math. Management Sci. **13–3** (1993), 371–406.

[7] W. Cook, CONCORDE solver for the TSP, Version 1.1. Computer software. `http://www.tsp.gatech.edu/concorde.html`.

[8] G.B. Dantzig, J.H. Ramser. *The truck dispatching problem.* Management Science **6:60** (1959).

[9] B. Gillet, J. Johnson. *Multi-terminal vehicle-dispatching algorithm.* Omega **4** (1976), 711–718.

[10] I. Giosa, I. Tansini, I. Viera. *New assignment algorithm for the multi-depot vehicle routing problem.* J. Oper. Res. Soc. **53** (2002), 283–292.

[11] B.L. Golden, T. L. Magnanti, H.Q. Nguyen. *Implementing vehicle routing algorithms.* Networks **7** (1977), 113–148.

[12] W. Hoeffding. *Probability inequalities for sums of bounded random variables.* J. Amer. Statist. Asso. **58 (301)** (1963), 13–30.

[13] T. Leighton, P.W. Shor. *Tight Bounds for minimax grid matching with applications to the average case analysis of algorithms.* Combinatorica **9** (1989), 161–187.

[14] A. Lim, F. Wang. *Multi-Depot Routing Problems: A one-Stage Approach.* IEEE transactions on automation science and engineering **4-2** (Oct. 2005).

[15] R. Motwani, P. Raghavan. *Randomized Algorithms.* Cambridge University Press 1995.

[16] J.F. Soumis, M. Desrochers. *A New Optimization Algorithm for the Vehicle Routing Problem With Time Windows by Column generation.* Networks **14** (1984), 545–565.

[17] J. M. Steele. *Growth rates of Euclidean minimal spanning trees with power weighted edges.* Ann. Probab. **16** (1988), 1767–1787.

[18] L. Tansini. *PhD Thesis.* Department of Computer Science, Chalmers Univeristy. To appear.

[19] A. Wren, A. Holliday. *Computer scheduling of vehicles from one or more depots to a number of delivery points.* Operations Research Quarterly, **23** (1972), 333–344.

[20] J. E. Yukich. *Probability Theory of Classical Euclidean Optimization Problems.*Lecture Notes in Mathematics **1675** (1998).

[21] X. Zhang, Y. Ye. COPL Computational Programming Optimization Library : Linear Programming. Computer software. `http://www.stanford.edu/~yyye/Col.html`.

[22] `http://www.top.sintef.no/vrp/benchmarks.html`.

Fields Institute Communications
Volume **55**, 2009

Trust-Tech-Based Global Optimization Methodology for Nonlinear Programming

Hsiao-Dong Chiang
School of Electrical and Computer Engineering
Cornell University
Ithaca, NY 14853 USA
chiang@ece.cornell.edu

Jeng-Huei Chen
School of Electrical and Computer Engineering
Cornell University
Ithaca, NY 14853 USA

Chandan K. Reddy
Department of Computer Sciences
Wayne State University
Detroit, MI 48202 USA
reddy@cs.wayne.edu

1 Introduction

Optimization technology has practical applications in almost every branch of science, business and technology. Indeed, a large variety of the quantitative issues such as the decision making, design, operation, planning, and scheduling are abundant in the practical systems of the sciences, engineering, and economics that can be perceived and modeled as either continuous or discrete optimization problems. Typically, the overall performance (or measure) of a system can be described by a multivariate function called the objective function. According to this generic description, one seeks the best solution for an optimization problem, often expressed by a real vector, in the solution space which satisfies all stated feasibility constraints and minimizes (or maximizes) the value of an objective function. The vector, if it exists, is termed the global optimal solution.

For most practical applications, the underlying objective functions are often nonlinear and depend on a large number of variables, making the task of searching the solution space to find the globally optimal solution very challenging. The primary challenge is that in addition to the high-dimension solution space, there are many local optimal solutions in the solution space (a local optimal solution is optimal in the local solution space, but not necessarily in the global solution space). The global optimal solution is just one such solution, yet both the global

2000 *Mathematics Subject Classification.* Primary 90xx.

optimal solution and the local optimal solutions share the same local properties. In general, the number of local optimal solutions is unknown and it can be quite large. Furthermore, the values of an objective function at local optimal solutions and at the global optimal solution may differ significantly. Hence, there are strong motivations to develop effective methods for finding the global optimal solution.

Most existing methods proposed for solving the global optimization problem can be categorized into three classes: the class of deterministic methods, the class of stochastic methods and the class of hybrid methods. The class of deterministic methods typically provides mathematical guarantees for convergence to a global minimum in finite steps for optimization problems involving a certain mathematical structure. This class of methods includes the branch and bound methods, cutting planes, decomposition based approaches, covering methods, interval methods, homotopy methods and generalized gradient methods. On the other hand, the class of stochastic methods requires sampling the objective function and perhaps computing their derivatives to provide asymptotic convergence guarantees for finding the global optimal solution. They are applicable to a wide range of optimization problems; however, the required number of samples to arrive at the global optimal solution can be prohibitive for high-dimension problems. Random search methods, genetic algorithms, simulated annealing, clustering methods and Bayesian methods all belong to the class of stochastic methods.

The only reliable way to find the global optimal solution of a nonlinear optimization problem is to first find all the local optimal solutions and then use them to find the global optimal solution. To this end, we have developed a new systematic method, which is deterministic in nature, to find all the local optimal solutions of the constrained nonlinear optimization problems. This method is based on the Transformation Under Stability-reTaining Equilibria Characterization, hence termed the *Trust-Tech method.*

The Trust-Tech method is distinguished from other methods in that it systematically and deterministically searches all the local optimal solutions of general constrained nonlinear optimization problems. In addition, a seamless interface between the Trust-Tech-based method and an effective local method for quickly finding a set of local optimal solutions is straightforward. An important feature of the Trust-Tech method is that it finds local optimal solutions in a tier-by-tier manner: starting from (or near) a local optimal solution to find the nearby first-tier local optimal solutions, then the second-tier local optimal solutions, and so on.

From a user's viewpoint, the Trust-Tech-based method can be applied to guide any existing computer package, designed for finding local optimal solutions used to locate a complete set of local optimal solutions. Thus, the identity of the global optimal solution can be identified. Furthermore, the Trust-Tech-based method can be programmed to interface with an existing computer package without changing its environments (including the graphical user interface and database). Hence, the Trust-Tech-based method imposes no new learning curve for the user to adopt the integrated Trust-Tech-based computer package.

Trust-Tech-based methodology addresses the following two important and challenging issues in the course of searching for the globally optimal solution:

(i) how to effectively move (escape) from a local optimal solution and move on toward another local optimal solution; and

(ii) how to avoid revisiting sub-regions that contain local optimal solutions which are already known.

In the past, significant efforts have been directed towards attempting to address these two issues, but without much success. Issue (i) is difficult to solve and the existing methods all encounter this difficulty. Issue (ii), related to computational efficiency during the course of a search, is also difficult to solve and the majority of the existing methods also encounter this difficulty. Issue (ii) is a common problem which degrades the performance of many existing methods when searching for the global optimal solution. The efforts spent in searching sub-regions that contain local optimal solutions which are already known are wasteful. This wastes computing resources without gaining new information regarding the location of the globally optimal solution.

2 Trust-Tech methodologies

Trust-Tech-based methods are dynamical methods for obtaining the global optimal solution of general constrained optimization problems. One local optimal solution is deterministically found by starting from an initial point. Then another local optimal solution is found by starting from the previous solution until all the local optimal solutions are found in a tier-by-tier manner. We believe that an effective approach to solving general optimization problems is to first find multiple, if not all local optimal solutions and then select the best solution among them.

We first consider the following unconstrained nonlinear programming problem:

$$\begin{aligned} &\min c(x) \\ &x \in R^n \end{aligned} \tag{2.1}$$

where $c \in C^2$ is a function bounded below and possesses only finite local optimal solutions. Our focus of solving this problem is to locate all or multiple local optimal solutions of $c(x)$.

Instead of directly solving the unconstrained optimization problem, described by Eq. (2.1), we consider the following gradient dynamical system:

$$\dot{x}(t) = -\nabla c(x) \tag{2.2}$$

where$x \in R^n$. It will be shown that each local optimal solution of problem (2.1) corresponds to a stable equilibrium point of the gradient system (2.2). This transformation allows us to locate each local optimal solution of problem (2.1) via each stable equilibrium point of the gradient system (2.2), a special class of the following general nonlinear dynamical system

$$\dot{x}(t) = f(x(t)) \tag{2.3}$$

where the state vector $x(t)$ belongs to the Euclidean space R^n and the function $f : R^n \to R^n$ satisfies the sufficient condition for the existence and uniqueness of solutions. A state vector $\hat{x}$ is called an equilibrium point of system (2.3) if $(f(\hat{x}) = 0)$. We say that an equilibrium point of (2.3) is *hyperbolic* if the Jacobian of $f(\cdot)$ at $\hat{x}$, denoted $J_f(\hat{x})$, has no eigenvalues with a zero real part. For a hyperbolic equilibrium point, it is an *(asymptotically) stable equilibrium point* if all the eigenvalues of its corresponding Jacobian have negative real parts; otherwise it is an *unstable equilibrium point.* If the Jacobian of the equilibrium point $\hat{x}$ has exactly one eigenvalue with a positive real part, we call it a *type-one equilibrium point.* Likewise, $\hat{x}$ is called a *type-k equilibrium point* if its corresponding Jacobian has exactly k eigenvalues with positive real parts.

The solution curve of (2.3) starting from x at $t = 0$ is called the system trajectory starting from x and is denoted by $\Phi_t(x)$. Let $\hat{x}$ be a hyperbolic equilibrium point. Its stable and unstable manifolds, $W^s(\hat{x})$ and $W^u(\hat{x})$, are defined as follows:

$$\begin{aligned} W^s(\hat{x}) &:= \{x \in R^n : \Phi_t(x) \to \hat{x} \text{ as } t \to \infty\} \\ W^u(\hat{x}) &:= \{x \in R^n : \mathrm{B}\Phi_t(x) \to \hat{x} \text{ as } t \to -\infty\}. \end{aligned}$$

Every trajectory in the stable manifold $W^s(\hat{x})$ converges to $\hat{x}$ as time goes to positive infinity while every trajectory in the stable manifold $W^u(\hat{x})$ converges to $\hat{x}$ as time goes to negative infinity. For a stable equilibrium point, it can be shown that there exists a number $\delta > 0$ such that $\|x_0 - \hat{x}\| < \delta$ implies $\Phi_t(x_0) \to \hat{x}$ as $(t \to \infty)$. If δ is arbitrarily large, then $\hat{x}$ is called a *global stable equilibrium point.* There are many physical systems containing multiple stable equilibrium points. The *stability region* (also called the *region of attraction*) is a useful concept for these kinds of systems. The stability region of a stable equilibrium point x_s is defined as:

$$A(x_s) := \{x \in R^n : \lim_{t \to \infty} \Phi_t(x) = x_s\}.$$

From a topological point of view, the stability region $A(x_s)$ is an open, invariant and connected set. The boundary of the stability region $A(x_s)$ is called the *stability boundary* of x_s and will be denoted by $\partial A(x_s)$.

We say a function $V : R^n \to R$ is an *energy function* for the system (2.3) if the following three conditions are satisfied:

(i) The derivative of the energy function $V(x)$ along any system trajectory $x(t)$ is non-positive, i.e.,

$$\dot{V}(x(t)) \leq 0.$$

(ii) If $x(t)$ is a non-trivial trajectory, i.e., $x(t)$ is not an equilibrium point, then, along the non-trivial trajectory $x(t)$ the set

$$\{t \in R : \dot{V}(x(t)) = 0\}$$

has measure zero in R.

(iii) That a trajectory $x(t)$ has a bounded value of $V(x(t))$ for $t \in R^+$ implies that the trajectory $x(t)$ is also bounded. Stating this in brief:

That $V(x(t))$ is bounded implies that $x(t)$ is also bounded.

Alone, property (i) states that the energy function is non-increasing along its trajectory, but does not imply that the energy function is strictly decreasing along its trajectory. There may exist a time interval $[t_1, t_2]$ such that $\dot{V}(x(t)) = 0$ for $t \in [t_1, t_2]$. However, when combined with property (ii), property (i) implies that the energy function is strictly decreasing along any system trajectory. Property (iii) states that the energy function must be a proper map along any system trajectory but need not be a proper map for the entire state space. Recall that a proper map is a function $f : X \to Y$ such that for each compact set $(D \in Y)$, the set $f^{-1}(D)$ is compact in X. Property (iii) which can be viewed as a 'dynamic' proper map is useful in the characterization of a stability boundary. From the above definition of an energy function, it is obvious that an energy function may not be a Lyapunov function.

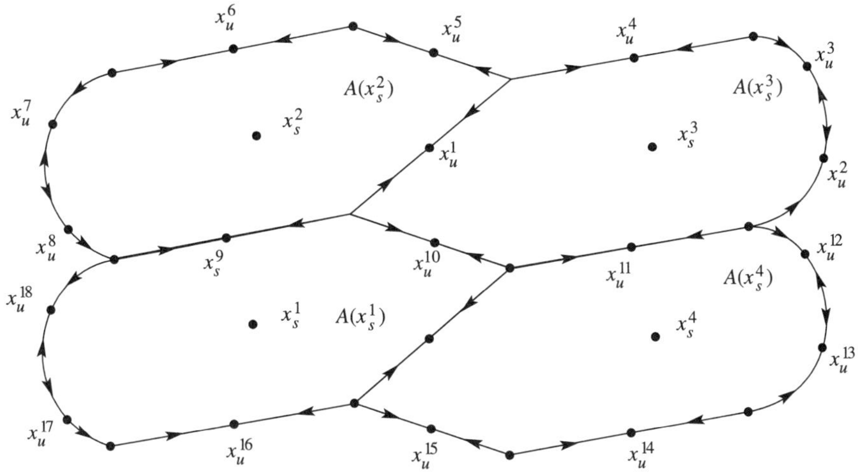

Figure 1 For a completely stable system, every trajectory converges to one of its equilibrium points, be it a stable or unstable equilibrium point

We next derive several geometrical and topological properties of the gradient system (2.2) which will be useful in our development of a systematic method for obtaining several local optimal solutions.

Proposition 2.1 (Completely stable [6]) *If there exist ε and δ such that $\|\nabla c(x)\| > \varepsilon$ unless $x \in B_\delta(\tilde{x})$, where $\tilde{x}$ is an equilibrium point; then the gradient system (2.2) is completely stable and $c(x)$ is an associated Lyapunov function.*

For a completely stable system, every trajectory converges to one of its equilibrium points. Hence, the state space is composed of the union of the closure of the stability region of each stable equilibrium point (see Figure 1 for an example). In other words, almost every trajectory converges to one of its stable equilibrium points and only those trajectories lying on the stability boundary converge to one of its unstable equilibrium points.

Theorem 2.2 (Characterization of stability boundary [3, 7]) *Suppose that all the equilibrium points of the gradient system (2.2) are hyperbolic. Let x_i, $i = 1, 2, \ldots$ be the equilibrium points on the stability boundary $\partial A(x_s)$ of a stable equilibrium point, say x_s. Then, the stability boundary is contained in the union of the stable manifolds of the equilibrium points on the stability boundary. In other words,*

$$\partial A(x_s) \subseteq \bigcup_{x_i \in \partial A} W^s(x_i).$$

It can be shown that an equilibrium point of a negative gradient system (2.2) is hyperbolic if and only if the corresponding critical point of $c(x)$ is nondegenerate.

Theorem 2.3 (Equilibrium points and local optimum [6]) *If $\bar{x}$ is a hyperbolic equilibrium point of gradient system (2.2), then*

1. *$\bar{x}$ is a stable equilibrium point of system (2.2) if and only if $c(x)$ has an isolated minimum at $\bar{x}$.*

2. *$\bar{x}$ is a source of system (2.2) if and only if $c(x)$ has an isolated maximum at $\bar{x}$.*

Theorem 2.3 characterizes the relationship between the optimal solutions of the unconstrained optimization problem (2.1) and the stable equilibrium points of its corresponding dynamical system (2.2). Hence, if x_s is a stable equilibrium point of (2.2), then it is a local optimal solution of the unconstrained optimization problem (2.1). Conversely, if x_s is a local optimal solution of (2.1), then it is a stable equilibrium point of the gradient system (2.2).

3 Decomposition points and the practical stability boundary

A new concept, called the practical stability region $A_p(x_s)$ was derived from the concept of the stability region $A(x_s)$ in [7] and is defined as

$$\text{practical stability region} = \int(\overline{A(x_s)}),$$

where $\overline{A(x_s)}$ represents the closure of $A(x_s)$ and int(:) represents the interior of the set. Hence, the practical stability region is the interior of the closure of the stability region. Its boundary is defined as the practical stability boundary and denoted as $\partial A_p(x_s)$. It has been shown that the practical stability region eliminates the complex portion of stability boundary lying inside the interior of $\overline{A(x_s)}$ and is more suitable for practical engineering applications [7].

Proposition 3.1 ([7]) *Consider the nonlinear dynamical system (2.3). An equilibrium point x_d is on the practical stability boundary $\partial A_p(x_s)$ of the stable equilibrium point x_s if $x_d \in \partial A(x_s)$ and $W^u(x_d) \cap (A(x_s))^c \neq \emptyset$.*

Proposition 3.2 ([7]) *Consider the nonlinear dynamical system (2.3). If $\tilde{x}$ is a type-k equilibrium point with $k > 1$ on the practical stability boundary $\partial A_p(x_s)$, then there exists a type-one equilibrium point $x_1 \in \partial A_p(x_s)$ such that $x \in \overline{W^s(x_1)}$.*

Motivated by the above two analytical results, we shall call a type-one equilibrium point lying on a practical stability boundary $\partial A_p(x_s)$ a *decomposition point* (with respect to x_s). Decomposition points and their stable manifolds will be used to characterize the structure of the practical stability boundary. The decomposition point x_d distinguishes itself from other type-one equilibrium points in the following way: as the Lyapunov (or energy) function value increases from $c(x_d) - \varepsilon$ to $c(x_d) + \varepsilon$, where ε is a small number, the number of the path components of $S_{c(x_d)}(x)$ will be decreased by one, i.e., $\#S_{c(x_d)+\varepsilon}(x) = \#S_{c(x_d)-\varepsilon}(x) - 1$. On the other hand, if x_1 is a type-one equilibrium point other than a decomposition point, the number of the path components remains the same as the Lyapunov function value increases from $c(x_d) - \varepsilon$ to $c(x_d) + \varepsilon$.

Proposition 3.3 (Another Characterization of practical stability boundary [7]) *Let σ_i, $i = 1, 2, \ldots$ be the decomposition points on the practical stability boundary $\partial A_p(x_s)$ of the stable equilibrium point of the gradient system (2.2). Then,*

$$\partial A_p(x_s) \subseteq \bigcup_{\sigma_i \in \partial A_q} \overline{W^s(\sigma_i)}$$

where $\partial A_p(x_s)$ is the practical stability boundary of x_s and σ_i is the decomposition point on the practical stability boundary of x_s. $\overline{W^s(\sigma_i)}$ represents the closure of stable manifold of decomposition point of σ_i.

Proposition 3.3 asserts that the practical stability boundary is contained in the union of the closure of the stable manifolds of all the decomposition points on the practical stability boundary. Thus, if all the decomposition points can be identified, then an explicit characterization of the practical stability boundary can be established via Proposition 3.3. Furthermore, we will use some dynamic information of the decomposition points to develop a systemic search method for obtaining multiple local optimal solutions. Before describing this procedure of locating multiple local optimal solutions, we need to justify that decomposition points can serve as the bridge linking two stable equilibrium points.

Theorem 3.4 (Decomposition point and stable equilibrium point [7, 8]) *Suppose that every equilibrium point of dynamical system (2.2) is hyperbolic and its stable and unstable manifold satisfies the transversality condition. If x_s^1 is a stable equilibrium point of dynamical system (2.2) and x_d is a decomposition point on its practical stability boundary, then there exists another stable equilibrium point x_s^2 to which the one dimensional unstable manifold of x_d converges.*

We next present analytical results asserting that all the quasi-gradient systems, a variation of the gradient system (2.2) of the following form, have similar equilibrium points of the same type.

$$\dot{x}(t) = -A(x)\nabla f(x).$$

The matrix $A(x)$ is positive definite. The gradient system (2.2) is a special system of the quasi-gradient system in which the coefficient matrix is an identity matrix. This nice property allows us to design a computationally efficient algorithm to find stable equilibrium points of system (2.2), thus finding local optimal solutions of system (2.1).

Theorem 3.5 (Type-persistence of equilibrium points) *Consider the gradient system (2.2); if $A_1(x) \neq A_2(x)$ are both positive definite for all $x \in U$, then*

1. *locations of the equilibrium points of $\dot{x}(t) = -A_1(x)\nabla f(x)$ are the same as the equilibrium points of $\dot{x}(t) = -A_2(x)\nabla f(x)$.*
2. *a type-k equilibrium point of $\dot{x}(t) = -A_1(x)\nabla f(x)$ is also a type-k equilibrium point of $\dot{x}(t) = -A_2(x)\nabla f(x)$.*

The above Theorem states that the static behaviors of any quasi-gradient system (2.2) are the same. However, their dynamic behaviors, such as stability regions, can be quite different. It can be shown that two quasi-gradient systems characterized by two positive definite matrices $A_1(x)$ and $A_2(x)$ have different stability regions, but their stability boundaries contain the same equilibrium points, provided that $A_1(x)$ and $A_2(x)$ are 'close' to each other.

The insights gained from the analytical characterizations derived in previous sections are useful to devise a mechanism for escaping from one stable equilibrium point to adjacent stable equilibrium points. The basic idea is roughly described as follows: the state space is the union of the closure of the stability regions of all stable equilibrium points. The two adjacent stability regions are separated by the intersection of their stability boundaries called a joint stability boundary. To identify an adjacent stable equilibrium point from a stable equilibrium point, it suffices to devise a mechanism to move across the joint stability boundary from one stability region to reach its adjacent stability region. We discuss two Trust-Tech based numerical methods for implementing such a mechanism in the next two sections.

4 Trust-Tech based DDP method

We present a Trust-Tech-based dynamic decomposition point method for locating another local optimal solution from a known local optimal solution.

Step 1: Construct a path moving away from the initial stable equilibrium point and moving toward its practical stability boundary.

Step 2: Identify the exit point at which the constructed path intersects with the practical stability boundary.

Step 3: Move along the practical stability boundary starting from the exit point and compute the decomposition point which separates the initial and corresponding stable equilibrium point whose stability boundary contains the joint stability boundary.

Step 4: Generate one point, starting from the decomposition point, which lies inside the stability region of the corresponding stable equilibrium point.

Step 5: Simulate the trajectory starting from the point generated at Step 4 which will converge to the corresponding stable equilibrium point, which is another local optimal solution.

We next describe numerical implementation of the above key steps. The task of identifying the exit point in Step 2 plays an important role in the overall Trust-Tech-based methodology. A numerical implementation is presented below.

Step 2: Identify the exit point. Starting from an initial stable equilibrium point x_s (i.e., a local optimal solution), we follow a search direction $\vec{d}$ and progressively sample a sequence of points $\{x_i\}$ along the search direction. By monitoring the value of the objective function $c(x)$ starting from the local optimal solution, it is clear that the value along the sequence will initially increase (because the sequence starts from a local optimal solution). The values along the sequence will continue to increase until it reaches a point at which the value of the objective function $c(x)$ decreases, provided the stability boundary exists. This point is termed as an approximated exit point x_{exit}. This approximated exit point is close to the exact exit point which is the intersection between the stability boundary and the curve connecting the sequence (see Figure 2).

Step 3: The stability-boundary following procedure to compute the decomposition point. This procedure seeks to construct a sequence of points along the stability boundary of x_s, starting from the exit point, leading to a point which is close to the corresponding dynamic decomposition point. The stability-boundary following procedure is described as follows.

We numerically integrate, starting from the initial condition x_{exit}, the gradient system (2.2) for a few time-steps. This will generate a point, say y_1 which is away from the stability boundary. From y_1, we move along the direction $\overline{y_1 x_s}$ and progressively sample a sequence of points with a small distance ε. This sampling process terminates when a change in the derivative of the objective function is detected. With a properly chosen ε, the point z_1, with a local maximal value of the objective function in the sequence of sampling points, is close to the stability boundary. We then repeat the same process, i.e., we numerically integrate, starting

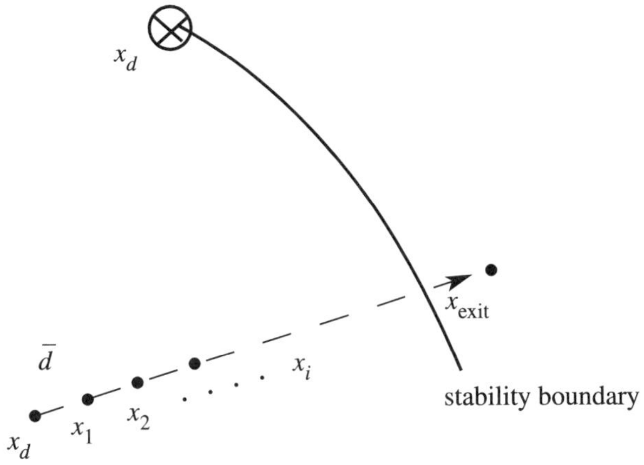

Figure 2 The computed exit point x_{exit} should be very close to the exact exit point

from the initial condition z_1, the gradient system (2.2) for a few time-steps. This will generate a point, say y_2, which is away from the stability boundary. From y_2, we move along the direction $\overline{y_2 x_s}$ and progressively sample a sequence of points with a small distance ε to find a point, say z_2, close to the stability boundary. This stability-boundary following procedure will generate a sequence of points $\{z_i\}$ which is close to the stability boundary. Since each point in the sequence $\{z_i\}$ is very close to the stability boundary, the behavior of the sequence $\{z_i\}$ will approximately follow the behavior of a trajectory located on the stability boundary.

Due to the structure of the stability boundary, where every trajectory on the stability boundary converges to a unique dynamical decomposition point, the point in the sequence $\{z_i\}$ that has the minimal norm value of the vector field is close to the decomposition point x_d. This point will be the output of the stability-boundary following procedure (see Figure 3). The purpose of this stability-boundary following procedure is to ensure that the search path for the decomposition point is confined to the stability boundary of x_s.

The point found by this procedure is termed the minimum gradient point (hereafter MGP). This point, among all the points constructed by the stability-boundary-following procedure, has the (first) minimum value of the norm of the vector field of the gradient system. From a geometrical viewpoint, the MGP is close to the desired decomposition point. By using the MGP as an initial guess and applying a robust nonlinear algebraic solver, one can obtain the dynamic decomposition point x_d.

Step 4: Locate a point inside the adjacent stability region. We locate a point inside the adjacent stability region by moving along the direction $\overline{x_s^1 x_d}$, outwards by 10% (i.e., $\overline{x_s^1 x_d} + 0.1 \cdot \overline{x_s^1 x_d}$). The end point will then locate inside the adjacent stability region. We use this point as an initial condition to numerically integrate the dynamical system (2.2). This numerical integration will generate a trajectory which converges to the stable equilibrium point x_s^2. Hence, another local optimal solution is located (see Figure 4).

Alternatively, one can monitor the ensuing trajectory until it reaches a point close to the targeted stable equilibrium point x_s^2 (i.e., the norm of the vector field

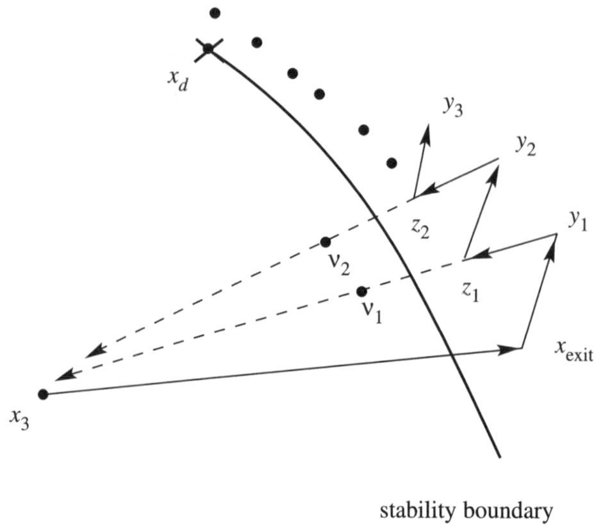

Figure 3 An illustration of the stability-boundary following procedure to compute the dynamic decomposition point

is small). One may then stop the numerical integration and resort to a local optimization solver. The ending point of the numerical integration serves as an initial condition for the solver to locate another local optimal solution x_s^2. The efficiency of searching another local optimal solution can be improved by using a hybrid of the proposed method and an efficient local optimizer.

The Trust-Tech-based DDP method has a nice feature of finding unique local optimal solutions. The theoretical basis is explained as follows. From Theorem 3.4, it follows that the unstable manifold of decomposition point x_d links the initial stable equilibrium point x_s^1 and the targeted stable equilibrium point x_s^2 via the unstable manifold. If one finds the same decomposition point from different exit points due to different selected search directions, it follows from Theorem 3.4 that the same stable equilibrium point (i.e., the same local optimal solution) will be found if one applies Step 4 and Step 5. Therefore, under this circumstance, these two steps are not necessary and the repetition of finding the same local optimal solution from two different search directions can be avoided.

5 Trust-Tech-based exit point method

The Trust-Tech-based DDP method can be time-consuming since the stability-boundary following procedure involves numerical integration, functional evaluation of the objective function (for closeness to the stability boundary) and a nonlinear algebraic solver (for locating the dynamic decomposition point). We, in this section, present a faster method, termed the Trust-Tech-based exit point method which does not compute the dynamic decomposition point. This method is faster in locating another local optimal solution compared to the Trust-Tech-based DDP method. But, this method, theoretically speaking, may suffer from the problem of possibly repeating a search of local optimal solutions found before.

The Trust-Tech-based exit point method is presented below:

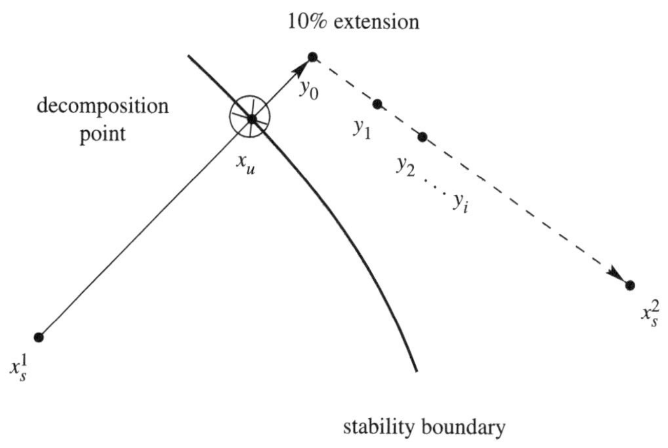

Figure 4 An illustration of the procedure of locating a point lying inside the stability region of the nearby stable equilibrium point (i.e., a local optimal solution)

Step 1: Construct a path moving away from the initial stable equilibrium point and moving toward its practical stability boundary.

Step 2: Identify the exit point at which the constructed path intersects with the practical stability boundary.

Step 3: Generate one point from the exit point which lies inside the stability region of the corresponding stable equilibrium point.

Step 4: Integrate the system trajectory starting from the point generated at Step 3. This trajectory will converge to the corresponding stable equilibrium point (which is another local optimal solution). Alternatively, apply a local optimization method to find the corresponding local optimal solution from a point in the trajectory.

The numerical implementation of Step 1 and Step 2 of the Trust-Tech-based exit point method is the same as that of the Trust-Tech-based decomposition point method. The numerical implementation of Step 3 is the same as Step 4 of the Trust-Tech-based decomposition point method (the minor difference is in the usage of the exit point instead of the decomposition point). The numerical implementation of Step 4 is the same as Step 5 of the Trust-Tech-based decomposition point method.

6 Numerical examples

We consider the following illustrative example

$$\min 2x_1 + 3x_2 + 1.5x_3 + 2x_4 - 0.5x_5$$

subject to the following constraints

$$\begin{aligned} x_1^2 + x_3 &= 1.25 \\ x_2^{1.5} + 1.5x_4 &= 3 \\ x_1 + x_3 &\leq 1.6 \\ 1.333x_2 + x_4 &\leq 3 \\ -x_3 - x_4 + x_5 &\leq 0 \\ x_1, x_2 &\geq 0. \end{aligned}$$

For an illustrative purpose, we show the construction of a unconstrained optimization problem using the penalty function method. We define the following objective function

$$c(x) = 2x_1 + 3x_2 + 1.5x_3 + 2x_4 - 0.5x_5.$$

The corresponding unconstrained optimization problem becomes:

$$\min f(x)$$

where

$$f(x) = c(x) + \sum_{i=1}^{6} \alpha_i g_i(x)$$

which is a combination of the objective function and the penalty functions $g_i(x)$ associated with the constraints $x \in R^5$, and α_i is the associated penalty coefficients. The penalty functions are built in the traditional way and the details are hence neglected. We build the corresponding gradient dynamical system as follows:

$$\dot{x} = \frac{\partial}{\partial x} f(x).$$

We next present the numerical results after applying the Trust-Tech-based DDP method to the test system. There are 16 stable equilibrium points (or equivalently, local optimal solutions) found in the numerical simulation. In addition, 15 dynamic decomposition points are found in our numerical study.

Table 1 The computed multiple local optimal solutions

x_s^1	x_s^2	x_s^3	x_s^4	x_s^5	x_s^6	x_s^7	x_s^8
1.1125	0.4680	1.1125	1.0922	−0.8620	0.4680	0.4680	−0.0542
2.0708	2.0709	1.2953	2.0397	2.0709	1.2953	2.0709	2.0709
−0.0004	1.0005	−0.0004	0.0441	0.0235	1.0005	1.0005	1.0109
0.0001	0.0001	1.0005	0.0446	0.0001	1.0005	0.0001	0.0001
0.0003	0.0004	0.0004	0.9559	0.0004	0.0004	1.0004	0.0004

x_s^9	x_s^{10}	x_s^{11}	x_s^{12}	x_s^{13}	x_s^{14}	x_s^{15}	x_s^{16}
1.1125	−0.8620	−0.8360	0.4680	−0.0542	−0.0542	−0.8620	−0.0542
1.2953	1.2953	2.0405	1.2953	1.2953	2.0709	1.2953	1.2953
−0.0004	0.0235	0.0683	1.0109	1.0109	1.0109	0.0235	1.0109
1.0005	1.0005	0.0434	1.0005	1.0005	0.0001	1.0005	1.0005
1.0003	0.0004	0.9570	0.0004	0.0004	1.0004	1.0004	1.0004

Table 2 The computed dynamic decomposition points

x_d^1	x_d^2	x_d^3	x_d^4	x_d^5	x_d^6	x_d^7	x_d^8
0.8562	1.1125	1.1015	0.0120	0.4680	0.4680	0.0609	1.1125
2.0709	1.7061	2.0540	2.0708	1.7061	2.0709	2.0709	1.2953
0.5002	−0.0004	0.0238	0.0606	1.0005	1.0005	1.0108	−0.0004
0.0001	0.4998	0.0243	0.0001	0.4998	0.0001	0.0001	1.0005
0.0004	0.0004	0.5238	0.0004	0.0004	0.4996	0.0004	0.4996

x_d^9	x_d^{10}	x_d^{11}	x_d^{12}	x_d^{13}	x_d^{14}	x_d^{15}
0.0120	0.0125	0.4680	0.0610	0.0120	0.0120	0.0610
1.2953	2.0418	1.2953	1.2953	1.2953	1.2953	1.2953
0.0606	0.0156	1.0005	1.0108	0.0606	0.0606	1.0108
1.0005	0.0417	1.0005	1.0005	1.0005	1.0005	1.0005
0.0004	0.9588	0.4996	0.0004	1.0004	1.0004	1.0004

The procedure of Trust-Tech-based DDP method for solving this test system can be illustrated in Table 1 in a tier-by-tier manner. It can be seen that four tiers of local optimal solutions have been found with the following:

tier-zero (starting local optimal solution): x_s^1
first-tier local optimal solutions: $x_s^2, x_s^3, x_s^4, x_s^5$
second-tier local optimal solutions: $x_s^6, x_s^7, x_s^8, x_s^9, x_s^{10}, x_s^{11}$
third-tier local optimal solutions: $x_s^{12}, x_s^{13}, x_s^{14}, x_s^{15}$
fourth-tier local optimal solutions: x_s^{16}

7 Some applications

We apply the Trust-Tech-based methods to several applications which are related to unconstrained optimization problems [23]–[25]. These problems arise in supervised machine learning using artificial neural networks which have numerous applications in various domains of science and engineering.

The task of training neural networks plays a vital role in deciding the network architecture and the performance of neural networks. Most popular training algorithms tend to be greedy and hence get stuck at a near-by local minimum of the error surface. To overcome this problem, some global methods (like multiple restarts, genetic algorithms, simulated annealing etc.) for efficient training make use of stochastic approaches in combination with local methods to obtain an effective set of training parameters.

In this section, we apply the Trust-Tech-based method to compute multiple tiers of local minimums for the error surface of the neural network training problem. The method computes multiple local optimal solutions surrounding the local optimal solution obtained by a global-local type training method in a deterministic and systematic manner. Our numerical studies on different machine learning datasets indicate that the Trust-Tech-based method outperforms several training methods available in the literature.

Numerical studies. Seven benchmark datasets taken from the UCI machine learning repository are available in the literature. Since the main focus of this

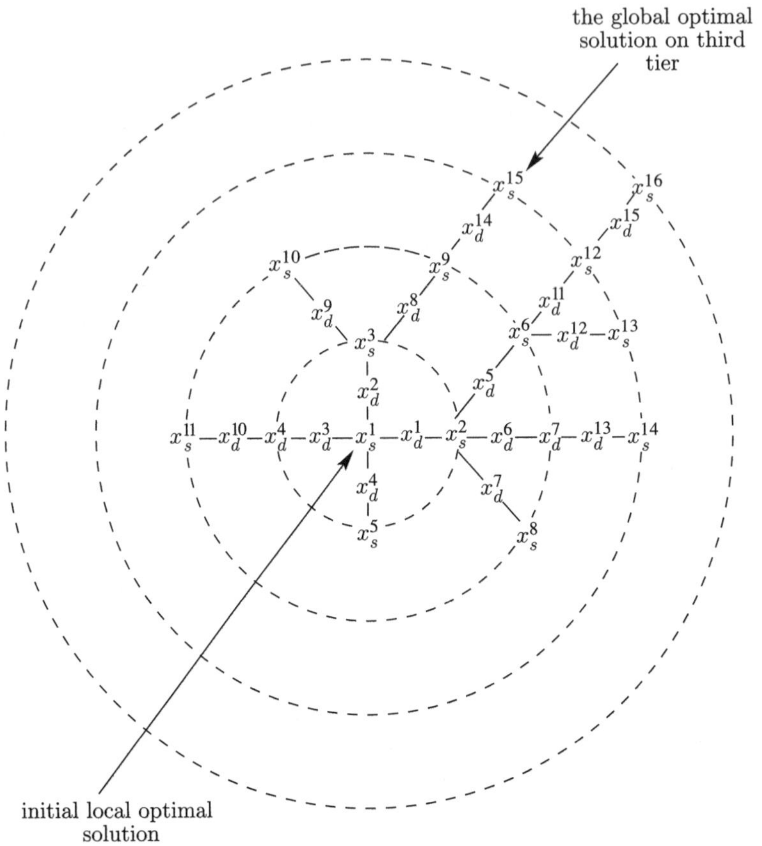

Figure 5 The numerical search procedure of the Trust-Tech-based method for finding multiple local optimal solutions

evaluation is the performance of Trust-Tech-based method based training algorithm, only simple experiments were conducted for choosing the optimal architecture of the neural network. The nodes in the hidden layer are added incrementally and the training error of the resulting neural network is computed. The final architecture is chosen when there is no significant improvement in the training error after a hidden node is added. Table 3 summarizes the datasets. It gives the number of samples, input features, and output classes along with the number of hidden nodes of the network architecture. These datasets have varying degrees of complexity in terms of sample size, output classes and the class overlaps. Here is the description of the datasets:

1. *Cancer:* This dataset contains data from cancer patients. It has 683 samples out of which 444 are benign cases and 239 are malignant cases. 9 attributes describing the tumor were used for classification.
2. *Diabetes:* This dataset gives information about patients who have some signs of diabetes according to World Health Organization criteria. Each sample has 8 real valued attributes. A total of 768 samples with 500 negative cases and 268 positive cases are available.

Table 3 Summary of benchmark datasets

Dataset (D)	Sample size (Q)	Input features (n)	Output classes (p)	Hidden nodes (H)	Search variables (n+2)k+1
Cancer	683	9	2	5	56
Diabetes	178	8	3	4	61
Image	2310	19	7	8	169
Ionosphere	351	34	2	9	325
Iris	150	4	3	3	19
Sonar	208	60	2	8	497
Wine	178	13	3	4	61

3. *Image:* This dataset contains images which were drawn randomly from a database of 7 outdoor images. The images were hand-segmented to create a classification for every pixel. There are 19 attributes that describe each instance (which is a 3×3 region) of a given image. The dataset contains a total of 2310 samples.
4. *Ionosphere:* This radar data was collected by a system consisting of a phased array of 16 high-frequency antennas with a total transmitted power on the order of 6.4 kilowatts. The targets were free electrons in the ionosphere. The dataset consists of 351 samples with 34 attributes. The classification task here is to separate good radar signals from bad ones.
5. *Iris:* This dataset contains 3 classes of 50 samples each, where each class refers to a type of iris plant. It is a relatively simple dataset where one class is linearly separable from the other two, but the other two have significant overlap and are not linearly separable from each other. The four attributes considered for classification are sepal length, sepal width, petal length and petal width. All attributes are measured in centimeters.
6. *Sonar:* This dataset is used for the classification of sonar signals. The task is to discriminate sonar signals bounced off a metal cylinder from those bounced off a roughly cylindrical rock. The dataset contains a total of 208 samples (111 for mines and 97 for rocks) which contain signals that were obtained from a variety of different aspect angles, spanning 90 degrees for the cylinder and 180 degrees for the rock. Each pattern is a set of 60 numbers in the range 0.0 to 1.0 that represents the energy within a particular frequency band, integrated over a certain period of time.
7. *Wine:* This dataset was obtained from the results of a chemical analysis of wines derived from three different cultivators. The analysis determined the quantities of 13 constituents found in each of the three types of wines (superior, normal, inferior): a total of 178 samples with the following distribution (59,71,48).

The popular back-propagation training algorithm is used in combination with the multiple random start (MRS) method to obtain multiple training solutions. Of these multiple local optimal solutions, the best was selected and then served as the initial solution for the Trust-Tech-based training method. The tier-one local optimal solutions of the initial solution are then computed by the Trust-Tech-based training method. The top five local optimal solutions are selected and the corresponding

Table 4 Training errors and testing errors of the combined MRS+BPmethod and of the Trust-Tech-BP training method are listed. The improvement in both training error and testing error is expressed in percentages

Dataset	Training error			Testing error		
	MRS+ BP	TRUST-TECH	Improvement %	MRS+ BP	TRUST TECH	Improvement %
Cancer	2.21	1.74	27.01	3.95	2.63	50.19
Diabetes	22.04	20.69	6.52	23.83	20.58	15.79
Image	9.37	8.04	16.54	11.08	9.74	13.76
Ionosphere	2.35	0.57	312.28	10.25	7.96	28.77
Iris	1.25	1.00	25.00	3.33	2.67	24.72
Sonar	1.56	0.72	116.67	19.17	12.98	47.69
Wine	4.56	3.58	27.37	14.94	6.73	121.99

tier-two local optimal solutions are then computed. Of these multiple solutions, the best one is selected and compared with the best obtained by the combined MRS method.

Error calculation. To demonstrate the generalization capability (and hence the robustness) of the Trust-Tech-based training method, ten-fold cross validation is performed on each dataset. This practice of cross validation effectively removes any bias in the dataset segmentation. The use of the validation dataset allows early stopping of the local method and prevents over-fitting to a particular dataset. Essentially, each dataset is partitioned into ten folds of approximately equal size. Let these folds be denoted by $T_1, T_2, \ldots, T_{10}$. Each time, the validation set will be T_i in which the target labels will be deleted. The test set is T_j for $j = (i+1)$ mod 10. The final mean square error (MSE) is the average of all the errors obtained across each of the ten folds. Usually, the training error is much lower than the test error because the network is modeled using the training data and this data will be more accurately classified compared to the unseen test data. All the network parameters including the architecture and the set of weights are obtained using the training data. Once the final model is fixed, the accuracy on the test data will provide an estimate for the generalization capability of the network model and the training algorithm.

Classification accuracy. The criteria of evaluation is given by the classification accuracy of the network mode:

$$\% \text{ accuracy} = \frac{\mathit{diff}\,(t(i), y(W, X))}{Q} \times 100$$

where *diff* gives the number of misclassified samples. Table 4 shows the improvements in the training error and the testing error using the Trust-Tech-based method. For effective implementation, only the best five tier-1 and corresponding tier-2 solutions were obtained using the Trust-Tech-based method. For some of the datasets, there had been considerable improvement in the classifier performance.

8 Trust-Tech methodology for constrained problems

Trust-Tech-based methodology for solving general constrained optimization problem has been developed [8, 16, 17]. We present in this section an overview of the methodology for solving the following constrained nonlinear optimization problem of the form:

$$\begin{array}{ll} & \textbf{Minimize} f(x) \\ \textbf{Subject to} & \begin{array}{ll} h_i(x) = 0, & i \in I = \{1, \dots, \ell\} \\ g_j(x) \le 0, & j \in J = \{1, \dots, s\}. \end{array} \end{array} \tag{8.1}$$

The constraint set of (8.1) defines the following

$$S = \{x \in R^n : h_i(x) = 0, \quad i \in I, g_j(x) \le 0, j \in J\}$$

which can be any closed subset of R^n and its structure can be very complex. The set S is usually non-convex and disconnected (i.e. composed of several (disjoint) connected feasible regions). The task of locating each connected feasible region of the set S is itself a difficult one.

Without loss of generality, we consider the following optimization problem with equality constraints:

$$\begin{array}{ll} & \textbf{Minimize} f(x) \\ \textbf{Subject to} & h_i(x) = 0, \quad i \in I = \{1, \dots, l\}. \end{array} \tag{8.2}$$

We define the following equality vector $H(x) = (h_1(x), \dots, h_m(x))^T = 0$. It can be shown that the constraint set (or the feasible set or region)

$$M = \{x \in R^n : H(x) := (h_1(x), \dots, h_m(x))^T = 0\} \tag{8.3}$$

is a smooth manifold. In general, the constraint set M can be very complicated with several disjoint path-connected feasible components. We believe that the only reliable way to find the global optimal solution of the constrained optimization problem is to first find all the local optimal solutions located in each feasible component and then find the global optimal solution. To this end, we develop the Trust-Tech methodology for systematically computing all the local optimal solutions of (8.1) in the following two phases.

Phase I: find all the feasible components by exploring the trajectories of one designed dynamical system.

Phase II: find all the local optimal solutions in each feasible component by exploring the trajectories of another designed dynamical system.

Phase I. In order to visit each path-connected feasible component of the constraint set M, Phase I is consisted of two steps:

Step 1.1: approach a path-connected feasible component of the optimization problem (8.2).

Step 1.2: escape from the path-connected feasible component and approach another path-connected feasible component of the optimization problem (8.2).

We propose to design a nonlinear dynamical system whose trajectories can be used to perform Step 1.1 and Step 1.2. The central idea of designing the nonlinear dynamical system is that every path-connected feasible component corresponds to a stable equilibrium manifold of the nonlinear dynamical system. In this way, the task of finding all the path-connected feasible components of the optimization problem (8.2) can be accomplished by finding all the stable equilibrium manifolds of the nonlinear dynamical system. One way to achieve this goal is to build the following quotient gradient system associated the constraint set characterized by $H(x)$

$$\dot{x} = -DH(x)^T H(x). \tag{8.4}$$

where $DH(x)$ represents the Jacobian matrix of the vector $H(x)$.

It can be shown that each feasible component of the optimization problem (8.2) can be located via each stable equilibrium manifold of the quotient gradient system (8.4). This analytical result provides a basis for the Trust-Tech methodology in searching for feasible components. To numerically implement Phase I of the methodology, we propose the following two tasks to be numerically implemented:

Step 1.1: approach a stable equilibrium manifold of the quotient gradient system (8.4).

Step 1.2: escape from the stable equilibrium manifold and approach another stable equilibrium manifold of the quotient gradient system (8.4).

Phase II. Phase II finds all the local optimal solutions in each feasible component based on the trajectories of another nonlinear dynamical system. In order to visit all the local optimal solutions located in each feasible component, Phase II is conceptually composed of two steps:

Step 2.1: starting from a point in the feasible component found in Phase I, approach a local optimal solution located in the feasible component of the optimization problem (8.2).

Step 2.2: escape from the local optimal solution and approach another local optimal solution of the same feasible component of the optimization problem (8.2).

We propose to design a nonlinear dynamical system whose trajectories can be used to perform Step 2.1 and Step 2.2. The central idea of designing the nonlinear dynamical system is that all the local optimal solutions of the optimization problem (8.2) correspond to all the stable equilibrium points of the nonlinear dynamical system; in particular every local optimal solution of the optimization problem (8.2) corresponds to a stable equilibrium point of the nonlinear dynamical system. To this end, we design the following projected gradient dynamical system

$$\dot{x} = -P_H(x(t))\nabla f(x(t)) \tag{8.5}$$

where $x(0) = x_0 \in M$ and the projection matrix, shown below

$$P_H(x) = (I - DH(x)^T(DH(x)DH(x)^T)DH(x)) \in R^{n\times n} \tag{8.6}$$

is a positive semi-definite matrix for every $x \in M$. Note that $P_H(x)\nabla f(x)$ is the orthogonal projection of $\nabla f(x)$ to the tangent space T_xM which means $P_H(x)\nabla f(x) \in T_xM$ for all $x \in M$. Note that every trajectory of (8.5) starting from $x_0 \in M_k$ stays in M_k. Therefore, M_k is an invariant set of (8.5).

One important issue when searching for the global optimal solution is how to effectively move (escape) from a local optimal solution and approach another local optimal solution. In the past, significant efforts have been attempted to address this issue, but without much success. With the goal of developing an effective method to overcome this issue, we have developed a computational method and incorporated it into Phase II of the Trust-Tech-based methodology. The basis of the method is

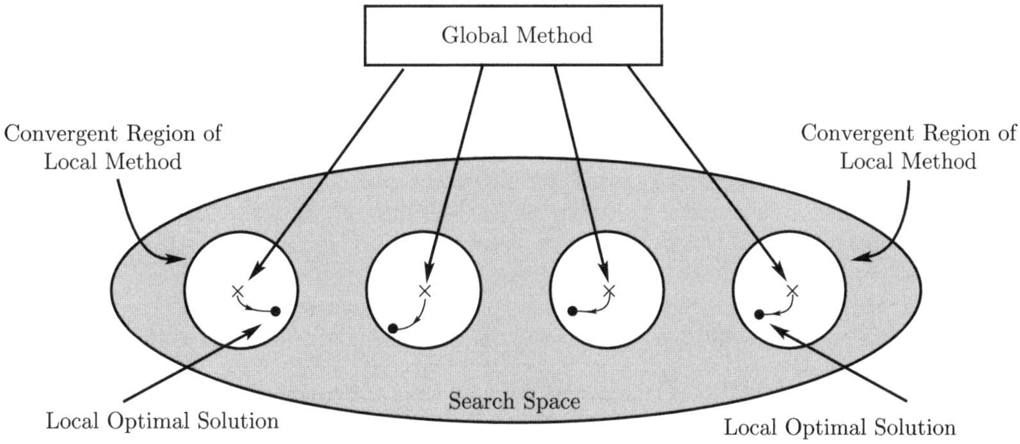

Figure 6 A global-type algorithm identifies several promising solutions as initial points which are sent to a local-type solution algorithm for fine-tuning

descried as follows. The problem of how to escape from a local optimal solution of the optimization problem (8.2) and approach another local optimal solution (in the same path-connected feasible component) is transformed into the problem of how to escape from the stability region of the corresponding stable equilibrium point (s.e.p.) of (8.5) and enter into the stability region of another corresponding s.e.p. of (8.5). Note that once a system trajectory lies inside the stability region of a s.e.p., the ensuing system trajectory will converge to the s.e.p. which is a local optimal solution of (8.2).

9 The hybrid trust-tech methodology

We next discuss a global/local type algorithm that can provide a trade-off between computational time and solution qualities. This type of architecture for algorithmic development is commonly adopted in the development of meta-heuristic algorithms [21]. However, the success of this type of architecture depends on its ability to effectively explore the entire search space. We use Figure 6 and Figure 7 to illustrate this point: a global-type algorithm identifies several promising solutions as initial points which are sent to a local-type solution algorithm for fine-tuning. Starting from these initial points, a local-type method computes a local optimal solution. The local type algorithm computes local optimal solutions which are only optimal in a local restrictive region. The problem is that the promising regions identified by global-type algorithms are not well-explored.

We believe that the strength of a global method is to identify promising regions in the entire search space where the identified promising regions contain the global optimal solution. While a global method does not have a fine-tuning capability, a local method only finds a local optimal solution. What is missing in the meta-heuristic architecture is the ability to locate all the local optimal solutions in the identified promising regions. It is likely that high-quality optimal solutions or even the global optimal solution lies in tier-one or a tier-two manner away from a local optimal solution found by a local-type algorithm (see Figure 8).

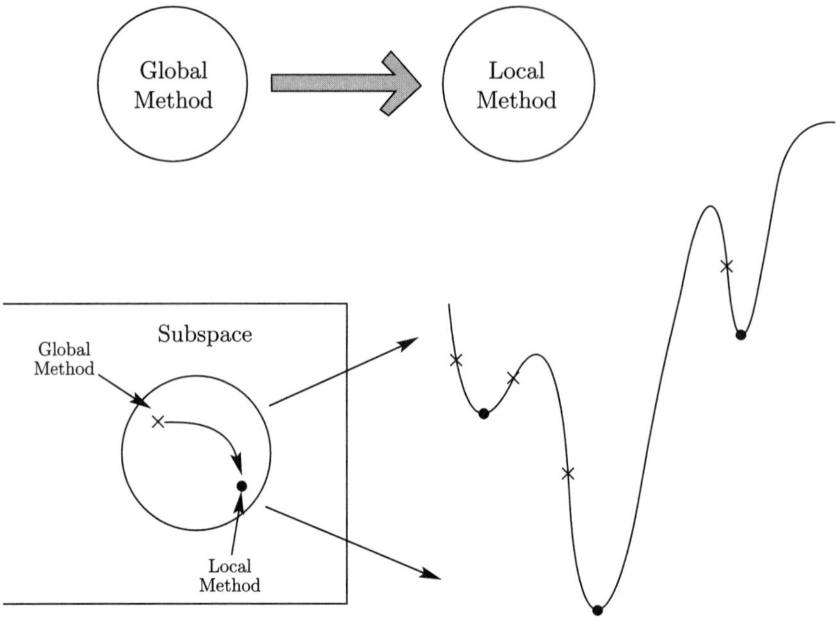

Figure 7 Depending on the initial points provided by a global method, the local method usually finds a nearby local optimal solution. Due to the local nature of a local method, the global optimal solution as indicated in Figure 6 may not be found by a local method

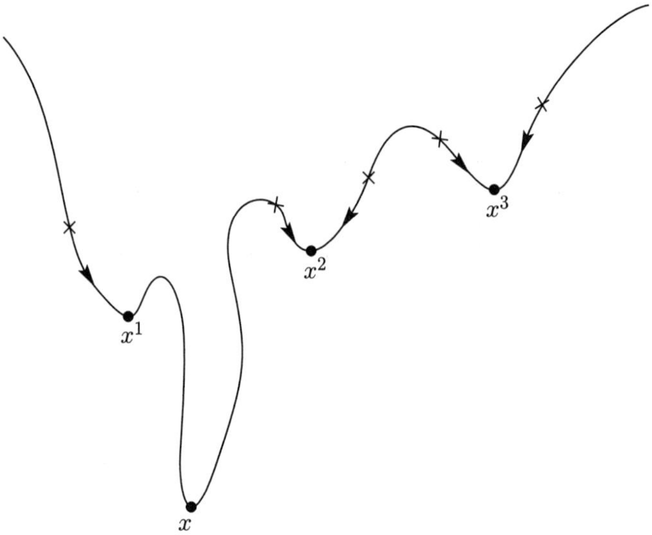

Figure 8 One problem with the global/local type algorithm is that the combined algorithm may not find high-quality local optimal solutions. For instance, the initial points provided by the global algorithm converge to local optimal solutions, by local algorithms; but the global optimal solution is missed

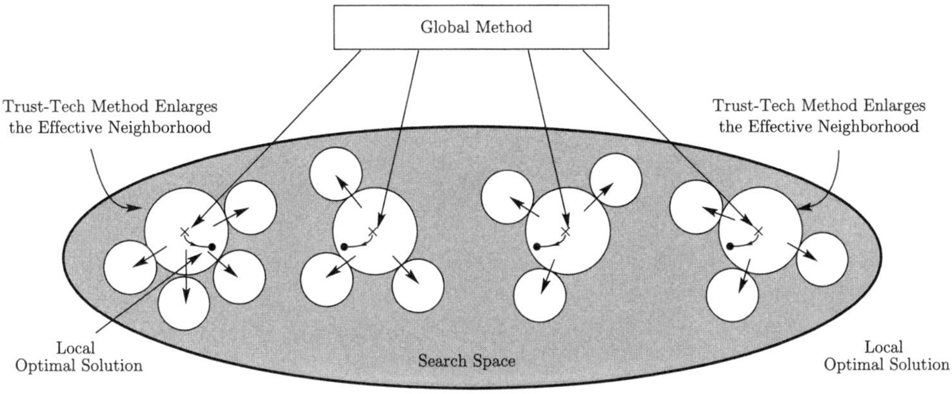

Figure 9 Trust-Tech-based methodology identifies and locates multiple local optimal solutions via enlarging the "effective neighborhood" of local methods

The Trust-Tech-based methodology is very effective in filling this gap. It is a systematic, deterministic methodology to effectively explore the surrounding neighborhood of a local optimal solution to find all the local optimal solutions in the neighborhood. This capability can be viewed as identifying and enlarging the "effective neighborhood" in which local type methods can find local optimal solutions. Figure 9 illustrates that the Trust-Tech-based methodology identifies and locates multiple local optimal solutions by enlarging the "effective neighborhood". The process of enlarging this effective neighborhood can be continued in the following ways

- for a pre-determined number of local optimal solutions, continue in a particular direction of the search space starting from the initial local optimal solution (single-direction search)
- for a pre-determined number of local optimal solutions, continue in multiple directions of the search space from the initial local optimal solution (multiple-direction search).
- for a pre-determined number of local optimal solutions, continue in 'all directions' of the search space from the initial local optimal solution (tier-by-tier search):

10 Concluding remarks

We have developed a two-phase Trust-Tech-based methodology to systematically compute all the local optimal solutions of constrained nonlinear optimization problems. The methodology consists of two distinct phases: Phase I, starting from an arbitrary starting point, systematically finds all the feasible regions which satisfy nonlinear equality and inequality constraints. The starting point of Phase I may be an infeasible point. Phase II then computes all the local optimal solutions in each feasible component found in phase I. To numerically implement Phase I, we have designed a non-hyperbolic gradient system whose trajectories can locate all the connected feasible components. By following the system trajectory from an arbitrary initial point, a feasible point lying in a feasible region can be obtained.

To account for the fact that the feasibility set is in general composed of several connected feasible components, Trust-Tech methodology designs in a mechanism which allows one move away from one feasible component and approachs another feasible component by following the system trajectories of the non-hyperbolic gradient system in reverse time. Once another feasible component is found, Phase II employs the trajectories of a projected gradient system to visit all the local optimal solutions in the feasible component. The complexity of the two-phase methodology depends on the number of local optimal solutions rather than on the system dimension, which equals the number of search variables.

From a computational viewpoint, it may well be due to the lack of computational methods for computing the complete set of decomposition points that the Trust-Tech-based methodology can only compute multiple local optimal solutions (in each feasible component) for general constrained optimization problems. In addition, due to the lack of computational methods for computing the complete set of unstable equilibrium manifolds, the Trust-Tech-based methodology can only compute multiple feasible components, instead of all of the feasible components for general constrained optimization problems.

Trust-Tech-based methodology is valuable and can be of significant importance in the following sense: since a sophisticated local-type method can only provide local optimal solutions in a restricted local region while a global-type algorithm can only provide sparse approximated solutions within reasonable computational efforts, the Trust-Tech-based methodology represents an enabling technology for:

- enhancing the functionality of local-type methods to effectively find optimal solutions in a broader region;
- enhancing the functionality of global-type methods to effectively find near-global optimal solutions;
- providing more flexibility in the integrating local-type methods and global-type algorithms to effectively find global optimal solutions.

References

[1] M.S. Bazaraa, Shetty, Nonlinear Programming: Theory and Algorithms, John Wiley and Sons, New York, 1990.

[2] A.A. Brown and M.C. Bartholomew-Biggs, "Some effective methods for unconstrained optimization based on the solution of systems of ordinary differential equations", Journal of Optimization Theory and Applications, **62**, pp. 211-224, 1989.

[3] H.D. Chiang, M.W. Hirsch, and F.F. Wu, "Stability region of nonlinear autonomous dynamical systems", IEEE Trans. Automatic Contr., **33**, pp. 16-27, 1988.

[4] H.D. Chiang, James S. Thorp, "Stability regions of nonlinear dynamical systems: A constructive methodology", IEEE Trans. Automatic Contr., **34**, pp. 1229-1241, 1989.

[5] H.D. Chiang, Chia-Chi Chu and G. Cauley, "Direct stability analysis of electric power systems using energy functions: Theory, applications, and perspective", Proceedings of the IEEE, **83**, No. 11, pp. 1497-1529, 1995.

[6] H.D. Chiang, Chia-Chi Chu, "A systematic search method for obtaining multiple local optimal solutions of nonlinear programming problems", IEEE Trans. Circuits and Systems, **43**, pp. 99-109, 1996.

[7] H.D. Chiang and L. Fekih-Ahmed, "Quasi-stability regions of nonlinear dynamical systems: Theory", IEEE Trans. Circuits and Systems, **43**, No.8, pp. 627-635, 1996.

[8] Hsiao-Dong Chiang and Jaewook Lee, "A Trust-Tech-based Hybrid Method for Computing High-quality Optimal Solutions: Method and Theory", (A Chapter of the IEEE Press Book) Modern Heuristic Optimization Techniques: Theory and application to Power Systems, IEEE Press Book, 2007.

[9] P. Fletcher, Practical Methods of Optimization, John Wiley and Sons, New York, 1987.

[10] C.A. Floudas and P.M. Pardalos, Recent advances in global optimization, Princeton Series in Computer Science, NJ, 1992.

[11] J. Guckenheimer and P. Homes, Nonlinear oscillations, dynamical systems, and bifurcations of vector fields, Applied Mathematical Sciences, **42**, Springer: New York, 1986.

[12] J. Guddat, F. Guerra Vazquez, and H.Th. Jongen, Parametric optimization: Singularities, path following and jumps, John Wiley and Sons, New York, 1990.

[13] R. Horst and H. Tuy, Global Optimization: Deterministic Approaches, Kluwer Academic Publishers, 1994.

[14] R. Horst and P.M. Pardalos, Handbook of Global Optimization, Kluwer Academic Publishers, 1994.

[15] R. Horst, P.M. Pardalos and N.V. Thoai, Introduction to Global Optimization, Kluwer Academic Publishers, 1995.

[16] Jaewook Lee, Path-following methods for global optimization, Ph.D. dissertation, Cornell Univ., NY, 1999.

[17] Jaewook Lee and Hsiao-Dong Chiang, "A dynamical trajectory-based methodology for systematically computing multiple optimal solutions of general nonlinear programming problems", IEEE Transactions on Automatic Control (Regular paper), **49**, Issue 6, pp. 888-899, June 2004

[18] D.A. Liotard, Algorithmic tools in the study of semi empirical potential surfaces, Int. J. Quantum Chem., **44**, pp.723-741, 1992.

[19] N. Mousseau, and G.T. Barkerma, Traveling through potential energy landscapes of disordered materials: The activation-relaxation technique, Physical Review E, **57**, 2, pp.2419-2424, 1998.

[20] P.M. Pardalos and J.B. Rosen, Constrained Global Optimization: Algorithms and Applications, Lecture Notes in Computer Science: Vol.268, Springer-Verlag, New York, 1987.

[21] Jean-Paul Penot, New Methods in Optimization and Their Industrial Uses, Birkhauser Verlag, Basel, 1989.

[22] Chandan K. Reddy, Hsiao-Dong Chiang and Bala Rajaratnam, "TRUST-TECH based expectation maximization for learning finite mixture models", IEEE Transactions on Pattern Analysis and Machine Intelligence, **30**, No. 7, July 2008, pp. 1146-1157.

[23] Chandan K. Reddy, Yao-Chung Weng and Hsiao-Dong Chiang, "Refining Motifs by improving Information Content Scores using Neighborhood Profile Search", (Invited paper) BMC Algorithms for Molecular Biology, Vol. 1, No. 23, November 2006

[24] Chandan K. Reddy and Hsiao-Dong Chiang, "A stability boundary based method for finding saddle points on potential energy surfaces", Journal of Computational Biology, **13**, No.3, pp. 745-766, April 2006

[25] H. Shashikala, N.K. Sancheti, and S.S. Keerthi, A new approach to global optimization using ideas from nonlinear stability theory, Int. Symp. Circuits Syst. (ISCAS'92), pp. 2817-2820, 1992.

[26] J.E. Sinclair and R. Fletcher, A new method of saddle-point location for the calculation of defect migration energies, J. Phy. Solid State Phys, **7**, pp. 864-870, 1974.

[27] Torn and A. Zilinskas, Global Optimization, Lecture Notes in Computer Science: Vol. 350, Springer-Verlag, New York, 1987.

[28] M.W. Hirsch, Differential Topology, Springer-Verlag: New York, Heidelberg, Berlin, 1976.

[29] S. Wiggins, Introduction to Applied Nonlinear Dynamical Systems and Chaos, Springer-Verlag: New York, 1990.

[30] D.A. Liotard, Algorithmic tools in the study of semiempirical potential surfaces, Int. J. Quantum Chem., **44**, pp. 723-741, 1992.

[31] N. Mousseau, and G.T. Barkerma, Traveling through potential energy landscapes of disordered materials: The activation-relaxation technique, Physical Review E, **57**, 2, pp. 2419-2424, 1998.

[32] J.E. Sinclair and R. Fletcher, A new method of saddle-point location for the calculation of defect migration energies, J. Phy. Solid State Phys, **7**, pp. 864-870, 1974.

Fields Institute Communications
Volume **55**, 2009

Global Optimization Issues in Parametric Programming and Control

Vivek Dua
Centre for Process Systems Engineering
Department of Chemical Engineering
University College London, London WC1E 7JE
v.dua@ucl.ac.uk

Kostas Kouramas
Centre for Process Systems Engineering
Department of Chemical Engineering
Imperial College London, London SW7 2AZ
k.kouramas@imperial.ac.uk

Stratos Pistikopoulos
Centre for Process Systems Engineering
Department of Chemical Engineering
Imperial College London, London SW7 2AZ
e.pistikopoulos@imperial.ac.uk

Abstract. A prime difficulty in applying optimization-based techniques for improved design, control and operation of various types of engineering systems arises from the unavoidable presence of variations in the problem parameters such as fluctuations in uncertain inputs and measurements, or variations in inherent system properties and characteristics. These variations readily translate to deviations from the prescribed optimal point, thus, either failing to exploit fully the benefits of the optimization based solution or requiring the repetitive solution of the problem for different values of the problem parameters. Parametric programming is a technique that determines computationally inexpensively the exact mapping of the optimal solution profile in the space of the system parameters, to overcome these limitations. In this chapter we present an overview of the work of our group on mathematical foundations of multi-parametric (mixed integer) linear, quadratic, non-linear and dynamic optimization problems that are commonly encountered in (i) optimization under uncertainty (ii) multi-level and multi-objective optimization and (iii) model-based on-line control and optimization.

2000 *Mathematics Subject Classification.* Primary 90C31, 90C29, 93C55; Secondary 90C05, 90C11, 90C20, 90C25, 90C26, 90C30.

Key words and phrases. Parametric Linear Programming; Parametric Quadratic Programming; Parametric Nonlinear Programming; Parametric Mixed-Integer Programming; Parametric Global Optimization; Model Based Control; Hybrid Control; Multi-Level Programming.

1 Introduction

Uncertainty and variability, typically characterised by varying parameters, are inherent characteristics of any process system - it is not at all surprising then that process models, the means for translating process related phenomena to some descriptive form (quantitative or qualitative) also involve elements of uncertainty. These varying parameters can be, for example, attributed to fluctuations in resources, technical characteristics, market requirements and prices, which can affect the feasibility and economics of a project. While the representation of the uncertainty is itself an important modelling question, the potential effect of variability on process decisions regarding process design and operations constitutes another challenging problem. Obviously the two problems are closely related: if an optimal decision is totally insensitive to the presence of uncertainty, acquiring a model for the description of the uncertainty is not really necessary. In this context, devising suitable mathematical techniques and algorithms through the application of which one could analyze and quantify if, how, what type of, and by how much, uncertainty affects decisions, becomes a major research goal.

A key approach to analyse the effect of variations and uncertainty in process systems engineering problems is parametric programming. In an optimization framework, where the objective is to minimize or maximize a performance criterion subject to a given set of constraints and where some of the parameters vary between lower and upper bounds, parametric programming is a technique for obtaining: (i) the objective function and the optimization parameters as a function of these parameters and (ii) the regions in the space of the parameters where these functions are valid. Another important area of application of parametric programming is in multi-objective optimization where multi-criteria problems can be reformulated as parametric programming problems and different (usually conflicting) optimal solutions, i.e. Pareto sets, can be obtained as parametric solutions.

For different process system problems, one obtains mathematical programs of different complexities. For example: (i) in process synthesis, design, planning and scheduling problems, one usually encounters algebraic equations which involve continuous and integer variables, (ii) model predictive control problems are represented by differential and algebraic equations which involve continuous variables and (iii) hybrid control problems, which simultaneously include dynamics and logical constraints, have differential and algebraic equations which involve continuous and integer variables. Theory, algorithms and computational tools for such mathematical programs have been developed and applied for the solution of many process systems engineering problems. In this paper an overview of recent developments in the area of parametric programming and control are presented; for a more detailed exposure to theory, algorithms and applications the reader may refer to the recently published books Pistikopoulos et al. (2007b,a). The rest of the paper is organised as follows. Theory and algorithms for multiparametric programs involving continuous optimization variables are presented in section 2 and solution techniques for multiparametric programs involving mixed integer and continuous variables are presented in section 3. Section 4 provides an overview of model based control via parametric programming and section 5 focusses on using fundamentals of parametric programming for solving bilevel programs to global optimality. Finally concluding remarks are presented in section 6.

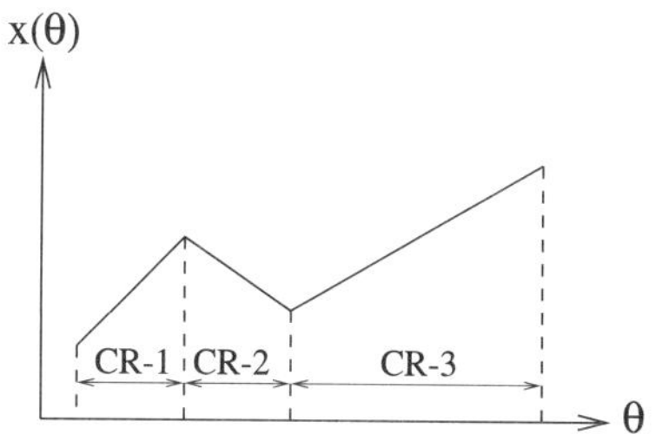

Figure 1 Parametric optimization

2 Multiparametric continuous programs

Consider a general parametric nonlinear programming problem of the following form:

$$\begin{aligned} & z(\theta) = \min_{x} \ f(x,\theta) \\ \text{s.t.} \quad & g_i(x,\theta) \leq \ 0, \quad \forall \ i = 1,...,p \\ & h_j(x,\theta) = \ 0, \quad \forall \ j = 1,...,q \\ & x \in X \subseteq \Re^n \\ & \theta \in \Theta \subseteq \Re^m \end{aligned} \tag{2.1}$$

where f, g and h are twice continuously differentiable in x and θ. Note that f is the performance criterion to be minimized, $g \leq 0$ and $h = 0$ are the constraints, x is the vector of optimization variables and θ is the vector of parameters. The objective is to obtain the optimal $x(\theta)$, which when substituted into $f(x,\theta)$ provides the optimal objective function value, $z(\theta)$, as a function of θ. See Figure 1, where $x(\theta)$ is plotted as a function of θ - the region where a particular functional relationship between $x(\theta)$ and θ holds is known as a Critical Region (CR). Note that in the figure there are three CRs and therefore $x(\theta)$ was obtained as a function of θ by solving only three optimization problems and therefore solving optimization problems for each and every value of θ has been avoided. In the figure θ is a scalar and CRs are given by intervals, in general when θ is a vector, a CR is given by a polyhedron.

The first-order KKT conditions for (2.1) are given as follows:

$$\begin{aligned} & \nabla f(x,\theta) + \sum_{i=1}^{p} \lambda_i \nabla g_i(x,\theta) + \sum_{j=1}^{q} \mu_j \nabla h_j(x,\theta) = 0 \\ & \lambda_i g_i(x,\theta) = \ 0, \quad \lambda_i \geq \ 0, \quad \forall \ i = 1,...,p \\ & h_j(x,\theta) = \ 0, \quad \forall \ j = 1,...,q, \end{aligned} \tag{2.2}$$

where λ is the vector of Lagrange multipliers. The main sensitivity result for (2.1) is given in the next theorem.

Theorem 2.1 *Basic Sensitivity Theorem (Fiacco, 1976): Let θ_0 be a vector of parameter values and (x_0, λ_0, μ_0) a KKT triple corresponding to (2.2), where λ_0 is nonnegative and x_0 is feasible in (2.1). Also assume that (i) strict complementary slackness (SCS) holds, (ii) the binding constraint gradients are linearly independent (LICQ: Linear Independence Constraint Qualification), and (iii) the second-order sufficiency conditions (SOSC) hold. Then, in neighborhood of θ_0, there exists a unique, once continuously differentiable function $[x(\theta), \lambda(\theta), \mu(\theta)]$ satisfying (2.2) with $[x(\theta_0), \lambda(\theta_0), \mu(\theta_0)] = (x_0, \lambda_0, \mu_0)$, where $x(\theta)$ is a unique isolated minimizer* *for (2.1), and*

$$\begin{pmatrix} \frac{dx(\theta_0)}{d\theta} \\ \frac{d\lambda(\theta_0)}{d\theta} \\ \frac{d\mu(\theta_0)}{d\theta} \end{pmatrix} = - \ (M_0)^{-1} N_0, \tag{2.3}$$

where,

$$M_0 = \begin{pmatrix} \nabla^2 L & \nabla g_1 & \cdots & \nabla g_p & \nabla h_1 & \cdots & \nabla h_q \\ -\lambda_1 \nabla^T g_1 & -g_1 & & & & & \\ \vdots & & \ddots & & & & \\ -\lambda_p \nabla^T g_p & & & -g_p & & & \\ \nabla^T h_1 & & & & & & \\ \vdots & & & & & & \\ \nabla^T h_q & & & & & & \end{pmatrix}$$

$$N_0 = (\nabla^2_{\theta x} L, -\lambda_1 \nabla^T_\theta g_1, \ldots, -\lambda_p \nabla^T_\theta g_p, \nabla^T_\theta h_1, \ldots, \nabla^T_\theta h_q)^T$$

and,

$$L(x, \lambda, \mu, \theta) = f(x, \theta) + \sum_{i=1}^{p} \lambda_i g_i(x, \theta) + \sum_{j=1}^{q} \mu_j h_j(x, \theta). \tag{2.4}$$

In this section three types of multiparametric continuous programs are discussed. These are multiparametric quadratic programs (Dua et al., 2002), multiparametric convex nonlinear programs (Dua and Pistikopoulos, 1999) and multiparametric nonconvex nonlinear programs (Dua et al., 2004).

2.1 Multiparametric quadratic programming. For a special case of (2.1) when the objective function is quadratic, the constraints are linear and the parameters are present on the right hand side of the constraints, we obtain a Multiparametric Quadratic Program (mp-QP) of the following form:

$$\begin{aligned} & z(\theta) = \min_x \ c^T x + \tfrac{1}{2} x^T Q x \\ \text{s.t.} \quad & Ax \le b + F\theta \\ & x \in X \subseteq \Re^n \\ & \theta \in \Theta \subseteq \Re^m \end{aligned} \tag{2.5}$$

where c is a constant vector of dimension n, Q is an $(n \times n)$ symmetric positive definite constant matrix, A is a $(p \times n)$ constant matrix, F is a $(p \times m)$ constant matrix, b is a constant vector of dimension p and X and Θ are compact polyhedral convex sets of dimensions n and m respectively. Note that a term of the form $\theta^T P x$ in the objective function can also be addressed in the above formulation, as it can be transformed into the form given in (2.5) by substituting $x = s - Q^{-1} P^T \theta$, where s is a vector of arbitrary variables of dimension n and P is a constant matrix of dimension $(m \times n)$. The main result for the solution of (2.5) is given in the next theorem.

Theorem 2.2 *(Dua et al., 2002) Let Q be symmetric and positive definite and the assumptions of theorem 2.1 hold, then x and λ are affine functions of θ.*

The algorithm starts by obtaining x and λ as an affine function of θ as follows. An application of Theorem 2.1 to (2.5) at $[x(\theta_Q), \theta_Q]$ gives the following result:

$$\begin{pmatrix} \frac{dx(\theta_Q)}{d\theta} \\ \frac{d\lambda(\theta_Q)}{d\theta} \end{pmatrix} = - \ (M_Q)^{-1} N_Q, \tag{2.6}$$

where,

$$M_Q = \begin{pmatrix} Q & A_{1.}^T & \cdots & A_{p.}^T \\ -\lambda_1 A_{1.} & -V_1 & & \\ \vdots & & \ddots & \\ -\lambda_p A_{p.} & & & -V_p \end{pmatrix}$$

$$N_Q = (Y, \lambda_1 F_{1.}, \ldots, \lambda_p F_{p.})^T$$

where $V_i = A_{i.}x(\theta_Q) - b_i - F_{i.}\theta_Q$ and Y is a null matrix of dimension $(n \times m)$.

The result presented in (2.6) can now be used to compute x and λ as an affine function of θ as follows:

$$\begin{bmatrix} x_Q(\theta) \\ \lambda_Q(\theta) \end{bmatrix} = -\ (M_Q)^{-1} N_Q(\theta - \theta_Q) + \begin{bmatrix} x(\theta_Q) \\ \lambda(\theta_Q) \end{bmatrix} \tag{2.7}$$

The space of θ where this solution, (2.7), remains optimal is defined as the critical region (CR^Q) and can be obtained by using feasibility and optimality conditions. Note that for convenience and simplicity in presentation, we use the notation CR to denote the set of points in space of θ that lie in CR as well as to denote the set of inequalities which define CR. Feasibility is ensured by substituting $x_Q(\theta)$ into the inactive inequalities given in (2.5), whereas optimality condition is given by $\tilde{\lambda}_Q(\theta) \geq 0$, where $\tilde{\lambda}_Q(\theta)$ corresponds to the vector of active inequalities, resulting in a set of parametric constraints. Let this set be represented by

$$CR^R = \{\breve{A}x_Q(\theta) \leq \breve{b} + \breve{F}\theta, \tilde{\lambda}_Q(\theta) \geq 0, CR^{IG}\} \tag{2.8}$$

where $\breve{A}, \breve{b}$ and $\breve{F}$ correspond to the inactive inequalities and CR^{IG} represents a set of linear inequalities defining an initial given region. From the parametric inequalities thus obtained, the redundant inequalities are removed and a compact representation of the CR^Q is obtained as follows:

$$CR^Q = \Delta\{CR^R\} \tag{2.9}$$

where Δ is an operator which removes redundant constraints - for a procedure to identify redundant constraints see Gal (1995). Note that a CR^Q is a polyhedral region. Once CR^Q for a solution, $[x(\theta_Q), \theta_Q]$, has been defined, the next step is to define the rest of the region, CR^{rest}, as follows:

$$CR^{rest} = CR^{IG} - CR^Q. \tag{2.10}$$

Note that in general CR^{rest} will be nonconvex. A procedure to obtain CR^{rest} as a union of convex regions has been proposed in Dua and Pistikopoulos (2000). Another set of parametric solutions in each of these regions is then obtained and corresponding CRs are obtained. The algorithm terminates when there are no more regions to be explored. Note that while defining the rest of the regions, some of the regions are split and hence the same optimal solution may be obtained in more than one regions. Therefore, the regions with the same optimal solution are united and a compact representation of the final solution is obtained.

2.1.1 *mp-QP algorithm.* The main steps of the algorithm are as follows.

Step 1: In a given region solve (2.5) by treating θ as a free variable to obtain a feasible point $[\theta_Q]$.

Step 2: Fix $\theta = \theta_Q$ and solve (2.5) to obtain $[x(\theta_Q), \lambda(\theta_Q)]$.

Step 3: Compute $[-(M_Q)^{-1}N_Q]$ from (2.6).

Step 4: Obtain $[x_Q(\theta), \lambda_Q(\theta)]$ from (2.7).

Step 5: Form a set of inequalities, CR^R, as described in (2.8).
Step 6: Remove redundant inequalities from this set of inequalities and define the corresponding CR^Q as given in (2.9).
Step 7: Define the rest of the region, CR^{rest} as given in (2.10).
Step 8: If no more regions to explore, go to next step, otherwise go to Step 1
Step 9: Collect all the solutions and unify the regions having the same solution to obtain a compact representation.

The algorithm has been fully automated (Pistikopoulos et al., 1999) and tested on a number of problems. Continuity and convexity properties of the optimal solution are summarized in the next theorem.

Theorem 2.3 *(Dua et al., 2002) Consider the mp-QP (2.5) and let Q be positive definite, Θ convex. Then the set of feasible parameters $\Theta_f \subseteq \Theta$ is convex, the optimizer $x(\theta) : \Theta_f \mapsto \Re^n$ is continuous and piecewise affine, and the optimal solution $z(\theta) : \Theta_f \mapsto \Re$ is continuous, convex and piecewise quadratic.*

Remark 2.4 Multiparametric Linear Program: Note that when Q is a null matrix, (2.5) reduces to a multiparametric linear program (mp-LP). This does not affect the solution procedure described above and the algorithm remains the same. This is because the results presented in the theorems are still valid as explained next. The results presented in theorem 2.1 continue to hold true and SOSC is valid inspite of the fact that Q is a null matrix as discussed on page 71 in (Fiacco, 1983). For mp-LPs x is an affine function of θ and λ remains constant in a CR as shown in chapter 4 in (Gal, 1995) and therefore the results of theorem 2.2 can be used. While the results of theorem 2.3 regarding Θ_f and $x(\theta)$ are still valid, $z(\theta)$ simplifies to a continuous, convex and piecewise linear function of θ as also shown in chapter 4 in (Gal, 1995). Note that for the solution of mp-LPs an algorithm based upon the simplex algorithm is described in Gal (1995).

2.2 Multiparametric nonlinear programming. Consider the following multiparametric convex nonlinear programming (mp-NLP) problem:

$$\begin{aligned} & z(\theta) = \min_x \ f(x) \\ \text{s.t.} \ & g_i(x) \leq b_i + F_i\theta, \ \ i = 1, \dots, p \\ & x \in X \subseteq \Re^n \\ & \theta \in \Theta \subseteq \Re^m. \end{aligned} \tag{2.11}$$

Continuity and convexity properties of the optimal solution are summarized in the next three theorems.

Theorem 2.5 *Continuity Property of the Objective Function Value (Fiacco, 1983): Let $z(\theta) = \inf_x\{f(x)|x \in X, g(x) \leq b + F\theta\}$. If (i) X is compact convex set in $\Re^n$, (ii) f and g are both continuous on $X \times \Re^n$, and (iii) each component of g is strictly convex on X for each θ, then, $z(\theta)$ is continuous on its effective domain.*

Theorem 2.6 *Convexity Property of the Solution Space (Fiacco and Kyparisis, 1986): If (i) g_i are jointly quasi-convex on X and θ, (ii) and X is convex, then the solution space $\Theta_f(\theta)$, $\Theta_f(\theta) = \{x \in X|g(x) \leq b + F\theta\}$, is convex.*

Theorem 2.7 *Convexity Property of the Objective Function Value (Fiacco and Kyparisis, 1986): If (i) f is convex on X, and (ii) the solution space $\Theta_f(\theta)$ is essentially convex, then $z(\theta)$ is convex on θ.*

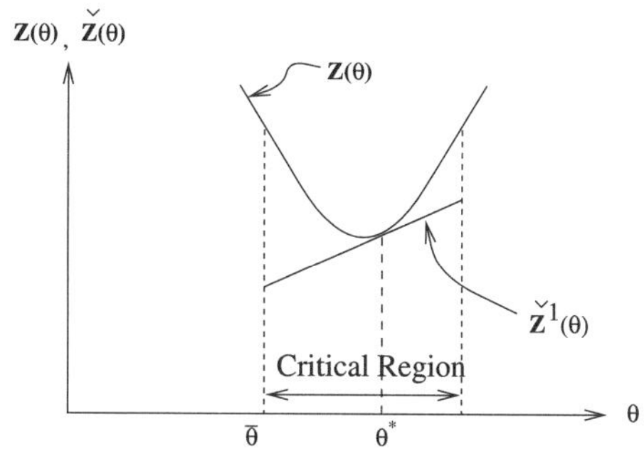

Figure 2 Maximum difference between $z(\theta)$ and $\check{z}(\theta)$ lies at the vertex of the CR

The solution procedure starts by creating an outer-approximation of (2.11) and then by formulating and solving the following multiparametric linear program:

$$\begin{aligned} \check{z}(\theta) &= \min_x \ f(x^*) + \nabla_x f(x^*)(x - x^*) \\ \text{s.t.} \quad & g(x^*) + \nabla_x g(x^*)(x - x^*) \leq b_i + F_i\theta, \ \ i = 1, \ldots, p \\ & x \in \Re^n \\ & \theta \in \Theta \subseteq \Re^m, \end{aligned} \tag{2.12}$$

where θ^* is an initial feasible point and x^* is the optimal solution of (2.11) for $\theta = \theta^*$. Note that the solution of (2.12) is given by a set of optimal solution profiles, $\check{z}(\theta)$, which are affine in θ and the corresponding polyhedral regions of optimality in Θ known as critical regions, CR, (remark 2.4). Also note that $\check{z}(\theta)$ is a piecewise linear, continuous and convex function of θ (Gal, 1995). Since f and g are convex, $\check{z}(\theta)$ provides an underestimator to $z(\theta)$. For a given CR the maximum difference between $z(\theta)$ and $\check{z}(\theta)$ will lie at one of the vertices of the CR - see Figure 2 for a graphical interpretation. At this vertex, $\theta = \bar{\theta}$, another mp-LP of the form (2.12) is formulated and solved to obtain another set of $\check{z}(\theta)$. The $\check{z}(\theta)$ obtained at two different θ, θ^* and $\bar{\theta}$, are then compared by using a procedure described in Acevedo and Pistikopoulos (1997b) and tighter of the two $\check{z}(\theta)$ is retained. This results in the subdivision of the initial CR - see Figure 3. This procedure of identifying $\bar{\theta}$ and obtaining $\check{z}(\theta)$ continues until the maximum difference between $z(\theta)$ and $\check{z}(\theta)$ is within a pre-specified tolerance, ϵ. The tolerance ϵ is then added to $\check{z}(\theta)$ to obtain a tight overestimator of $z(\theta)$. The final solution of (2.11) is given by linear parametric profiles, that are within ϵ of $z(\theta)$, and the corresponding regions of optimality.

2.2.1 *mp-NLP algorithm.* The basic steps of the mp-NLP algorithm are as follows.

Step 1: For a given region of θ create outer-approximation of the mp-NLP problem (2.11) and solve the resulting mp-LP problem (2.12) to obtain a set of parametric profiles, $\check{z}(\theta)$, and the corresponding critical regions CR.

Step 2: Identify the point where the difference between the solution of mp-NLP, $z(\theta)$, and the solution of the mp-LP, $\check{z}(\theta)$ is maximum. If $z(\theta) - \check{z}(\theta) \geq \epsilon$ in some region CR, return to Step 1 with this CR and the point where the difference is maximum as the new point for outer-approximation.

Step 3: If $z(\theta) - \check{z}(\theta) \leq \epsilon$ for some region of θ, then the solution in that region is given by $\check{z}(\theta) + \epsilon$.

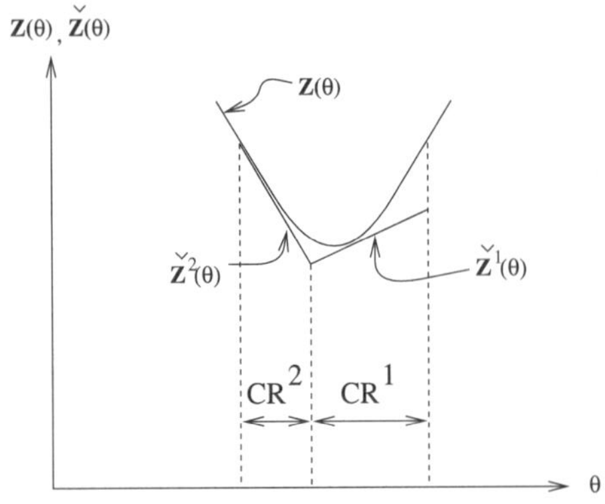

Figure 3 Comparison of $\check{z}(\theta)$

2.3 Multiparametric global optimization. Consider the following multiparametric Global Optimization (mp-GO) problem:

$$\begin{array}{rl} & z(\theta) = \min\limits_{x} \ f(x) \\ \text{s.t.} & g_i(x) \le b_i + F_i\theta, \ \ i = 1, \dots, p \\ & x_j^L \le x_j \le x_j^U, \ \ j = 1, \dots, n \\ & x \in \Re^n \\ & \theta \in \Theta \subseteq \Re^m, \end{array} \tag{2.13}$$

where f and g in general are nonconvex functions of x, θ is a vector of parameters and Θ is a compact and polyhedral convex set, b_i is the ith row of p dimensional constant column vector b and F_i is the ith row of constant matrix F of dimension $p \times m$.

The central idea behind the Branch and Bound (B&B) tree type algorithm for the solution of (2.13) is to create convex parametric underestimators and overestimators, denoted by $\check{z}(\theta)$ and $\hat{z}(\theta)$ respectively, of $z(\theta)$ and then branch and bound on x until the difference between $\hat{z}(\theta)$ and $\check{z}(\theta)$ is within a certain pre-specified tolerance, ϵ. The difference between $\hat{z}(\theta)$ and $\check{z}(\theta)$ will be referred to as "Global Parametric Gap". Note that for the case when $\hat{z}(\theta)$ and $\check{z}(\theta)$ are affine in θ the global parametric gap is checked by using a comparison procedure proposed by Acevedo and Pistikopoulos (1997b). The parametric underestimators are obtained by creating convex underestimators of f and g and then formulating and solving problem (2.11) as described in section 2.2. The solution of (2.11) is given by linear parametric profiles and the corresponding critical regions. Convex parametric overestimators can also be similarly obtained by creating convex overestimators of f and g and solving the resulting multiparametric convex nonlinear program as described in section 2.2. The final solution will again be given by linear parametric profiles and the corresponding critical regions. Let the solution of the underestimating and the overestimating subproblems be given by $\check{x}(\theta)$ and $\hat{x}(\theta)$ respectively. If the global parametric gap for a CR is greater than ϵ, for some x_j, the interval $[x_j^L, x_j^U]$ is partitioned into, say two, smaller intervals: $[x_j^L, x_j^{\otimes}]$ and $[x_j^{\otimes}, x_j^U]$ where $x_j^L < x_j^{\otimes} < x_j^U$ - this is known as the branching of the B&B tree. In the partitioned intervals the tighter underestimators and overestimators are obtained and the corresponding underestimating and overestimating subproblems are solved.

The overestimators are updated by taking the minimum of all the upper bounds obtained in the CR so far. For a given CR, the partitioned intervals where the solution of the underestimating subproblem is greater than or within a certain tolerance of the upper bound are removed from further consideration - this is known as bounding or fathoming. The remaining partitioned intervals are further partitioned into smaller intervals and this procedure continues until all the intervals except the ones where the global solution lies and is within a certain tolerance of the upper bound, have been fathomed. Note that the partitioning or branching may take place for a different x_j at each iteration.

2.3.1 *Multiparametric overestimators.* In some cases it is not required to formulate and solve an overestimating subproblem and instead a parametric overestimator can be obtained merely from the solution of the underestimating subproblem - these cases are discussed in detail in Dua et al. (2004). Note that four different ways of obtaining parametric overestimators are discussed and the key features of each of the parametric overestimators are also presented in Dua et al. (2004). The relative merits of different types of parametric overestimators can be weighed in terms of three attributes: (i) ease of obtaining, (ii) tightness and (iii) functional description, i.e., linear or nonlinear. These issues are briefly reviewed next.

Parametric overestimator of type-1, which is denoted by $\hat{z}(\theta)_{O1}$, can be (easily) obtained by substituting the solution of the underestimating subproblem into the original problem (2.13) - there is no need to resolve the original problem (2.13). One disadvantage is that, in general, it may lead to nonlinear and nonconvex functional description of $\hat{z}(\theta)_{O1}$ and corresponding critical regions. In such cases comparison of $\hat{z}(\theta)_{O1}$ to $\check{z}(\theta)$ is nontrivial. For some specific examples when: (i) $m = 1$, the comparison is easier and (ii) $\hat{z}(\theta)_{O1}$ simplifies to affine expressions in θ, the comparison is much simpler and can be achieved by the procedure of Acevedo and Pistikopoulos (1997b).

For $\hat{z}(\theta)_{O2}$, the overestimator of the second type consider that the only nonconvexity present in the formulation is due to a bilinear term of the type x_1x_2. An underestimator of x_1x_2 can be obtained by replacing it by an additional variable $\check{w}$ and introducing the following linear constraints (McCormick, 1976):

$$\begin{aligned} \check{w} &\geq x_1^L x_2 + x_2^L x_1 - x_1^L x_2^L \\ \check{w} &\geq x_1^U x_2 + x_2^U x_1 - x_1^U x_2^U . \end{aligned} \tag{2.14}$$

$\hat{z}(\theta)_{O2}$ which is based upon the work of Floudas and co-workers relies on the following lemma.

Lemma 2.8 *(Androulakis et al., 1995) The maximum separation between x_1x_2 and $\check{w}$ inside the rectangle $[x_1^L, x_1^U] \times [x_2^L, x_2^U]$ is equal to $\delta_{12} = \frac{(x_1^U - x_1^L)(x_2^U - x_2^L)}{4}$.*

An overestimator of x_1x_2 can therefore be obtained by replacing x_1x_2 by $\hat{w}$ where

$$\begin{aligned} \hat{w} &\geq x_1^L x_2 + x_2^L x_1 - x_1^L x_2^L + \delta_{12} \\ \hat{w} &\geq x_1^U x_2 + x_2^U x_1 - x_1^U x_2^U + \delta_{12}. \end{aligned} \tag{2.15}$$

Lemma 2.9 *(Dua et al., 2004) If $f(x) = f^c(x) + \sum_{j=1}^{J}\sum_{j'=1}^{J} a_{jj'}x_jx_{j'}$, where $j \neq j', f^c(x)$ is a convex function of x, $a_{jj'}$ is constant and positive for all j and j' (without loss of generality, we assume $a_{jj'} > 0$; for some $a_{jj'} < 0$ the following results can be accordingly modified) and g is convex then $\hat{z}_{O2}(\theta) = \check{z}(\theta) +$*

$\sum_{j=1}^{J}\sum_{j'=1}^{J} a_{jj'}\delta_{jj'}$. *Note that terms of the form* $a_{jj'}x_jx_{j'}, j = j'$ *are convex and a part of the convex function* $f^c(x)$.

$\hat{z}(\theta)_{O2}$ also has the advantage that it can be obtained with negligible effort after the underestimating subproblem has been solved (lemma 2.9). This feature holds true for the case when the only nonconvexity is due to the bilinear terms and these terms are present only in the objective function. For the case when bilinear terms are also present in the constraints there are some cases where $\hat{z}(\theta)_{O2}$ can be obtained more efficiently at modest extra effort by shifting δ_{12} on the right hand side of the constraints and solving the underestimating subproblem for a slightly expanded range of θ. Since the underestimator is affine in θ, $\hat{z}(\theta)_{O2}$ is also affine in θ and the comparison can be carried out (Acevedo and Pistikopoulos, 1997b). The extension of these ideas for the case when general nonconvex terms are present is also possible along the work of Floudas and co-workers.

Unlike for the case of $\hat{z}(\theta)_{O1}$ and many cases of $\hat{z}(\theta)_{O2}$, the overestimator of the third type, $\hat{z}(\theta)_{O3}$, requires solving a parametric optimization problem. All the nonconvex terms are replaced by the corresponding overestimators. For example, the bilinear terms are replaced by $\hat{w}$ where

$$\begin{aligned} \hat{w} &\geq x_1^L x_2 + x_2^U x_1 - x_1^L x_2^U \\ \hat{w} &\geq x_1^U x_2 + x_2^L x_1 - x_1^U x_2^L. \end{aligned} \tag{2.16}$$

The key advantage is that this type of formulation is completely general and does not have limitations regarding the presence of only bilinear terms and restrictions regarding nonconvex terms only in the objective function. The solution is affine in θ and can be compared to $\check{z}(\theta)$. Note the following lemma.

Lemma 2.10 *(Dua et al., 2004) The overestimator in (2.15), which corresponds to* $\hat{z}_{O2}(\theta)$, *is tighter than the overestimator in (2.16), which corresponds to* $\hat{z}_{O3}(\theta)$, *in 87.5% of the area of the rectangle* $[x_1^L, x_1^U] \times [x_2^L, x_2^U]$ *provided that the rectangle is not empty.*

The fourth type of overestimator, $\hat{z}(\theta)_{O4}$, also requires solving a parametric optimization problem. It is tighter than $\hat{z}(\theta)_{O3}$ but is limited to the case when only bilinear terms are present in the objective function because it relies on maximization of the auxiliary variables, $\hat{w}$, which replace the bilinear terms, where

$$\begin{aligned} \hat{w} &\leq x_1^L x_2 + x_2^U x_1 - x_1^L x_2^U \\ \hat{w} &\leq x_1^U x_2 + x_2^L x_1 - x_1^U x_2^L. \end{aligned} \tag{2.17}$$

2.3.2 *mp-GO algorithm.* Based upon the above developments an algorithm for the solution of (2.13) is presented as follows.

Step 1: Initialize the current upper bound as $\bar{z}(\theta) = \infty$, a region of θ, CR, a space of continuous variables x - determined by the lower and upper bounds x^L and x^U respectively, and tolerances, ϵ_1 and ϵ_2, where ϵ_1 is specified for the global parametric gap and ϵ_2 for the solution of parametric convex programs as described in section 2.2.

Step 2: For a given region of θ, CR, and the corresponding space of x,

1. formulate and solve (2.11) as described in section 2.2 and obtain the parametric underestimators, $\check{z}(\theta)$.

2. obtain the parametric overestimators, $\hat{z}(\theta)$, by using one of the methods described in sections 2.3.1.

Step 3: Compare $\hat{z}(\theta)$ to $\bar{z}(\theta)$, as described in Acevedo and Pistikopoulos (1997b), and update the current upper bound $\bar{z}(\theta) = \min(\hat{z}(\theta), \bar{z}(\theta))$, in the corresponding regions of θ.

Step 4: Compare $\check{z}(\theta)$ to $\bar{z}(\theta)$, as described in Acevedo and Pistikopoulos (1997b), and in the spaces of x where,

1. $\check{z}(\theta) \leq \bar{z}(\theta) - \epsilon_1$, branch on x by subdividing the bounds on x and go to Step 2,
2. $\check{z}(\theta) \geq \bar{z}(\theta) - \epsilon_2$, fathom those spaces of x only for the corresponding regions of θ.

Step 5: In the spaces of x where (2.11) is infeasible, fathom those spaces and the corresponding regions of θ.

Step 6: If no more spaces of x and regions of θ to explore, terminate, otherwise go to Step 2.

Step 7: The final solution is given by $\bar{z}(\theta)$.

3 Multiparametric mixed-integer programs

In this section algorithms for the solution of (i) multiparametric mixed integer linear programs (Dua and Pistikopoulos, 2000), (ii) multiparametric mixed integer quadratic programs (Dua et al., 2002), (iii) multiparametric mixed integer convex nonlinear programs (Dua and Pistikopoulos, 1999), and (iv) multiparametric mixed integer nonconvex nonlinear programs (Dua et al., 2004) are presented.

3.1 Multiparametric mixed-integer linear programming. Consider a multiparametric Mixed Integer Linear Programming (mp-MILP) problem of the following form:

$$\begin{aligned} & z(\theta) = \min_{x,y} c^T x + d^T y \\ \text{s.t.} \quad & Ax + Ey \leq b + F\theta \\ & \theta_k^L \leq \theta_k \leq \theta_k^U, \quad k = 1, \ldots, m \\ & x \in \Re^n \\ & y \in \{0,1\}^l \\ & \theta \in \Theta \subseteq \Re^m, \end{aligned} \tag{3.1}$$

where x is a vector of continuous variables, y is vector of 0-1 binary variables, θ is a vector of parameters, θ^L and θ^U are the vectors of lower and upper bounds on θ, A is an $(p \times n)$ constant matrix, E is an $(p \times l)$ constant matrix, F is an $(p \times m)$ constant matrix, b, c and d are constant vectors of dimension p, n and l respectively.

Note that Acevedo and Pistikopoulos (1997b) presented an algorithm for solving (3.1) based upon B&B principles. The algorithm described in this section is based upon decomposing (3.1) into an mp-LP and an MILP subproblem. The solution of the mp-LP, which is obtained by fixing the vector of binary variables, provides a parametric upper bound, whereas the solution of the MILP, which is obtained by treating θ as a free variable, provides a new integer vector. The parametric solutions corresponding to two different integer solutions are then compared, using a procedure proposed by Acevedo and Pistikopoulos (1997b), in order to keep as tight upper bounds as possible. The steps of the algorithm are described in detail in the following sections.

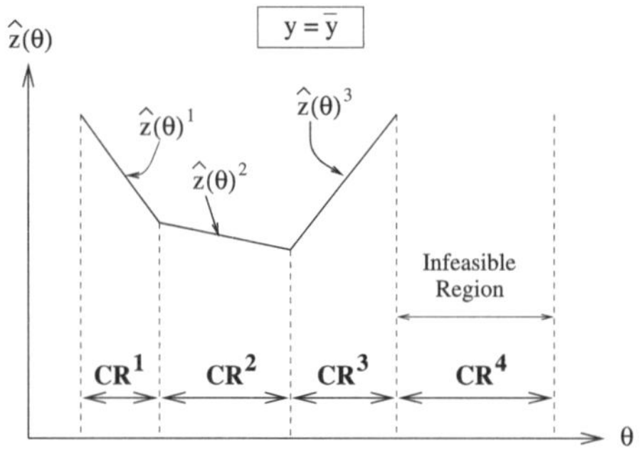

Figure 4 Parametric solution - multiparametric LP subproblem for $y = \bar{y}$

3.1.1 *Initialization.* An initial feasible y is obtained by solving the following MILP:

$$\begin{aligned} & z = \min_{x,y,\theta} c^T x + d^T y \\ \text{s.t.} \quad & Ax + Ey \leq b + F\theta \\ & \theta_k^L \leq \theta_k \leq \theta_k^U, \ \ k = 1, \ldots, m \\ & x \in \Re^n \\ & y \in \{0,1\}^l, \end{aligned} \tag{3.2}$$

where θ is treated as a variable to find a starting feasible solution. Let the solution of (3.2) be given by $y = \bar{y}$.

3.1.2 *Multiparametric LP subproblem.* Fix $y = \bar{y}$ in (3.1) to obtain a multi-parametric LP problem of the following form:

$$\begin{aligned} & \hat{z}(\theta) = \min_{x} c^T x + d^T \bar{y} \\ \text{s.t.} \quad & Ax + E\bar{y} \leq b + F\theta \\ & \theta_k^L \leq \theta_k \leq \theta_k^U, \ \ k = 1, \ldots, m \\ & x \in \Re^n. \end{aligned} \tag{3.3}$$

The solution of (3.3) is given by a set of linear parametric profiles, $\hat{z}(\theta)^i$, where $\hat{z}(\theta)$ is convex, and corresponding critical regions, CR^i (remark 2.4). The parametric solution has been graphically depicted in Figure 4, where $\hat{z}(\theta)^1, \hat{z}(\theta)^2$ and $\hat{z}(\theta)^3$ represent solution in the regions CR^1, CR^2 and CR^3 respectively.

The region, CR^4 in Figure 4, where no solution is found, represents an infeasible region for the current integer solution, $\bar{y}$. Note that unlike in the case of a single parameter where the infeasible region is simply given by intervals, for the multi-parametric case an infeasible region can be defined by convex regions, which are obtained by systematically subdividing the initial region of θ as described in Dua and Pistikopoulos (2000). The final solution of the multiparametric LP subproblem in (3.3) which represents a parametric upper bound on the final solution is given by (i) a set of parametric profiles, $\hat{z}(\theta)^i$, and the corresponding critical regions, CR^i, and (ii) a set of infeasible regions where $\hat{z}(\theta)^i = \infty$.

3.1.3 *MILP subproblem.* For each critical region, CR^i, obtained from the solution of the multiparametric LP subproblem in (3.3), an MILP subproblem is formulated as follows:

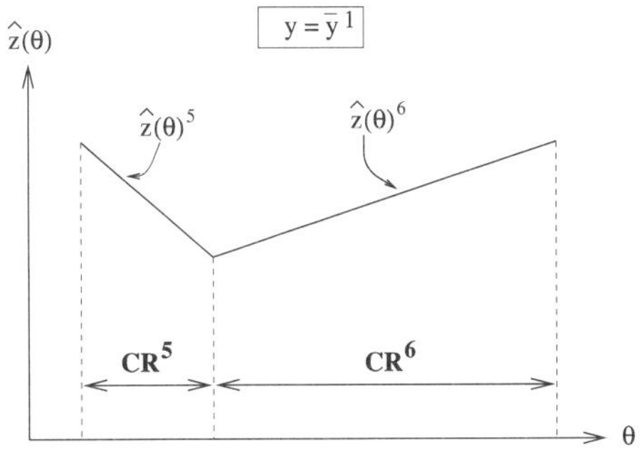

Figure 5 Parametric solution - multiparametric LP subproblem for $y = \bar{y}^1$

$$
\begin{array}{ll}
 & z = \min\limits_{x,y,\theta} c^T x + d^T y \\
\text{s.t.} & Ax + Ey \le b + F\theta \\
 & c^T x + d^T y \le \hat{z}(\theta)^i \\
 & \sum_{j \in J^{ik}} y_j^{ik} - \sum_{j \in L^{ik}} y_j^{ik} \le |J^{ik}| - 1, \;\; k = 1, \dots, K^i \\
 & \theta \in CR^i \\
 & x \in \Re^n \\
 & y \in \{0,1\}^l,
\end{array}
\tag{3.4}
$$

where θ is treated as a variable and $\theta \in \text{CR}^i$ indicates that θ is bounded by the set of inequalities which define CR^i; $J^{ik} = (j|y_j^{ik} = 1)$ and $L^{ik} = (j|y_j^{ik} = 0)$, and $|J^{ik}|$ is the cardinality of J^{ik} and K^i is the number of integer solutions that have already been analysed in CR^i. Note that the inequality, $c^T x + d^T y \le \hat{z}(\theta)^i$, excludes integer solutions with higher values than the current upper bound, $\hat{z}(\theta)^i$; the inequality, $\sum_{j \in J^{ik}} y_j^{ik} - \sum_{j \in L^{ik}} y_j^{ik} \le |J^{ik}| - 1$, corresponds to integer cuts prohibiting previous integer solutions from appearing again. The integer solution, $y = \bar{y}^1$, and the corresponding CRs, obtained from the solution of (3.4), are then recycled back to the multiparametric LP subproblem - to obtain another set of parametric profiles, as graphically depicted in Figure 5, where $\hat{z}(\theta)^5$ and $\hat{z}(\theta)^6$ represent solution in the regions CR^5 and CR^6, respectively (for simplicity in the graphical presentation the same integer solution, $y = \bar{y}^1$, is shown for all critical regions). Note that, in general one may obtain different integer solutions in different critical regions.

If there is no feasible solution to the MILP subproblem (3.4) in a CR^i, that region is excluded from further consideration and the current upper bound in that region represents the final solution. Note also that the integer solution obtained from the solution of (3.4) is guaranteed to appear in the final solution, since it represents the minimum of the objective function at the point, in θ, obtained from the solution of (3.4). The final solution of the MILP subproblem is given by a set of integer solutions and their corresponding CR^is.

3.1.4 *Comparison of parametric solutions.* The set of parametric solutions corresponding to an integer solution, $y = \bar{y}$, which represents the current upper bound are then compared to the parametric solutions corresponding to another integer solution, $y = \bar{y}^1$, in the corresponding CRs in order to obtain the lower of the two parametric solutions and update the upper bound. This is achieved by employing the procedure proposed by Acevedo and Pistikopoulos (1997b). The result of the

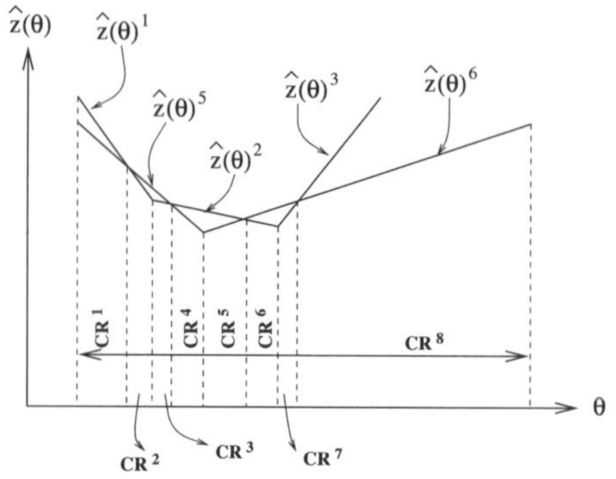

Figure 6 Comparison of two parametric solutions

comparison procedure is graphically depicted in Figure 6, where $\hat{z}(\theta)^1$ represents the current upper bound in CR^2, $\hat{z}(\theta)^2$ represents the current upper bound in CR^3 and CR^6, $\hat{z}(\theta)^3$ is the current upper bound in CR^7, $\hat{z}(\theta)^5$ the current upper bound in CR^1 and CR^4, and $\hat{z}(\theta)^6$ represents the current upper bound in CR^5 and CR^8.

3.1.5 *Multiparametric MILP algorithm.* Based upon the above theoretical developments, the steps of the proposed algorithm can be stated as follows:

Step 0: (Initialization) Define an initial region of θ, CR, with best upper bound $\hat{z}^*(\theta) = \infty$, and an initial integer solution $\bar{y}$.

Step 1: (Multiparametric LP Problem) For each region with a new integer solution, $\bar{y}$:

- **(a)** Solve multiparametric LP subproblem (3.3) to obtain a set of parametric upper bounds $\hat{z}(\theta)$ and corresponding critical regions CR.
- **(b)** If $\hat{z}(\theta) \leq \hat{z}^*(\theta)$ for some region of θ, update the best upper bound function, $\hat{z}^*(\theta)$, and the corresponding integer solutions, y^* (as described in Section 3.1.4).
- **(c)** If an infeasibility is found in some region CR, go to Step 2.

Step 2: (Master Subproblem) For each region CR, formulate and solve the MILP master problem in (3.4) by (i) treating θ as a variable bounded in the region CR, (ii) introducing an integer cut, and (iii) introducing a parametric cut, $c^T x + d^T y \leq \hat{z}^*(\theta)$. Return to Step 1 with new integer solutions and corresponding CRs.

Step 3: (Convergence) The algorithm terminates in a region where the solution of the MILP subproblem is infeasible. The final solution is given by the current upper bounds $\hat{z}^*(\theta)$ in the corresponding CRs.

Note that this algorithm has been implemented in a computational environment (Pistikopoulos et al., 2000).

3.2 Multiparametric mixed-integer quadratic programming: An enclosure of the solution.

3.2.1 *Mathematical formulation.* Consider an mp-MIQP problem of the following form:

$$\begin{aligned} z(\theta) = & \min_{x,y} c^T x + \tfrac{1}{2} x^T Q x + d^T y \\ \text{s.t.} \quad & Ax + Ey \leq b + F\theta \\ & x \in X \subseteq \Re^n \\ & y \in \{0,1\}^l \\ & \theta \in \Theta \subseteq \Re^m . \end{aligned} \tag{3.5}$$

The procedure proposed in this section is based upon decomposing (3.5) into a multi-parametric quadratic program (mp-QP) and a mixed-integer nonlinear program (MINLP). The solution of the mp-QP, which is obtained by fixing the vector of binary variables, provides a parametric upper bound, whereas the solution of the MINLP, which is obtained by treating θ as a free variable, provides a new integer vector. The parametric solutions corresponding to two different integer solutions are then retained to keep as tight upper bounds as possible. The steps of the procedure are described in detail in the following sections.

3.2.2 *Initialization.* An initial feasible y is obtained by solving the following MIQP:

$$\begin{aligned} & \min_{x,y,\theta} c^T x + \tfrac{1}{2} x^T Q x + d^T y \\ \text{s.t.} \quad & Ax - F\theta + Ey \leq b \\ & x \in X \subseteq \Re^n \\ & y \in \{0,1\}^l \\ & \theta \in \Theta \subseteq \Re^m , \end{aligned} \tag{3.6}$$

where θ is treated as a vector of free variables to find a starting feasible solution. Let the solution of (3.6) be given by $y = \bar{y}^1$.

3.2.3 *Multiparametric QP subproblem.* Fix $y = \bar{y}^1$ in (3.5) to obtain an mp-QP problem of the following form:

$$\begin{aligned} \hat{z}(\theta) = & \, d^T \bar{y} + \min_{x} c^T x + \tfrac{1}{2} x^T Q x \\ \text{s.t.} \quad & Ax \leq (b - E\bar{y}^1) + F\theta \\ & x \in X \subseteq \Re^n \\ & \theta \in \Theta \subseteq \Re^m . \end{aligned} \tag{3.7}$$

The solution of (3.7) is given by a set of linear parametric profiles, $\hat{x}(\theta)^i$, and the corresponding critical regions, CR^i (Section 2.1). Note that the optimal objective function, $\hat{z}(\theta)$, is continuous, convex and piecewise quadratic (theorem 2.3). This solution represents a parametric upper bound to the solution of (3.5). Let $\hat{z}(\theta)$ in CR^i be denoted by $\hat{z}(\theta)^i$. Note that CR^i also includes the regions where $\bar{y}$ is infeasible and in these regions $\hat{z}(\theta)^i = \infty$. These infeasible regions can be characterized by a procedure outlined in Dua and Pistikopoulos (2000).

3.2.4 *MINLP subproblem.* For each critical region, CR^i, obtained from the solution of the mp-QP subproblem in (3.7), an MINLP subproblem is formulated as follows:

$$\begin{aligned} z = & \min_{x,y,\theta} c^T x + \tfrac{1}{2} x^T Q x + d^T y \\ \text{s.t.} \quad & Ax - F\theta + Ey \leq b \\ & c^T x + \tfrac{1}{2} x^T Q x - \hat{z}(\theta)^{ik} + d^T y \leq 0, \;\; k = 1, \ldots, K^i \\ & \sum_{j \in J^{ik}} y_j^{ik} - \sum_{j \in T^{ik}} y_j^{ik} \leq |J^{ik}| - 1, \;\; k = 1, \ldots, K^i \\ & \theta \in CR^i \\ & x \in X \subseteq \Re^n \\ & y \in \{0,1\}^l, \end{aligned} \tag{3.8}$$

where θ is treated as a vector of free variables and $\theta \in \mathrm{CR}^i$ indicates that θ is bounded by the set of inequalities which define CR^i; $J^{ik} = (j|y_j^{ik} = 1)$ and $T^{ik} = (j|y_j^{ik} = 0)$, and $|J^{ik}|$ is the cardinality of J^{ik} and K^i is the number of integer solutions that have already been analysed in CR^i. Note that K^i also corresponds to the number of iterations that have taken place in CR^i. It may also be noted that the set of inequalities, $c^T x + \frac{1}{2} x^T Q x - \hat{z}(\theta)^{ik} + d^T y \leq 0$, known as parametric cuts, excludes integer solutions with higher values than the current upper bound, $\hat{z}(\theta)^{ik}$, where $\hat{z}(\theta)^{ik}$ denotes $\hat{z}(\theta)^i$ for a given k; the set of inequalities, $\sum_{j \in J^{ik}} y_j^{ik} - \sum_{j \in T^{ik}} y_j^{ik} \leq |J^{ik}| - 1$, corresponds to integer cuts prohibiting previous integer solutions from appearing again. The integer solution, $y = \bar{y}^2$, obtained by solving (3.8), and the corresponding CRs are returned back to the mp-QP subproblem, (3.7), to obtain another set of parametric profiles.

Note that (3.8), in general, is a non-convex MINLP, where the nonconvexity is due to $\hat{z}(\theta)^{ik}$, which is present in the second set of constraints. This problem can be solved, for example, by using the algorithms described in Adjiman et al. (2000). Since solving this problem by using global optimization can be computationally intensive, some tests are suggested to alleviate this problem in Dua et al. (2002).

If there is no feasible solution to the MINLP subproblem, (3.8), in a CR, that region is excluded from further consideration and the current upper bounds in that region represent the final solution. The final solution of the MINLP subproblem is given by a set of integer solutions and the CRs in which they were identified.

3.2.5 *Envelope of parametric solutions.* The set of parametric solutions corresponding to the new integer solution, $y = \bar{y}^2$, are included into the set of parametric current upper bounds, given by the parametric solutions corresponding to $\bar{y}^1$, in the corresponding CRs. Note that a comparison of the parametric (quadratic) solutions corresponding to two different integer solution is not performed. If this comparison were made it would result in the lower of the two profiles as an upper bound. Instead of the comparison, all the parametric profiles corresponding to all the integer solutions obtained in a CR are retained and the best solution is obtained at a particular vector point θ, on-line, by taking the minimum of the values obtained by function evaluation of parametric profiles for a given θ. Also note that a rigorous procedure for comparing nonlinear profiles, similar to the one for linear profiles (Acevedo and Pistikopoulos, 1997b), can be developed and will result, in general, in nonlinear and nonconvex CRs, further complicating the solution of (3.8).

3.2.6 *Redundant parametric solutions.* The algorithm in a CR^i terminates with a number of integer solutions, $\bar{y}^{ik}$, and the corresponding objective function profiles, $\hat{z}(\theta)^{ik}$. As discussed in the previous section the final parametric solution in CR^i is given by a lower envelope of the profiles $\hat{z}(\theta)^{ik}$, $k = 1, \ldots, K^i$. In this section

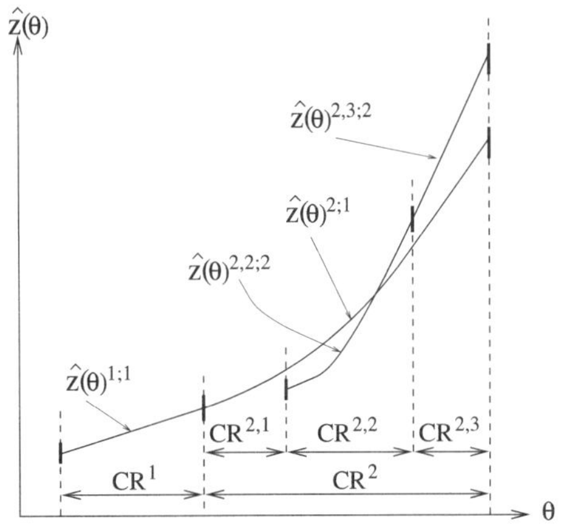

Figure 7 Redundant profiles

we argue that, for a CR^i, some of the profiles from the set $\hat{z}(\theta)^{ik}$, $k = 1, \dots, K^i$ could be eliminated without affecting the final parametric solution. Consider a hypothetical example depicted in Figure 7. In the first iteration, an integer solution $\bar{y}^1$ is identified. The parametric solution corresponding to $\bar{y}^1$ subdivides the region of θ into two critical regions, CR^1 and CR^2. The optimal objective function profiles corresponding to $\bar{y}^1$ are denoted by $\hat{z}(\theta)^{1;1}$ and $\hat{z}(\theta)^{2;1}$ in CR^1 and CR^2 respectively. In the second iteration, MINLPs of the form given in (3.8) are solved in CR^1 and CR^2. In CR^1 the MINLP is infeasible and hence the algorithm terminates in CR^1 with $\hat{z}(\theta)^{1;1}$ as the final solution. In CR^2 an integer solution $\bar{y}^2$ is identified, the optimal objective function profiles corresponding to $\bar{y}^2$ are denoted by $\hat{z}(\theta)^{2,1;2}$, $\hat{z}(\theta)^{2,2;2}$ and $\hat{z}(\theta)^{2,3;2}$, which further subdivide CR^2 into CR^{2,1}, CR^{2,2} and CR^{2,3}; $\bar{y}^2$ is infeasible in CR^{2,1} and hence $\hat{z}(\theta)^{2,1;2} = \infty$. In the third iteration, the MINLPs are infeasible in CR^{2,1}, CR^{2,2} and CR^{2,3} and hence the algorithm terminates in CR^{2,1} with $\hat{z}(\theta)^{2;1}$, in CR^{2,2} with both $\hat{z}(\theta)^{2;1}$ and $\hat{z}(\theta)^{2,2;2}$ and in CR^{2,3} with both $\hat{z}(\theta)^{2;1}$ and $\hat{z}(\theta)^{2,3;2}$ as the final solutions. It is clear from Figure 7 that $\hat{z}(\theta)^{2,3;2}$ is a redundant solution in CR^{2,3} and can be removed without affecting the final solution. In a general case, the redundant profiles can be identified by recursively solving MINLPs of the form given in (3.8) - see Dua et al. (2002) for details.

3.2.7 *Enclosure of the solution of multiparametric MIQP.* Based upon the above discussion, the steps of the proposed solution procedure can be summarized as follows:

Step 0: (Initialization) Define an initial region of θ, CR, and an initial integer solution $\bar{y}$.

Step 1: (mp-QP Subproblem) For each region with a new integer solution, $\bar{y}$:

- **(a)** Solve mp-QP subproblem (3.7) to obtain a set of parametric upper bounds $\hat{z}(\theta)$ and corresponding critical regions CR.
- **(b)** If an infeasibility is found in some region CR, go to Step 2.

Step 2: (Master Subproblem) For each region CR, formulate and solve the MINLP master problem in (3.8) by (i) treating θ as a vector of free variables bounded in the region CR, (ii) introducing integer cuts, and (iii) introducing parametric cuts. Return to Step 1 with new integer solutions and corresponding CRs.

Step 3: (Convergence) The algorithm terminates in a region where the solution of the MINLP subproblem is infeasible. The solution is given by an envelope of the current upper bounds $\hat{z}(\theta)$, integer solutions, $\bar{y}$, and the corresponding CRs.

Step 4: (Redundancy Check) For all the CRs remove redundant parametric solutions as described in Section 3.2.6.

3.3 Multiparametric mixed-integer nonlinear programming. Consider multiparametric mixed integer nonlinear programs (mp-MINLP) of the following from:

$$\begin{aligned} z(\theta) = \min_{y,x} \; & d^T y + f(x) \\ \text{s.t.} \quad & E\,y + g(x) \leq b + F\,\theta \\ & \theta_k^L \leq \theta_k \leq \theta_k^U, \; k = 1, \ldots, m \\ & x \in \Re^n \\ & y \in \{0,1\}^l \\ & \theta \in \Re^m, \end{aligned} \tag{3.9}$$

where θ is a vector of uncertain parameters and F is a constant matrix. The solution of problem (3.9) is approached by decomposing it into (i) a Primal and (ii) a Master subproblem, representing valid parametric upper and lower bounds respectively. While Primal Subproblem is obtained by fixing the integer variables, the Master Subproblem identifies new integer variables. The algorithm starts by finding an initial feasible $y = \bar{y}$, similar to that in sections 3.1.1 and 3.2.2 for mp-MILP and mp-MIQP. Fixing $y = \bar{y}$ in formulation (3.9) results in an mp-NLP that is solved by using the algorithm described in section 2.2. The solution of the Primal Subproblem in the ith critical region, CR^i, is given by $\hat{z}(\theta)^i$. This denotes an upper bound on the solution of (3.9). The master subproblem is formulated as follows.

3.3.1 *Master subproblem.* The following MINLP Master Subproblem is formulated and solved to identify the next integer solution:

$$\begin{aligned} z = \min_{y,x,\theta} \; & d^T y + f(x) \\ \text{s.t.} \quad & Ey + g(x) \leq b + F\theta \\ & d^T y + f(x) \leq \hat{z}(\theta)^i \\ & \textstyle\sum_{j \in J^{ik}} y_j^{ik} - \sum_{j \in L^{ik}} y_j^{ik} \leq |J^{ik}| - 1, \quad k = 1, \ldots, K^i \\ & \theta \in CR^i \\ & x \in \Re^n \\ & y \in \{0,1\}^l, \end{aligned} \tag{3.10}$$

where θ is treated as a variable and $\theta \in CR^i$ indicates that θ is bounded by the set of inequalities which define CR^i; $J^{ik} = (j|y_j^{ik} = 1)$ and $L^{ik} = (j|y_j^{ik} = 0)$, and $|J^{ik}|$ is the cardinality of J^{ik} and K^i is the number of integer solutions that have already been analysed in CR^i. Note that the inequality, $d^T y + f(x) \leq \hat{z}(\theta)^i$, constrains the objective function, $d^T y + f(x)$, to take a value which is lower than the current upper bound $\hat{z}(\theta)^i$ and the inequality, $\sum_{j \in J^{ik}} y_j^{ik} - \sum_{j \in L^{ik}} y_j^{ik} \leq |J^{ik}| - 1$, corresponds to integer cuts which eliminate the integer solutions which have already been explored in the Primal Subproblem. Note that (3.10) is solved for each region, CR^i, obtained from the solution of Primal Subproblem. The integer solution obtained from the solution of (3.10) is then returned back to the Primal Subproblem (2.11). The algorithm terminates in a region, CR^i, when there is no feasible solution to the deterministic Master Subproblem (3.10) in that region. Note that master subproblems based upon Outer-Approximation and Generalized Benders Decomposition principles can also be formulated to obtain candidate integer solutions as described in Papalexandri and Dimkou (1998) and Dua and Pistikopoulos (1999).

3.4 Multiparametric MINLP algorithm. Based upon the above theoretical developments the main steps of the algorithm can be summarized as follows:

Step 0: (Initialization). Define an initial region of θ, CR, with the best upper bound $\hat{z}(\theta) = \infty$, an initial integer structure $\bar{y}$, a feasible point for the linearization and a tolerance ϵ for the approximation of mp-NLP by linear profiles.

Step 1: (Primal Subproblem) For each region with a new integer structure, $\bar{y}$, solve the resulting mp-NLP problem as described in Section 2.2.

Step 2: (Master Subproblem) For each region CR, formulate and solve a deterministic MINLP Master Subproblem (3.10) with (i) θ as a variable bounded in the region CR, (ii) an integer cut, and (iii) a parametric cut $d^T y + f(x) \leq \hat{z}(\theta)$. Return to Step 1 with the new integer solutions and the corresponding CRs.

Step 3: (Convergence) The algorithm terminates when the solution of the deterministic MINLP is infeasible for each region CR. The final solution is given by the current upper bounds $\hat{z}(\theta)$ in the corresponding CRs.

3.5 Multiparametric mixed-integer global optimization. Consider the following multiparametric mixed-integer nonlinear programming problem:

$$\begin{aligned} z(\theta) = & \min_{x,y} \; f(x,y) \\ \text{s.t.} \; & g(x,y) \leq b + F\theta \\ & x_j^L \leq x_j \leq x_j^U, \; j = 1, \ldots, n \\ & y \in \{0,1\}^l \\ & x \in \Re^n \\ & \theta \in \Theta \subseteq \Re^m, \end{aligned} \tag{3.11}$$

where f and g are nonconvex functions of x and y. The basic idea of the algorithm for the solution of (3.11) is to obtain parametric solution for y fixed at integer values and then use this solution as the current solution to cut-off suboptimal integer solutions and identify the ones that are better than the current one. The better integer solutions that are identified then become the current solution and this procedure continues until all the optimal solutions and the corresponding regions of optimality, in the space of θ, have been identified. An initial integer solution is identified by solving the following problem:

$$\begin{aligned} & \min_{x,y,\theta} \; f(x,y) \\ \text{s.t.} \; & g(x,y) - F\theta \leq b \\ & x_j^L \leq x_j \leq x_j^U, \; j = 1, \ldots, n \\ & y \in \{0,1\}^l \\ & x \in \Re^n \\ & \theta \in \Theta \subseteq \Re^m, \end{aligned} \tag{3.12}$$

where θ is treated as a vector of free variables. See Adjiman et al. (2000) for an algorithm for the solution of (3.12). Let the solution of (3.12) be given by $y = \tilde{y}$.

The algorithm starts by fixing $y = \tilde{y}$ in (3.11) to obtain the following nonconvex mp-NLP:

$$\begin{aligned} z_{\tilde{y}}(\theta) = \min_{x} \ & f(x,\tilde{y}) \\ \text{s.t.} \ & g(x,\tilde{y}) \le b + F\theta \\ & x_j^L \le x_j \le x_j^U, \ j = 1,\ldots,n \\ & x \in \Re^n \\ & \theta \in \Theta \subseteq \Re^m. \end{aligned} \tag{3.13}$$

The solution of (3.13) is then approached by creating convex underestimators and overestimators of nonconvex functions $f(x,\tilde{y})$ and $g(x,\tilde{y})$ which converge to the global optima by branching on the space of x - similar to solution of (2.13). The ith solution of (3.13), $z_{\tilde{y}}(\theta)^i$, which is given by a linear parametric profile valid in its region of optimality, CR^i represents a current upper bound in CR^i. Another optimal vector of integer variables is then obtained in each region CR^i by formulating the following problem:

$$\begin{aligned} \min_{x,y,\theta} \ & f(x,y) \\ \text{s.t.} \ & g(x,y) \le b + F\theta \\ & f(x,y) \le \hat{z}(\theta)^l \\ & \textstyle\sum_{t \in T^{ik}} y_t^{ik} - \sum_{t \in P^{ik}} y_t^{ik} \le |T^{ik}| - 1, \ k = 1,\ldots,K^i \\ & x_j^L \le x_j \le x_j^U, \ j = 1,\ldots,n \\ & y \in \{0,1\}^l \\ & x \in \Re^n \\ & \theta \in \mathrm{CR}^i \in \Theta \subseteq \Re^m, \end{aligned} \tag{3.14}$$

where θ is treated as a vector of free variables and $\theta \in \mathrm{CR}^i$ indicates that θ is bounded in the region given by CR^i, $T^{ik} = (t|y_t^{ik} = 1)$ and $P^{ik} = (t|y_t^{ik} = 0)$, and $|T^{ik}|$ is the cardinality of T^{ik} and K^i is the number of integer solutions that have already been analysed in CR^i. The formulation in (3.14) is a nonconvex problem which is solved to global optimality (Adjiman et al., 2000) by branching on x, y and θ and treating θ as a vector of free variables. The solution of (3.14) identifies the next set of optimal integer variables by introducing the constraint, $f(x,y) \le \hat{z}(\theta)^i$, to restrict the objective function to take values which are less than the current upper bound and introducing the integer cut, $y \ne \tilde{y}$, which is given by the constraint $\sum_{t \in T^{ik}} y_t^{ik} - \sum_{t \in P^{ik}} y_t^{ik} \le |T^{ik}| - 1$, to eliminate the integer solutions that have already been analysed.

The integer vector obtained from the solution of (3.14) is then returned back to (3.13) to obtain another set of parametric profiles. Parametric solutions corresponding to two integer solutions are then compared and a lower envelope of the parametric solutions is retained by using the comparison procedure (Acevedo and Pistikopoulos, 1997a) to update the current upper bound $z_{\tilde{y}}(\theta)$. The algorithm proceeds in this way until there is no feasible solution to (3.14) in each region in the space of θ. The final solution is given by the current upper bound $z_{\tilde{y}}(\theta)$.

4 Multiparametric model based control

4.1 Linear model predictive control. Model Predictive Control (MPC) has been widely adopted by industry to address on-line optimization problems with input and output constraints. MPC is based on the so called *receding horizon* philosophy - a sequence of future control actions is chosen according to a prediction of the future evolution of the system and applied to the plant until new measurements

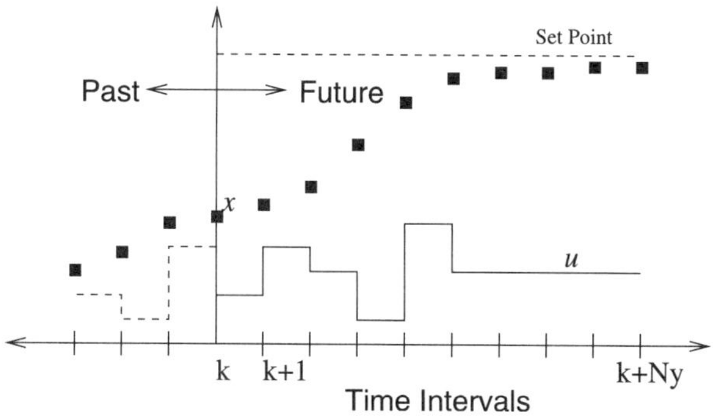

Figure 8 Model predictive control

are available. Then, a new sequence is determined which replaces the previous one. Each sequence is evaluated by means of an optimization procedure which takes into account two objectives, optimize the tracking performance; and protect the system from possible constraint violations.

Consider the following state-space representation of a given process model (Pistikopoulos et al., 2002):

$$\begin{cases} x(t+1) &= Ax(t) + Bu(t) \\ y(t) &= Cx(t), \end{cases} \tag{4.1}$$

subject to the following constraints:

$$\begin{array}{c} y_{min} \leq y(t) \leq y_{max} \\ u_{min} \leq u(t) \leq u_{max}, \end{array} \tag{4.2}$$

where $x(t) \in \Re^n$, $u(t) \in \Re^m$, and $y(t) \in \Re^p$ are the state, input, and output vectors respectively, subscripts min and max denote lower and upper bounds respectively and (A, B) is stabilizable. Model based control problems for regulating to the origin can then be posed as the following optimization problems:

$$\begin{aligned} \min_U \quad & J(U, x(t)) = x^T_{t+N_y|t} P x_{t+N_y|t} + \sum_{k=0}^{N_y-1} [x^T_{t+k|t} Q x_{t+k|t} + u^T_{t+k} R u_{t+k}] \\ \text{s.t.} \quad & y_{min} \leq y_{t+k|t} \leq y_{max}, \ k = 1, \ldots, N_c \\ & u_{min} \leq u_{t+k} \leq u_{max}, \ k = 0, 1, \ldots, N_c \\ & x_{t|t} = x(t) \\ & x_{t+k+1|t} = Ax_{t+k|t} + Bu_{t+k}, \ k \geq 0 \\ & y_{t+k|t} = Cx_{t+k|t}, \ k \geq 0 \end{aligned} \tag{4.3}$$

where $U \triangleq \{u_t, \ldots, u_{t+N_u-1}\}$, $Q = Q^T \succeq 0$, $R = R^T \succ 0$, $P \succeq 0$, $N_y \geq N_u$ and the superscript T denotes the transpose of the corresponding vector or matrix. The basic idea of an MPC implementation is shown in Figure 8 where at the current time interval, k, the optimization problem (4.3) is solved to minimize the state and control deviations from the set point by implementing the opitmal values of the control variables. Note that only the first control element is implemented and this sequence is repeated at the next time intervals, for the new state measurements or estimates, until the desired or set point values are obtained.

The problem (4.3) is solved repetitively at each time t for the current measurement $x(t)$ and the vector of predicted state variables, $x_{t+1|t}, \ldots, x_{t+k|t}$ at time $t+1, \ldots, t+k$ respectively and corresponding control actions $u_t, \ldots, u_{t+k-1}$ is obtained. In the following paragraphs, a parametric programming approach which avoids a repetitive solution of (4.3) is presented. First, we do some algebraic manipulations to recast (4.3) in a form suitable for using and developing some new parametric programming concepts. By making the following substitution in (4.3):

$$x_{t+k|t} \quad = \quad A^k x(t) + \sum_{j=0}^{k-1} A^j B u_{t+k-1-j} \tag{4.4}$$

the objective $J(U, x(t))$ can be formulated as the following Quadratic Programming (QP) problem:

$$\begin{aligned} \min_U \quad & \tfrac{1}{2} U^T H U + x^T(t) F U + \tfrac{1}{2} x^T(t) Y x(t) \\ \text{s.t.} \quad & GU \leq W + Ex(t) \end{aligned} \tag{4.5}$$

where $U \triangleq [u_t^T, \ldots, u_{t+N_u-1}^T]^T \in \Re^s$, $s \triangleq mN_u$, is the vector of optimization variables, $H = H^T \succ 0$, and H, F, Y, G, W, E are obtained from Q, R and (4.3)–(4.4). The QP problem (4.5) can now be formulated as the following Multi-parametric Quadratic Program (mp-QP):

$$\begin{aligned} \mu(x) = \quad & \min_z \quad \tfrac{1}{2} z^T H z \\ & \text{s.t.} \quad Gz \leq W + Sx(t), \end{aligned} \tag{4.6}$$

where $z \triangleq U + H^{-1}F^T x(t)$, $z \in \Re^s$, represents the vector of optimization variables, $S \triangleq E + GH^{-1}F^T$ and x represents the vector of parameters. The main advantage of writing (4.3) in the form given in (4.6) is that z (and therefore U) can be obtained as an affine function of x for the complete feasible space of x by using the algortihm described in section 2.1.

This approach provides a significant advancement in the solution and real time implementation of model based control problems. Since its application results in a complete set of control variables as a function of state-variables and the corresponding regions of validity, which are computed off-line. Therefore during on-line optimization, no optimizer needs to be called and instead for the current state of the plant, the region, CR^0, where the value of the state variables is valid, can be identified by substituting the value of these state variables into the inequalities which define the regions. Then, the corresponding control variables can be computed by using a function evaluation of the corresponding affine function (see Figure 9). Figure 10 demonstrates how advanced controllers can be implemented on a simple hardware.

4.2 Example. Consider the following state-space representation:

$$\left\{ \begin{aligned} x(t+1) &= \begin{bmatrix} 0.7326 & -0.0861 \\ 0.1722 & 0.9909 \end{bmatrix} x(t) + \begin{bmatrix} 0.0609 \\ 0.0064 \end{bmatrix} u(t) \\ y(t) &= \begin{bmatrix} 0 & 1.4142 \end{bmatrix} x(t). \end{aligned} \right. \tag{4.7}$$

The constraints on input are as follows:

$$-2 \leq u(t) \leq 2 \tag{4.8}$$

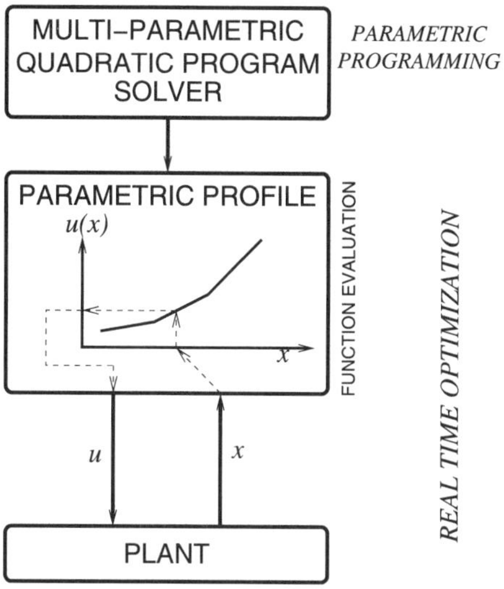

Figure 9 Real time optimization via parametric programming

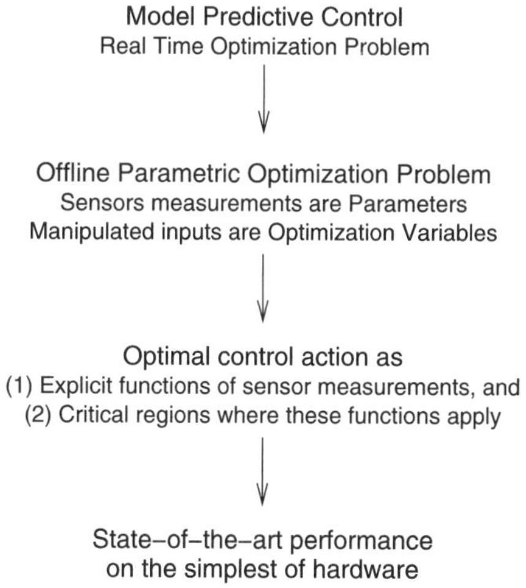

Figure 10 Achieving state-of-the-art control performance on simple hardware

The corresponding optimization problem of the form (4.3) for regulating to the origin is given as follows:

$$\begin{aligned} \min_{u_t,u_{t+1}} \quad & x'_{t+2|t}Px_{t+2|t} + \sum_{k=0}^{1} x'_{t+k|t}x_{t+k|t} + .01u^2_{t+k} \\ \text{s.t.} \quad & -2 \le u_{t+k} \le 2, \;\; k = 0,1 \\ & x_{t|t} = x(t) \end{aligned} \tag{4.9}$$

where P, Q and R are the weighting matrices, $N_u = N_y = N_c = 2$. The closed-loop response from the initial condition $x(0) = [1\ 1]'$ is shown in Figure 11. The

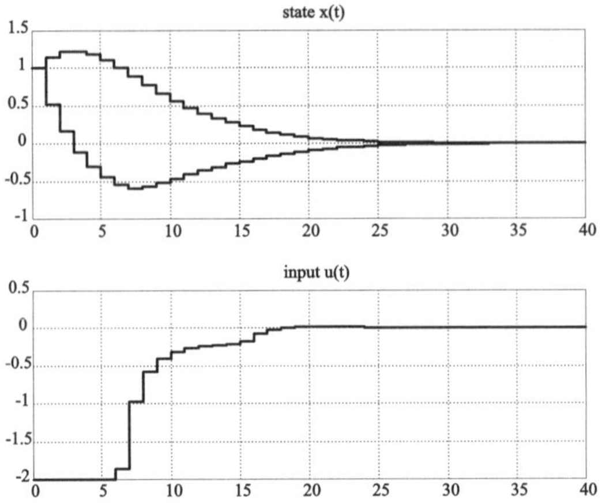

Figure 11 Closed loop response

same problem is now solved by using the parametric programming approach. The corresponding mp-QP problem of the form (4.6) has the following constant vectors and matrices.

$$H = \begin{bmatrix} 0.0196 & 0.0063 \\ 0.0063 & 0.0199 \end{bmatrix}, F = \begin{bmatrix} 0.1259 & 0.0679 \\ 0.0922 & -0.0924 \end{bmatrix},$$

$$G = \begin{bmatrix} 1 & 0 \\ -1 & 0 \\ 0 & 1 \\ 0 & -1 \end{bmatrix}, W = \begin{bmatrix} 2 \\ 2 \\ 2 \\ 2 \end{bmatrix}, E = \begin{bmatrix} 0 & 0 \\ 0 & 0 \\ 0 & 0 \\ 0 & 0 \end{bmatrix}$$

The solution of the mp-QP problem as computed by using the algorithm is provided in Table 1 and is depicted in Figure 12. Note that the CRs 2,4 and 7,8 in Table 1 are combined together and a compact convex representation is obtained. To illustrate how on-line optimization reduces to a function evaluation problem, consider the starting point $x(0) = [1\ 1]'$. This point is substituted into the constraints defining the CRs in Table 1 and it satisfies only the constraints of $CR^{7,8}$ (see also Figure 12). The control action corresponding to $CR^{7,8}$ from Table 1 is $u^{7,8} = -2$, which is obtained without any further optimization calculations and it is same as the one obtained from the closed loop response depicted in Figure 11.

For the case when the model of the system involves uncertain parameters explicit robust control law can be derived as described in (Sakizlis et al., 2004). Systems described by continuous time linear differential equations are considered in Sakizlis et al. (2005). For application of parametric control to biomedical, process and automotive systems refer to (Dua et al., 2006, 2007).

4.3 Hybrid systems. Hybrid systems can be defined as systems comprising a number of interconnected continuous subsystems where the interconnections are determined by logical or discrete switchings. Each subsystem is governed by a unique set of differential and/or algebraic equations. In this section we focus on piecewise affine (PWA) systems (Bemporad and Morari, 1999). PWA systems are defined by partitioning the state and input space into polyhedral regions and

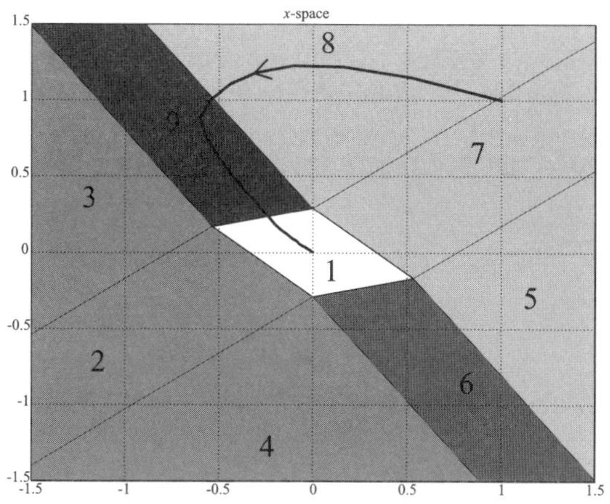

Figure 12 Polyhedral partition of the state-space

Table 1 Parametric solution of the numerical example

Region#	Region	u
1	$\begin{bmatrix} -5.9220 & -6.8883 \\ 5.9220 & 6.8883 \\ -1.5379 & 6.8291 \\ 1.5379 & -6.8291 \end{bmatrix} x \leq \begin{bmatrix} 2.0000 \\ 2.0000 \\ 2.0000 \\ 2.0000 \end{bmatrix}$	$\begin{bmatrix} -5.9220 & -6.8883 \end{bmatrix} x$
2,4	$\begin{bmatrix} -3.4155 & 4.6452 \\ 0.1044 & 0.1215 \\ 0.1259 & 0.0922 \end{bmatrix} x \leq \begin{bmatrix} 2.6341 \\ -0.0353 \\ -0.0267 \end{bmatrix}$	2.0000
3	$\begin{bmatrix} 0.0679 & -0.0924 \\ 0.1259 & 0.0922 \end{bmatrix} x \leq \begin{bmatrix} -0.0524 \\ -0.0519 \end{bmatrix}$	2.0000
5	$\begin{bmatrix} -0.1259 & -0.0922 \\ -0.0679 & 0.0924 \end{bmatrix} x \leq \begin{bmatrix} -0.0519 \\ -0.0524 \end{bmatrix}$	$-$ 2.0000
6	$\begin{bmatrix} -6.4159 & -4.6953 \\ -0.0275 & 0.1220 \\ 6.4159 & 4.6953 \end{bmatrix} x \leq \begin{bmatrix} 1.3577 \\ -0.0357 \\ 2.6423 \end{bmatrix}$	$\begin{bmatrix} -6.4159 & -4.6953 \end{bmatrix} x + 0.6423$
7,8	$\begin{bmatrix} 3.4155 & -4.6452 \\ -0.1044 & -0.1215 \\ -0.1259 & -0.0922 \end{bmatrix} x \leq \begin{bmatrix} 2.6341 \\ -0.0353 \\ -0.0267 \end{bmatrix}$	$-$ 2.0000
9	$\begin{bmatrix} 6.4159 & 4.6953 \\ 0.0275 & -0.1220 \\ -6.4159 & -4.6953 \end{bmatrix} x \leq \begin{bmatrix} 1.3577 \\ -0.0357 \\ 2.6423 \end{bmatrix}$	$\begin{bmatrix} -6.4159 & -4.6953 \end{bmatrix} x - 0.6423$

associating with each region a different linear state update equation

$$x(t+1) = A^i x(t) + B^i u(t) + f^i \tag{4.10}$$

$$\text{if} \begin{bmatrix} x(t) \\ u(t) \end{bmatrix} \in P^i$$

where $i = 1, ..., s$, $x \in \Re^{n_c} \times \{0,1\}^{n_l}, u \in \Re^{m_c} \times \{0,1\}^{m_l}, \{P_i\}_{i=1}^s$ is a polyhedral partition of the set of the state and input space $P \subset \Re^{n+m}, n \triangleq n_c + n_l, m \triangleq m_c + m_l$. P is assumed to be closed and bounded and $x_c \in R^{n_c}$ and $u_c \in \Re^{m_c}$ denote the continuous components of the state and input vector, respectively; $x_l \in \{0,1\}^{n_l}$ and $u_l \in \{0,1\}^{m_l}$ similarly denote the binary components.

Note that PWA models are not suitable for recasting analysis/synthesis problems into more compact optimization problems. For this purpose the Mixed Logical Dynamical (MLD) framework (Bemporad and Morari, 1999) is used. The general MLD form of a hybrid system is:

$$x(t+1) = Ax(t) + B_1 u(t) + B_2 \delta(t) + B_3 z(t) \tag{4.11}$$

$$y(t) = Cx(t) + D_1 u(t) + D_2 \delta(t) + D_3 z(t) \tag{4.12}$$

$$E_2 \delta(t) + E_3 z(t) \leq E_1 u(t) + E_4 x(t) + E_5 \tag{4.13}$$

where $x = \begin{bmatrix} x_c^T & x_l^T \end{bmatrix}^T \in R^{n_c} \times \{0,1\}^{n_l}$ are the continuous and binary states, $u = \begin{bmatrix} u_c^T & u_l^T \end{bmatrix}^T \in R^{m_c} \times \{0,1\}^{m_l}$ are the inputs, $y = \begin{bmatrix} y_c^T & y_l^T \end{bmatrix}^T \in R^{p_c} \times \{0,1\}^{p_l}$ the outputs, and $\delta \in \{0,1\}^{r_l}$, $z \in R^{r_c}$ represent auxiliary binary and continuous variables respectively. All constraints on the states, the inputs, the z and δ variables are summarized in the inequalities (4.13). Note that, although the description (4.11)-(4.12)-(4.13) seems to be linear, nonlinearity is hidden in the integrality constraints over the binary variables. MLD systems are a versatile framework to model various classes of systems. For a detailed description of such capabilities we defer the reader to Morari et al. (2003).

4.4 Predictive control of MLD systems. Let t be the current time, and $x(t)$ the current state. Consider the following optimal control problem

$$\min_{\{v_0^{T-1}\}} J(v_0^{T-1}, x(t))$$

$$\triangleq \sum_{k=0}^{T-1} \|v(k)\|_{Q_1}^2 + \|\delta(k|t)\|_{Q_2}^2 + \|z(k|t)\|_{Q_3}^2 + \|x(k|t)\|_{Q_4}^2 + \|y(k|t)\|_{Q_5}^2 \tag{4.14}$$

$$\begin{aligned} \text{s.t. } & x(k+1|t) = Ax(k|t) + B_1 v(k) + B_2 \delta(k|t) + B_3 z(k|t) \\ & y(k|t) = Cx(k|t) + D_1 v(k) + D_2 \delta(k|t) + D_3 z(k|t) \\ & E_2 \delta(k|t) + E_3 z(k|t) \leq E_1 v(k) + E_4 x(k|t) + E_5 \end{aligned} \tag{4.15}$$

where $v_0^{T-1} \triangleq [v^T(0), ..., v^T(T-1)]^T$, $Q_1 = Q_1^T > 0$, $Q_2 = Q_2^T \geq 0$, $Q_3 = Q_3^T \geq 0$, $Q_4 = Q_4^T > 0$ and $Q_5 = Q_5^T \geq 0$. $x(k|t) \triangleq x(t+k, x(t), v_0^{k-1})$ is the state predicted at time $t+k$ resulting from the input $u(t+k) = v(k)$ to (4.11-4.13) starting from $x(0|t) = x(t)$. $\delta(k|t)$, $z(k|t)$ and $y(k|t)$ are similarly defined. Assume for the moment that the optimal solution $\{v_t^*(k)\}_{k=0,\ldots,T-1}$ exists. According to the *receding horizon* philosophy mentioned above, set

$$u(t) = v_t^*(0), \tag{4.16}$$

disregard the subsequent optimal inputs $v_t^*(1), \ldots, v_t^*(T-1)$, and repeat the whole optimization procedure at time $t+1$. Note that (4.14-4.15) is a Mixed Integer Quadratic Program (MIQP). This problem can be formulated as a Mixed Integer Linear Program (MILP) if 1 norm instead of the 2 norm is considered in the objective function. The repetitive solution of the MIQP or MILP can be avoided by formulating (4.14-4.15) as a multiparametric program and solving it to obtain the control variables as a set of explicit functions of the current state of the system and the regions in the space of the state variables where the explicit functions remain

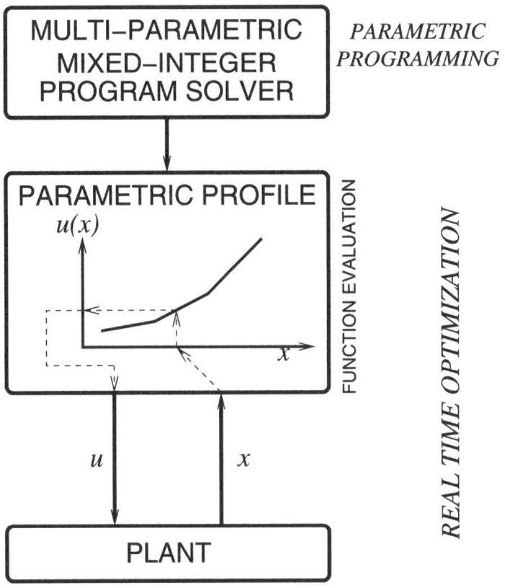

Figure 13 Real time optimization of hybrid systems via parametric programming

valid (Bemporad et al., 2000; Sakizlis et al., 2002). This is achieved by recasting (4.14-4.15) in a compact form as follows:

$$\begin{aligned} J(x(t)) &= \min_{\pi_c,\pi_d} \pi_c^T Q_c \pi_c + \phi^T \pi_d \\ \text{s.t.} \quad G_c\pi_c + G_d\pi_d &\le S + Fx(t) \end{aligned} \tag{4.17}$$

where π_c and π_d are continuous and discrete variables of (4.14-4.15), Q_c, ϕ^T, G_c, G_d, S, F are constant matrices and vectors of appropriate dimensions and Q_c is symmetric and positive definite. $x(t)$ is the state at the current time t. The objective is to obtain π_c and π_d as a function of $x(t)$ without exhaustively enumerating the entire space of $x(t)$. This can be achieved by using the algorithm for Multiparametric Mixed Integer Linear Programs (mp-MILP) described in section 3.1. This reduces the real time hybrid system control problem to a function evaluation problem (Figure 13).

5 Multiparametric programming for bilevel programming problems

A global optimization approach to a bilevel programming problem which refers to an optimization problem that is constrained by another problem is presented Faisca et al. (2007). Using parametric programming techniques, the proposed approach transforms the bilevel problem into a family of single optimization problems, which can be solved to global optimality for linear-linear, linear-quadratic, quadratic-linear, and quadratic-quadratic bilevel models.

Many industrial situations involve several groups which are inter-connected in a hierarchical structure. Each group may be an individual or an agency which has independent, perhaps conflicting, objectives. These situations can be mathematically modeled into a multi-level programming problem (Clark and Westerberg 1983, 1990). When there are two decision makers, this can be formulated as a BiLevel

Programming Problem (BLPP) which is generally of the following form:

$$\begin{aligned} &\min_x F(x,y) \\ s.t.\ &G(x,y) \le 0 \\ &\min_y \ f(x,y) \\ s.t.\ &g(x,y) \le 0 \end{aligned} \tag{5.1}$$

where $x \in X$ and $y \in Y$ and X and Y are compact and polyhedral regions. $F(x,y)$ is called a higher level decision problem, namely *a leader*'s problem or an outer problem and $f(x,y)$ is called a lower level decision problem, *a follower*'s problem or an inner problem.

As can be seen in (5.1), a BLPP refers to an optimization problem that is constrained by another optimization problem. The two problems are connected in such a way that the leader's problem sets parameters influencing the follower's problem and the leader's problem, in turn, is affected by the outcome of the follower's problem.

To compare the two problems in terms of scope of information, a follower makes decisions using only its local information while a leader does so using the complete information including the follower's possible reaction to the leader's decision. This is an important feature which can address complicated industrial situations which may be difficult to model using other modeling methodologies. Therefore the BLPP has been applied in extensive and diverse areas. A bibliography and good reviews of works on BLPPs can be found in Vicente and Calamai (1994).

In the literature, research on solving BLPPs has mainly focused on the linear case in the following two major ways. One aims to transform the original problem into a single optimization problem by employing the Karush-Kuhn-Tucker optimality condition of the lower level problem (Visweswaran et al., 1996). The other utilizes enumeration techniques based on the fact that an optimal solution to a bilevel problem is a basic feasible solution of the linear constraints involved at the lower and upper level and consequently must occur at an extreme point of the feasible set (Bard and Moore, 1990). Both methodologies require many additional variables and complex computations which involve many iterations. The algorithms and applications which have been presented so far can be found in Bard(1998) and test problems can be found in the recent book by Floudas et al.(1999). Many references on complexity issues of BLPPs can be found in Gümüs and Floudas (2001) and Floudas et al. (2001).

In order to fully utilize the potential of BLPPs which is such an important modeling framework, an efficient solution methodology is of critical importance. Many approaches have been proposed but most of them do not allow problems to be solved to global optimality because the solution space of the BLPP is non-convex (Visweswaran et al., 1996). A novel global optimization methodology to solve the bilevel programming problem is presented here.

5.1 Theory and algorithm. The following terminologies relating to BLPPs in the literature are worth noting again. First, the relaxed feasible set of BLPP, Ω is defined as follows:

$$\Omega = \{(x,y) : G(x,y) \le 0, g(x,y) \le 0\}.$$

If there are no $(x, y) \in \Omega$, then the BLPP is infeasible. This paper hereafter assumes that there exist solutions satisfying the above set.

The feasible set of the inner problem, $S(x)$ is defined for every $x \in X$ as :

$$S(x) = \{y | g(x, y) \leq 0\}.$$

The rational reaction set of the inner problem, $RR(x)$ is defined for every $x \in X$ as:

$$RR(x) = \{y \in \arg\min f(x, y) | y \in S(x)\}.$$

The BLPP feasible set (the induced, or inducible region), IR is defined as follows:

$$IR = \{(x, y) : (x, y) \in \Omega, y \in RR(x)\}.$$

The challenge associated with solving BLPPs is mainly caused by the fact that the BLPP feasible set (inducible region) becomes non-convex (Visweswaran et al., 1996). However, this difficulty may be avoided if the inducible region can be obtained as a set of explicit functions of variable x of the outer problem. The issue is then how to compute the functions with the corresponding conditions which validate the functions.

As an alternative method to obtain the explicit expression of y as a function of x, this paper introduces parametric programming techniques which have been mainly used for computing the solutions of problems involving uncertain parameters. This is a novel idea because there is little research to our knowledge employing parametric programming techniques in solving BLPP although it is widely known that the feasible set of the inner problem is *parametric* in terms of the decision variables of the outer problem.

At first, the proposed methodology will be illustrated for the case of a bilevel linear programming problem which is of the following form:

$$\begin{aligned} \min_x F(x, y) &= c_1^T x_1 + d_1^T y \\ s.t.\ \ A_1 x + B_1 y &\leq b_1 \\ \min_y f(x, y) &= \ c_2^T x + d_2^T y \\ s.t.\ \ A_2 x + B_2 y &\leq b_2 \end{aligned} \tag{5.2}$$

where $x \in X \subseteq R^{n1}, y \in Y \subseteq R^{n2}$, $c_1, c_2 \in R^{n1}, d_1, d_2 \in R^{n2}, b_1 \in R^j, b_2 \in R^{j'}, A_1 \in R^{j \times n1}, A_2 \in R^{j' \times n1}, B_1 \in R^{j \times n2}$, and $B_2 \in R^{j \times n2}$.

In order to obtain the explicit expression of IR, consider the inner problem of the above BLPP:

$$\begin{aligned} \min_y \ & c_2^T x + d_2^T y \\ s.t.\ \ & A_2 x + B_2 y \leq b_2 \end{aligned} \tag{5.3}$$

(5.3) can be transformed into the following Multiparametric Linear Programming (mp-LP) problem by assuming variables x as parameters:

$$\begin{aligned} \min_y \ & d_2^T y + c_2^T x \\ s.t.\ \ & B_2 y \leq b_2 - A_2 x. \end{aligned} \tag{5.4}$$

By solving (5.4) using the parametric programming techniques, a complete profile of optimal solutions of (5.3) can be obtained as an explicit function of variables of the outer problem with the corresponding valid boundary conditions, which is as follows:

$$y = \begin{cases} \xi_1(x) = m_1 + n_1 x & \text{if } H_1 x \leq h_1, \\ \xi_2(x) = m_2 + n_2 x & \text{if } H_2 x \leq h_2, \\ \vdots & \vdots \\ \xi_k(x) = m_k + n_k x & \text{if } H_k x \leq h_k, \end{cases} \tag{5.5}$$

where k denotes the number of computed parametric solutions, H_k is a constant matrix, h_k is a constant vector, and the corresponding valid boundary condition of the kth parametric solution, $H_k x \leq h_k$ is called a critical region, CR^k (2.1)..

Using the explicit expression of the inducible region in the form of (5.5), a bilevel programming problem (5.2) can be transformed into a family of single optimization problems which are as follows:

$$\begin{aligned} & z_1 = \min_x c_1^T x + d_1^T \xi_1(x), \\ s.t. \quad & A_1 x + B_1 \xi_1(x) \leq b_1, \\ & H_1 x \quad \leq h_1, \\ \\ & z_2 = \min_x c_2^T x + d_2^T \xi_2(x), \\ s.t. \quad & A_2 x + B_2 \xi_2(x) \leq b_2, \\ & H_2 x \quad \leq h_2, \\ & \vdots \\ & z_k = \min_x c_k^T x + d_k^T \xi_k(x), \\ s.t. \quad & A_k x + B_k \xi_k(x) \leq b_k, \\ & H_k x \quad \leq h_k. \end{aligned} \tag{5.6}$$

Solutions of the above single LP problems correspond to the local optimal solutions of the original BLPP because parametric programming techniques search all solution spaces of the parametric problem. Therefore the original bilevel programming problems are solved to global optimality by solving the above single linear programming problems.

The proposed methodology can be also applied for the case of bilevel quadratic programming problems. If the inner problem has a quadratic objective function, we can formulate the problem as a multi-parametric Quadratic Programming (mp-QP) problem which corresponds to the following general class of mp-QP problem :

$$\begin{aligned} z(x) = \min_y c^T y + ct^T x + \frac{1}{2} y^T Q y \\ s.t. \ A'y \leq b' + F'x \\ x_{min} \leq x \leq x_{max} \end{aligned} \tag{5.7}$$

where Q is an $(n \times n)$ symmetric positive definite constant matrix, A' and F' are constant matrices and b' is a constant vector. The remaining procedure is the same as the one for the linear case.

In the next section, a numerical example will be solved to illustrate the proposed methodology.

5.2 Numerical example. This problem is taken from Bard (1983) and Visweswaran et al.(1996):

$$
\begin{aligned}
& \min_x F(x,y) = x + y \\
s.t. \quad & -x \le 0 \\
& \min_y f(x,y) = -5x - y \\
s.t. \quad & -x - 0.5y \le -2 \\
& -0.25x + y \le 2 \\
& x + 0.5y \le 8 \\
& x - 2y \le 2 \\
& -y \le 0
\end{aligned}
$$

According to the proposed methodology, the inner problem is transformed into the following mp-LP problem:

$$
\begin{aligned}
& \min_y -y - 5x \\
s.t. \quad & \begin{bmatrix} -0.5 \\ 1 \\ 0.5 \\ -2 \\ -1 \end{bmatrix} y \le \begin{bmatrix} -2 \\ 2 \\ 8 \\ 2 \\ 0 \end{bmatrix} + \begin{bmatrix} 1 \\ 0.25 \\ -1 \\ -1 \\ 0 \end{bmatrix} x
\end{aligned}
\tag{5.8}
$$

The application of parametric programming techniques yields the following optimal parametric solution of (5.8):

$$
y = \begin{cases} 0.25x + 2 & \text{if } 0.8889 \le x \le 6.22 \\ -2x + 16 & \text{if } 6.22 \le x \le 6.8 \end{cases}
$$

As a result, the leader's problem corresponds to the following two LP problems:

$$
\begin{aligned}
& \min_x 1.25x + 2 \\
& s.t \ 0.8889 \le x \le 6.22
\end{aligned}
\tag{5.9}
$$

$$
\begin{aligned}
& \min_x -x + 16 \\
& s.t \ 6.22 \le x \le 6.8
\end{aligned}
\tag{5.10}
$$

The optimal solution of problem (5.9) is 3.111 when (x,y) is (0.8889, 2.222) and the optimal solution of problem (5.10) is 9.2 when (x,y) is (6.8, 2.4). Therefore the global optimum is 3.111 which is the same as that in the previous studies.

6 Concluding remarks

An overview of theory and applications of parametric programming and control has been presented. Some topics within this area have been discussed in detail and illustrative examples have also been presented. A bird's eye view of the remaining topics and relevant references have been provided. The main fundamental contribution of this work is to be able to obtain the solution vector as an explicit function of the parameters which is guaranteed to be optimal and is obtained without exhaustively enumerating the entire space of the parameters. Due to the generic nature of the techniques presented here, the optimization variables and parameters can be defined according to the particular application under consideration.

References

Acevedo, J., Pistikopoulos, E. N., 1997a. A hybrid parametric/ stochastic programming approach for mixed-integer linear problems under uncertainty. Industrial and Engineering Chemistry Research 36, 2262–2270.

Acevedo, J., Pistikopoulos, E. N., 1997b. A multiparametric programming approach for linear process engineering problems under uncertainty. Industrial and Engineering Chemistry Research 36, 717–728.

Adjiman, C. S., Androulakis, I. P., Floudas, C. A., 2000. Global optimization of mixed-integer nonlinear problems. American Institute of Chemical Engineers Journal 46, 1769–1797.

Androulakis, I. P., Maranas, C. D., Floudas, C. A., 1995. α-BB: A global optimization method for general constrained nonconvex problems. J. Global Optimization 7, 337–363.

Bard, J., 1983. An algorithm for solving the general bilevel programming problem. Math. Oper. Res. 8, 260–272.

Bemporad, A., Borrelli, F., Morari, M., 2000. Piecewise linear optimal controllers for hybrid systems. In: Proc. American Control Conf.

Bemporad, A., Morari, M., 1999. Control of systems integrating logic, dynamics, and constraints. Automatica 35, 407–427.

Dua, P., III, F. J. D., Pistikopoulos, E. N., 2006. Model-based blood glucose control for type i diabetes. IEEE Transactions on Biomedical Engineering 53, 1478.

Dua, P., Kouramas, K., Dua, V., Pistikopoulos, E. N., 2007. Mpc on a chip - recent advances on the application of multi-parametric model based control. Computers & Chemical Engineering .

Dua, V., Bozinis, N. A., Pistikopoulos, E. N., 2002. A multiparametric programming approach for mixed-integer quadratic engineering problems. Computers & Chemical Engineering 26, 715–733.

Dua, V., Papalexandri, K. P., Pistikopoulos, E. N., 2004. Global optimization issues in multiparametric continuous and mixed-integer optimization problems. Journal of Global Optimization 30, 59–89.

Dua, V., Pistikopoulos, E. N., 1999. Algorithms for the solution of multiparametric mixed-integer nonlinear optimization problems. Industrial and Engineering Chemistry Research 38, 3976–3987.

Dua, V., Pistikopoulos, E. N., 2000. An algorithm for the solution of multiparametric mixed integer linear programming problems. Annals of Operations Research 99, 123–139.

Faisca, N. P., Dua, V., M., P., Saraiva, Rustem, B., Pistikopoulos, E. N., 2007. Parametric global optimisation for bilevel programming. Journal of Global Optimization 38, 609–623.

Fiacco, A. V., 1976. Sensitivity analysis for nonlinear programming using penalty methods. Mathematical Programming 10, 287–311.

Fiacco, A. V., 1983. Introduction to Sensitivity and Stability Analysis in Nonlinear Programming. Academic Press, New York.

Fiacco, A. V., Kyparisis, J., 1986. Convexity and concavity properties of the optimal value function in parametric nonlinear programming. J. of Optim. Theory and Appl. 48, 95–126.

Floudas, C.A., Z. G., Ierapetritou, M., 2001. Global optimization in design under uncertainty: Feasibility test and flexibility index problems. Industrial and Engineering Chemistry Research 40, 4267–4282.

Gal, T., 1995. Postoptimal Analyses, Parametric Programming, and Related Topics. de Gruyter, New York.

Gümüs, Z., Floudas, C., 2001. Global optimization of nonlinear bilevel programming problems. J. of G. Opti. 20, 1–31.

McCormick, G. P., 1976. Computability of global solutions to factorable nonconvex programs: Part i - convex underestimating problems. Math. Progr. 10, 147–175.

Morari, M., Baotic, M., Borrelli, F., 2003. Hybrid systems modeling and control. European Journal of Control 9, 177–189.

Papalexandri, K. P., Dimkou, T. I., 1998. A parametric mixed integer optimization algorithm for multi-objective engineering problems involving discrete decisions. Industrial and Engineering Chemistry Research 37 (5), 1866–1882.

Pistikopoulos, E. N., Bozinis, N. A., Dua, V., August 1999. POP: A MATLAB (© The Math Works, Inc.) implementation of multi-parametric quadratic programming algorithm. Centre for Process Systems Engineering, Imperial College.

Pistikopoulos, E. N., Bozinis, N. A., Dua, V., April 2000. A MATLAB (© The Math Works, Inc.) implementation of multi-parametric mixed-integer linear programming algorithm. Centre for Process Systems Engineering, Imperial College.

Pistikopoulos, E. N., Dua, V., Bozinis, N. A., Bemporad, A., Morari, M., 2002. On-line optimization via off-line parametric optimization tools. Computers & Chemical Engineering 26, 175–185.

Pistikopoulos, E. N., Georgiadis, M. C., Dua, V. (Eds.), 2007a. Multi-parametric model based control - theory and applications. Wiley-VCH.

Pistikopoulos, E. N., Georgiadis, M. C., Dua, V. (Eds.), 2007b. Multi-parametric programming - theory, algorithms and applications. Wiley-VCH.

Sakizlis, V., , Perkins, J. D., Pistikopoulos, E. N., 2005. Explicit solution to optimal control problems for constrained continuous-time linear systems. IEE Proceedings - Control Theory and Applications 152, 443–452.

Sakizlis, V., Dua, V., Kakalis, N., Perkins, J. D., Pistikopoulos, E. N., 2004. Design of robust model-based controllers via parametric programming. Automatica 40, 189–201.

Sakizlis, V., Dua, V., Perkins, J. D., Pistikopoulos, E. N., 2002. The explicit control law for hybrid systems via parametric programming, proceedings of the 2002 American Control Conference, Anchorage.

Fields Institute Communications
Volume **55**, 2009

An Overview of Advances in Global Optimization During 2003–2008

C.A. Floudas
Department of Chemical Engineering
Princeton University
Princeton, NJ 08544-5263 USA
`floudas@titan.princeton.edu`

C.E. Gounaris
Department of Chemical Engineering
Princeton University
Princeton, NJ 08544-5263 USA
`gounaris@princeton.edu`

Abstract. This chapter presents an overview of the research progress in deterministic global optimization during the years 2003-2008, and a brief account of our recent research contributions. The review part covers the areas of twice continuously differentiable nonlinear optimization, mixed-integer nonlinear optimization, optimization with differential-algebraic models, semi-infinite programming, optimization with grey box/non-factorable models, and bilevel nonlinear optimization. In the research contributions part, we discuss the convex and concave envelopes for trilinear monomials, convex underestimators for trigonometric functions, p-αBB and g-αBB global optimization approaches, tight convex underestimators for $\mathcal{C}^2$-continuous functions, optimal functional forms of convex underestimators, and the convexity of products of univariate functions.

1 Introduction

Global optimization addresses the computation and characterization of global solutions to nonconvex continuous, mixed-integer, differential-algebraic, bilevel, and non-factorable problems. Given an objective function f that is to be minimized and a set of equality and inequality constraints S, the main task of *Deterministic Global Optimization* is to determine (with theoretical guarantees) an epsilon global minimum of the objective function f subject to the set of constraints S. Besides

2000 *Mathematics Subject Classification.* Primary 90C26; Secondary 90C90.
Key words and phrases. Global optimization; NLP; MINLP; GDP; DAE; SIP; Nonfactorable models; Bilevel optimization; Nonconvexities; Convex envelopes; Convex underestimation; αBB theory.

this, global optimization focuses also on the important issues of how to determine lower and upper bounds on the global minimum of the objective function f that are valid for the whole feasible region S, and how to determine an ensemble of good quality local solutions in the vicinity of the global solution. Other objectives may involve the enclosure of all solutions of a set of equality and inequality constraints S, or proving that a constrained nonlinear problem is feasible or infeasible.

Global optimization has found an increased number of applications not only in Chemical Engineering but also across all branches of engineering, applied sciences, and sciences. Complex problems, like the ones arising in refinery pooling, azeotropic distillation and phase and chemical equilibrium, have been tackled by global optimization approaches. Furthermore, many interesting mathematical problems, such as the enclosure of all solutions of systems of nonlinear equations or the parameter estimation in nonlinear algebraic models, can be expressed with global optimization formulations. As a result, we have experienced significant interest in new theoretical advances, algorithmic and implementation related investigations, and their application to important scientific problems.

During the last five years, 2003-2008, many research contributions in the area of global optimization were presented at various conferences around the world – notably including the two more recent editions of the dedicated conference series "Frontiers in Global Optimization", which were held on the Greek islands of Santorini (in 2003) and Myconos (in 2007).

Furthermore, several outstanding books have been published addressing different facets of global optimization. These include the textbooks by Zabinsky (2003) and Zhigljavsky and Zilinskas (2007) on stochastic approaches, Hansen and Walster (2004) on interval methods, Price *et al.* (2005) on differential evolution algorithms, as well as the collection of essays edited by Audet *et al.* (2005). An essay on parallel computing in global optimization was presented by D'Apuzzo *et al.* (2006). In addition to the above, collections of contributed papers were also published. These include the books by Floudas and Pardalos (2003), Liberti and Maculan (2006), that includes implementation discussions and commercial software presentation, and Törn and Zilinskas (2007) on global optimization algorithms of various kinds.

The introduction of many algorithms and the need for their computational comparison led to the development of standardized collections of benchmark problems in local and global optimization. Such collections of problems can be found in the publications by Casado *et al.* (2003) and Ali *et al.* (2005), as well as the benchmark suite compiled by Shcherbina *et al.* (2003). The latter was used by Neumaier *et al.* (2005) to perform a computational comparison between eight global optimization solvers with the help of the COCONUT environment Schichl (2004).

A second edition of the Encyclopedia of Optimization was recently published Floudas and Pardalos (2008). This Encyclopedia constitutes an effort to systematically compile results in the field and present them in an orderly and carefully designed fashion. Providing extensive references for each entry, it introduces the reader to a complete set of topics that show the spectrum of research, the richness of ideas, and the breadth of applications that has come from the field of Optimization.

Finally, a number of review papers discussed the various advances in the field. A survey paper by Neumaier (2004) discussed constrained global optimization and continuous constraint satisfaction problems with a particular emphasis on the use of complete techniques that provably find all solutions, while Floudas *et al.* (2005) provided a thorough review of developments that covered the years 1998-2003.

This chapter presents an overview of the research progress in deterministic global optimization during the years 2003-2008, and a brief account of our recent research contributions. The review part covers the following classes of mathematical problems: (i) twice continuously differentiable nonlinear optimization, NLPs; (ii) mixed-integer nonlinear optimization, MINLPs; (iii) differential-algebraic systems, DAEs; (iv) semi-infinite programming, SIP; (v) grey-box and nonfactorable problems; and (vi) bilevel nonlinear optimization. After presenting all the advances in the aforementioned classes, we will focus on a select number of contributions from the Computer-Aided Systems Laboratory at Princeton University. These contributions are on: (a) explicit facets of convex and concave envelopes for trilinear monomials Meyer and Floudas (2003, 2004); (b) convex underestimators for trigonometric functions Caratzoulas and Floudas (2005); (c) convex underestimators by piecewise quadratic perturbation Meyer and Floudas (2005b); (d) the generalized αBB global optimization approach Akrotirianakis and Floudas (2004a,b); (e) tight convex underestimators for $\mathcal{C}^2$-continuous functions Gounaris and Floudas (2008a,b); (f) functional forms of convex underestimators Floudas and Kreinovich (2007a,b); and (g) convexity of products of univariate functions Gounaris and Floudas (2008c).

2 Twice continuously differentiable NLPs

In the first part of this section, we will review the advances in convex envelopes and underestimators, convexification techniques for twice continuously differentiable NLPs, and results regarding concave, bilinear and fractional models. We will subsequently cover general theoretical and algorithmic advances in the area of C^2 NLPs, including results that are focused on particular applications. Special focus will be given on advances for the global optimization of computational biology and phase equilibrium problems.

2.1 Convex envelopes. Liberti and Pantelides (2003) proposed a nonlinear continuous and differentiable convex envelope for monomials of odd degree, derived its linear relaxation, and compared their results with other relaxation methods. Meyer and Floudas (2003) studied trilinear monomials with positive or negative domains, derived explicit expressions for the facets of the convex and concave envelopes and showed that these outperform the previously proposed relaxations based on arithmetic intervals or recursive arithmetic intervals. In a later publication Meyer and Floudas (2004), they presented the results for the case of mixed-sign domains.

Tardella (2003) studied the class of functions whose convex envelope on a polyhedron coincides with the convex envelope based on the polyhedron vertices, and proved important conditions for a vertex polyhedral convex envelope. Meyer and Floudas (2005a) described the structure of the polyhedral convex envelopes of edge-concave functions over polyhedral domains using geometric arguments. They developed an algorithm for computing the facets of the convex envelope over hyperrectangles in R^3 and derived sufficient conditions for the convex envelope of a sum of edge-concave functions to be equivalent to the sum of the convex envelopes of these functions. Tardella (2008) studied vertex polyhedral convex envelopes, derived sufficient conditions for their existence, and characterized their sum decomposability. This work effectively extends and unifies several results previously obtained for special cases of this problem.

2.2 Convex underestimators and relaxations. Akrotirianakis and Floudas (2004a) introduced a new class of convex underestimators for twice continuously differentiable NLPs, studied their theoretical properties, and proved that the resulting convex relaxation is improved compared to the αBB one. Furthermore, Akrotirianakis and Floudas (2004b) presented computational results of the new class of convex underestimators embedded in a branch-and-bound framework for box-constrained NLPs. They also proposed a hybrid global optimization method that includes the random-linkage stochastic approach with the aim at improving the computational performance. Caratzoulas and Floudas (2005) proposed novel convex underestimators for trigonometric functions which are trigonometric functions themselves. The underestimation method can be applied to one-dimensional as well as multi-dimensional problems involving trigonometric polynomials, since the product of trigonometric functions can always be decomposed into the sum of sin and cos functions with arguments that are linear combinations of the problem variables.

Meyer and Floudas (2005b) proposed two new classes of convex underestimators for general C^2 NLPs which combine the αBB underestimators within a piecewise quadratic perturbation, derived properties for the smoothness of the convex underestimators, and showed the improvements over the classical αBB convex underestimators for box-constrained optimization problems. Sherali *et al.* (2005) proposed a new cutting plane methodology that is based on the construction of a partial convex hull representation for a given 0-1 mixed-integer programming problem by using the reformulation-linearization technique (RLT). The cuts are generated by projecting the extended space of the RLT formulation into the original space, and the authors investigated several variable selection rules for performing this convexification in a computationally efficient manner. Gounaris and Floudas (2008a) developed tight convex underestimators for univariate C^2-continuous functions of arbitrary structure. These are based on a piecewise application of the αBB underestimators and it is theoretically proven that a finite number of pieces is sufficient for the method to yield the -a priori unknown- convex envelope of the function. The methodology was extended to handle multivariate functions Gounaris and Floudas (2008b), through appropriate projections of the function's epigraph into select one-dimensional spaces. Orthonormal transformations were also employed to improve the quality of the underestimation.

2.3 Convexification techniques. Björk *et al.* (2003) studied convexifications for signomial terms, introduced properties of power convex functions, compared the effect of the convexification schemes for heat exchanger network problems, and studied quasi-convex convexifications. Li *et al.* (2005) discussed nonconvex minimization problems that can be equivalently transformed into convex ones, and thus be globally solved by use of only local techniques. They derived sufficient conditions for identifying such classes of problems, which are called hidden convex minimization problems. Hidden convexity was also the focus in the work by Wu *et al.* (2007). General convexification and concavification methods were proposed for some classes of monotone Wu *et al.* (2005a) and non-monotone Wu *et al.* (2005b) optimization problems, which convert them into the better structured classes of concave minimization, d.c. programming, and reverse convex programming problems.

Zlobec (2005) showed that the Liu-Floudas convexification Liu and Floudas (1993) can be possible for every Lipschitz continuously differentiable function, practically extending the applicability of the GOP algorithm Floudas and Visweswaran (1990) to almost every smooth program occuring in practice. Zlobec (2006) also worked on the characterization of convexifiable functions by developing necessary and sufficient conditions for the decomposition of a function into a sum of a convex and a quadratic concave function. Westerlund (2006) reviewed the work on power convex transformations that are applicable to signomial programs, the optimal exponents to be used in such an approach, the tight underestimation of the resulting program via piecewise linear approximations, and the computational framework developed for the global optimization of signomial programs. Lundell *et al.* (2008) presented a method for the optimal selection of the variables that need to be transformed in such a framework. The technique minimizes the number of transformations. Li *et al.* (2008) applied the reciprocal transformation for the convexification of posynomial programs with positive variables.

2.4 Concave, bilinear and fractional models. Ryoo and Sahinidis (2003) studied linear and generalized linear multiplicative models, applied the recursive arithmetic interval approach for the derivation of lower bounds, introduced greedy heuristics for a branch-and-reduce approach, and applied it to benchmark problems and randomly generated problems. Goyal and Ierapetritou (2003a) introduced an approach for the systematic evaluation of the infeasible domains using a simplicial outer approximation framework that is applicable to concave or quasiconvex constraints.

Benson (2004) showed how fractional programs can be reformulated so as to be solvable by parametric convex programming and concave minimization methods. Tuy *et al.* (2004) proposed a new approach for optimizing polynomial fractional functions under polynomial constraints. The approach is based on reformulation into a monotone optimization problem. Tsai (2005) proposed a method to solve fractional programs via convexification transformation and piecewise underestimation. Wang *et al.* (2005) developed a branch and bound algorithm for the global optimization of sums of several linear franctional functions over a polytope, that is based on successively refined linear relaxations. Liberti (2005) developed an automatic symbolic reformulation procedure that can uncover linearity that is embedded in bilinear programs and thus simplify such formulations. Linderoth (2005) proposed a branch-and-bound algorithm for solving nonconvex quadratically-constrained quadratic programs. The novelty of their algorithm lies in the fact that branching is done by partitioning the feasible region into the Cartesian product of two-dimensional triangles and rectangles. Explicit relaxations of the bilinear function over such triangles and rectangles are derived and used. Chinchuluun *et al.* (2005) developed a new approach for solving concave quadratic programming problems which is based on the improved global optimality conditions of Dur *et al.* (1998) and Hiriart-Urruty and Ledyav (1996).

Beck *et al.* (2006) derived an efficient global optimization algorithm for the regularized total least squares problem and applied it on problems arising from the inverse Laplace transform and image processing. Jiao *et al.* (2006) proposed solving generalized linear fractional problems via a branch and bound algorithm that is based on the construction of a linear relaxation. Benson (2006, 2007a) studied fractional programs that involve ratios of convex terms and presented a new branch and bound algorithm that requires solving a sequence of convex optimization problems. He also focused on linear sum-of-ratios problems and used a simplicial branch and bound duality-bounds algorithm to globally solve them Benson (2007b). Nahapetyan and Pardalos (2007) presented a continuous relaxation for the Concave Piecewise Linear Network Flow Problem (CPLNFP), which has a bilinear objective function and network constraints. Ding and Al-Khayyal (2007) presented two linear cutting plane algorithms that refine methods for solving disjoint bilinear programs.

Shen and Yuan (2007) proposed a branch and bound approach for the global optimization problem of the sum of generalized polynomial fractional functions under generalized polynomial constraints. Yamamoto and Konno (2007) proposed an efficient algorithm for solving convex-convex quadratic fractional programs that combines the classical Dinkelbach approach, the integer programming approach for solving nonconvex quadratic programs, and a standard NLP solver. Nie *et al.* (2008) discussed the global minimization of rational functions via sum of squares relaxations. Nahapetyan and Pardalos (2008) suggested a bilinear reduction of the original MILP formulation of the capacitated multi-item dynamic pricing problem, and proposed a heuristic algorithm to solve effectively the equivalent bilinear problem.

2.5 General theoretical and algorithmic developments. Lucia and Feng (2003) extended their terrain methodology Lucia and Feng (2002) to multivariable problems and integral curve bifurcations associated with valleys and ridges, showed that the terrain methods are superior to arc homotopy continuation in the presence of parametric disconnectedness, and studied examples for the location of all azeotropes, retrograde flash calculations, and CSTR problems. Casado *et al.* (2003) proposed new interval analysis support functions. These enable one to develop more powerful bounding, selection, and rejection criteria, which result in significant coputational savings. They applied their enhancements on a wide set of multiextremal test functions, and reported that the new algorithm works almost two times faster than a traditional interval analysis global optimization method. A natural n-dimensional extension to these algorithms was later presented by Martinez *et al.* (2004). Zilinskas and Bogle (2003) studied the evaluation of ranges of functions through balanced random interval arithmetic, investigated the hypothesis on the normal distribution of the centers and radii of the evaluated balanced random intervals through several computational studies, and concluded that this hypothesis is incorrect.

Zhu and Fu (2003) developed a sequential convexification method (SCM) for continuous global optimization problems that is based on a simple transformation to convert the objective function into an auxiliary function with gradually fewer number of local minimizers. The method requires use of only local techniques and with probability one converges assymptotically to a global minimizer. Gao (2003, 2004) presented detailed applications of the canonical dual transformation method in quadratic and d.c. programming. Based on the same method, Gao (2005) also presented a duality theory for solving concave minimization problems subject to nonlinear inequality constraints, and polynomial programming problems subject to box constraints Gao (2007). Sun *et al.* (2005) considered four classes of augmented Lagrangian functions and presented new results on the existence of their saddle points. The findings provide new insights to the role played by augmented

Lagrangians in local duality theory of constrained nonconvex optimization. Borradaile and Van Hentenryck (2005) worked on how to derive safe linear relaxations to account for numerical errors due to the underlying floating-point environment of the calculations. They derived two classes of safe linear estimators for univariate functions, studied their tightness, and combined them to derive estimators for the multivariate case.

Floudas and Jongen (2005) considered the minimization of smooth functions in the Euclidean space with a finite number of stationary points having moderate asymptotic behavior at infinity, and emphasized the crucial role of saddle points. They showed that, in general, any two local minima can be connected via an alternating sequence of local minima and transition points of first order. Shen (2005) developed a linearization method for the global optimization of generalized geometric programs (GGPs). Through a sequence of linear programming problems, the method is proven to be convergent to the global optimum of the original nonconvex problem. Wang and Liang (2005) applied a combination of a branch-and-bound method with a cutting-plane method to globally solve GGPs after their tranformation to reverse convex programming problems. Li and Tsai (2005), Tsai and Lin (2006), and Tsai *et al.* (2007) studied generalized GGPs in the presence of non-positive variables, proposed a technique for treating these variables, and provided convexification rules for certain signomial terms. Tsai and Lin (2007) applied this work and developed an algorithm to find all solutions of systems of nonlinear equations with free variables.

Marcovecchio *et al.* (2006) presented a new algorithm for nonconvex NLPs that is based on the solution of two problems, called the reformulated and the main problem. The former is a suitable reformulation of the original and involves only convex and univariate concave temrs, while the latter is a valid convex relaxation. Parpas *et al.* (2006) proposed a stochastic algorithm for the global optimization of nonconvex functions over a polytope. Their method follows the trajectory of an appropriately defined stochastic differential equation. They also studied the global optimization of robust chance constrained problems Parpas *et al.* (2008), assuming that the probability measure with which the constraints are evaluated is known only through its moments. Hirsch *et al.* (2007) extended the GRASP algorithm Feo and Resende (1989, 1995) of discrete optimization to continuous problems, and applied this stochastic heuristic to two hard global optimization problems. Li *et al.* (2007) proposed a class of general transformation methods to convert a nonconvex problem to an equivalent one for which, under certain assumptions, the existence of a local saddle point of its Lagrangian function can be guarranteed.

Floudas and Kreinovich (2007a,b) showed that natural symmetry requirements enable one to analytically solve the problem of finding the optimal functional forms for convex underestimators for twice continuously differentiable functions, and identified the original Adjiman *et al.* (1998a,b) and the generalized Akrotirianakis and Floudas (2004a,b) αBB method as the only optimal schemes that satisfy such a requirement. Gounaris and Floudas (2008c) developed sufficient convexity conditions for products of univariate functions. The results also provide insight on why transformations already proposed in the literature, like the exponential Maranas and Floudas (1997) or the reciprocal Li *et al.* (2008), actually work in the convexification of posynomial programs. A novel global optimization method based on an augmented Lagrangian framework was introduced by Birgin *et al.* (2008). They utilize the αBB method to ϵ-globally solve the subproblems that are generated at each outer iteration.

2.6 Computational biology. Klepeis and Floudas (2003a) introduced a deterministic global optimization approach, αBB coupled with torsional angle dynamics, for the protein structure prediction given restraints predicted from the identification of α-helices and β-sheets. Klepeis and Floudas (2003b) proposed Astro-Fold, a first principles framework for the protein structure prediction. They described the global optimization and mixed-integer optimization advances, and presented a variety of test systems including several blind protein predictions. Klepeis *et al.* (2003a) introduced a new class of hybrid global optimization methods denoted as integrated hybrids for the oligopeptide structure prediction. Klepeis *et al.* (2003b) proposed new alternating hybrid global optimization methods, studied and developed their distributed computing algorithms, and applied them to the structure prediction of met-enkaphalin and mellitin. These two classes of hybrid global optimization approaches combine the αBB for the generation of rigorous lower bounds with the modified genetic algorithm, CSA, for the upper bounding calculations. Reviews of the global optimization activities in the areas of protein folding and peptide docking can be found in Floudas (2005, 2007) and Floudas *et al.* (2006).

Doye *et al.* (2004) developed new transformation techniques to improve the global optimization of models that include Morse potentials and, thus, aid the study of biomolecules. Schafroth and Floudas (2004) studied the protein-peptide interactions via deterministic global optimization, atomistic-level modeling, and several solvation methods that include the area accessible to the solvent, the volume accessible to the solvent, and the Poisson-Boltzmann method, and reported excellent agreement on the binding motifs. Moloi and Ali (2005) proposed an algorithm for minimizing potential energy fuctions that is based on the differential evolution algorithm by Storn and Price (1997), and tested it on two different types of potentials. Gattupalli and Lucia (2008) applied a multi-scale global optimization methodology, based on terrain methods and funneling algorithms, to find the minimum energy molecular conformations of pure n-alkanes. The results are indicative of typical fuel oils and useful in understanding the waxing of petroleum fuels.

2.7 Phase equilibriium. Scurto *et al.* (2003) developed a methodology to predict the behavior of high-pressure solid-fluid equilibrium systems with cosolvents, where the likelihood of formation of more than two phases is great. Lin and Stadtherr (2004a) applied their interval-Newton procedure, enhanced by a new LP bounding strategy, on various phase equilibrium problems. Nichita *et al.* (2004) used the tunneling global optimization method to directly minimize the Gibbs free energy in multiphase equilibria calculations. Nichita *et al.* (2006) used the same method to solve the nonconvex phase stability problem that results from the tangent plane distance criterion in terms of the Helmholtz free energy. Henderson *et al.* (2004) formulated an optimization problem based on a slight modification of the Gibbs tangent plane criterion, and used a simulated annealing algorithm to solve it. Freitas *et al.* (2004) studied a number of binary mixtures by solving the aforementioned formulation. They used a novel and robust algorithm that does not require use of derivatives and is insensitive to the quality of initial guesses. Lucia *et al.* (2005) incoroporated new ideas within the terrain methods to study the phase stability and equilibrium of n-alkanes. Srinivas and Rangaiah (2006) evaluated their implementation of the random tunneling algorithm on a number of medium sized problems, including VL,LL, and VLL phase equilibrium problems. Srinivas and Rangaiah (2007) also focused on solving phase equilibrium problems with differential evolution and tabu search methods. They reported better performance when the methods are combined and a tabu list is used in the generation step of differential evolution. Mitsos and Barton (2007) reinterpreted the Gibbs tangent plane stability criterion via a Lagrangian duality approach, and performed case studies based on the NRTL and UNIQUAC activity coefficient model.

2.8 Other application-oriented developments. Lin and Stadtherr (2004b) proposed a linear programming bounding scheme that enhances the performance of the interval-Newton method, and applied the new technique on parameter estimation problems. Ulas *et al.* (2005) studied the uncertainties in thermodynamic models that result from parameter estimation. Their results identified errors in published thermodynamic data collections and can improve the performance of batch distillation controllers. Singer *et al.* (2006) presented a method that can find the best possible least-squares fit of experimental data by a nonlinear kinetic model. Amaral *et al.* (2008) developed an RLT-based algorithm for the global optimization of a nonconvex problem that arises in total least squares with inequality constraints, and in the correction of infeasible linear systems of equalities.

Bringas *et al.* (2007) worked on the optimal design of groundwater remediation networks that use selective membranes for the valorization of anionic pollutants. They globally solved select instances of their superstructure using a Lagrangean decomposition algorithm. Maringer and Parpas (2008) proposed an extension of the Markowitz mean variance model for the global optimization of the higher moments in portfolio selection, which they solved using two stochastic algorithms based on differential evolution. Kallrath (2008) studied the simultaneous packing of circles and polygons in rectangles, using global optimization approaches.

3 Mixed-integer nonlinear optimization

This section lists recent publications in areas of mathematical programming that involve discrete components. The entries are classified into theoretical and implementation contributions for MINLP, formulation of MINLP models and studies of particular problems, and advances in the area of generalized disjunctive programming (GDP). As this classification is not mutually exclusive, some publications could have appeared in more than one subsections.

3.1 MINLP: Theoretical and algorithmic developments. Zhu and Kuno (2003) proposed a hybrid global optimization method for nonconvex MINLPs which combines convex quadratic underestimation techniques with a revised form of the generalized Benders decomposition, suggested its convergence properties, and illustrated it via a two variable problem. Goyal and Ierapetritou (2003b) studied MINLP models where the objective function is convex, and the constraints are convex, concave or quasi-concave, introduced the simplicial approximation of the convex hull of the feasible region, proposed algorithmic procedures and illustrated them via small benchmark problems.

Yan *et al.* (2004) extended the line-up competition algorithm to handle integer variables, and tested its performance with several nonconvex MINLP problems published in the literature. Dua *et al.* (2004) studied issues concerning the solution of parametric nonconvex programs, proposed a branch and bound algortihm, and discussed the need for defining a suitable overestimating subproblem. Tawarmalani and Sahinidis (2004) developed a branch and bound framework for the global optimization of continuous, integer, and mixed-integer nonlinear programs. It involves novel linear relaxation schemes, a Lagrangian/linear duality-based theory for domain and range reduction, as well as branching strategies that guarrantee finiteness of the solution sequence for certain classes of problems. They also discuss implementation issues and present computational results with a variety of benchmark problems. Kesavan *et al.* (2004) studied separable MINLP models with nonconvex functions, proposed two decomposition algorithms based on alternating sequences of relaxed master problems, two nonlinear programming problems, and outer approximation, showed that the first algorithm yields the global solution while the second provides a rigorous bound on the global solution, and presented computational results on several benchmark problems and heat exchanger network problems.

Chaovalitwongse *et al.* (2004) considered the reduction of multi-quadratic 0-1 programming problems to linear mixed 0-1 programming problems, with a reformulation that limits the number of required continuous variables. Prokopyev *et al.* (2005a,b) considered single- and multiple-ratio unconstrained hyperbolic 0-1 programming problems, proved that checking for the uniqueness of their solution is NP-hard, and discussed the complexity of local search and the approximability of multiple-ratio problems. Prokopyev *et al.* (2005b) studied also their MILP reformulations, and presented a GRASP-based heuristic for solving cardinality constrained problems. Tawarmalani and Sahinidis (2005) introduced a polyhedral branch-and-cut approach in global optimization. Their algorithm exploits convexity in order to generate the polyhedral cuts and relaxations for multivariate nonconvex problems. Computational results with a collection of benchmark NLP and MINLP problems were presented. Munawar and Gudi (2005) proposed a hybrid method that combines stochastic and deterministic approaches to improve the convergence rate of differential evolution (DE) algorithms, and utilized a nonlinear transformation proposed in the literature so as to more effectively take into account the integrality restrictions of the integer variables in an MINLP formulation. In the context of linear and polynomial MIP problems, Adams and Sherali (2005) used Lagrange interpolating polynomials to generalize the level-1 RLT constraints for a binary variable to the general discrete case.

Zhu and Kuno (2006) developed a cutting-plane branch-and-cut algorithm for the solution of convex MINLPs. The method utilizes the lift-and-project technology, which accelerates the branching process by generating cuts through the solution of a transformed projection problem. The computational results indicate that the method can address large-scale problems. Luo *et al.* (2007) developed an improved particle swarm optimization (PSO) algorithm for solving nonconvex MINLPs. The algorithm includes a transformation procedure where mixed variables are partitioned and reduced variables are identified through analyzing and tearing equality constraints. In addition, the discrete variables are updated independently according to a given criterion and not simultaneously with the continuous ones. Young *et al.* (2007) proposed the implementation of information theory to the mutation stage of a genetic algorithm so as to refresh the premature population. The novel algorithm, called information-guided genetic algorithm (IGA), can address general MINLPs and does not require any additional number of variables and constraints or knowledge of an initial feasible point. Computational results on a variety of problems show superior convergence and global minimum discovery rates than traditional genetic algorithm methods. Pardalos *et al.* (2008) used global equilibrium search techniques for the solution of benchmark unconstrained binary quadratic optimization problems, studied their computational efficiency, and reported better performance when compared with multistart tabu search methods. Based on the canonical dual transformation Gao (2004), Fang *et al.* (2008) constructed a pair of canonical dual problems for the 0-1 quadratic programming problem, showed that no duality gap exists, derived optimality conditions of both the local and global type, and developed an algorithm for large scale problems.

3.2 MINLP: Formulations and applications. Ostrovsky *et al.* (2003) revisited their molecular design reduced dimension branch-and-bound algorithm by studying further the branching functions concept and the special tree function representation, proposed the sweep method for the construction of the linear underestimators, investigated the problem size dependency on the algorithmic performance, and showed that the computational effort increases almost linearly. Vaia and Sahinidis (2003) studied the simultaneous parameter estimation and model structure identification in infrared spectroscopy, proposed two methods out of which the second corresponds to a single nonconvex MINLP model, presented a branch-and-bound approach which is based on a relaxation of terms that are logarithmic, bilinear, and multilinear depending on the determinant of the covariance matrix, and presented comparative computational results. Sinha *et al.* (2003) studied the systematic design of cleaning solvent blends for lithographic printing, modelled it as a nonconvex MINLP problem, introduced an interval analysis based global optimization approach with modifications on the upper bounding calculation and the local feasibility test which are solved via SQP, and an interval-based domain reduction algorithm, and presented computational results for the design of aqueous blanket wash blends. Sahinidis *et al.* (2003) revisited the design of alternative refrigerants problem, introduced an integer formulation for previously described structural constraints, proposed new structural constraints between one-bonded and higher-bonded groups in the absence of rings and new clique constraints for rings, applied a branch-and-reduce global optimization algorithm with a modification so as to generate all feasible integer solutions, and generated new compounds for refrigerants.

Kallrath (2003) studied and solved a nonconvex product portfolio problem via an approximate MILP formulation of the objective function and exact linear relations for the constraints, modelled it as a nonconvex MINLP problem for the optimization of the number and size of batch process units, analyzed the sources of nonconvexity consisting of concave functions and trilinear products, investigated the piecewise linear approximation of the objective function, the use of a local MINLP solver, SBB, and a global optimization solver, Baron, and reported that for the large instances weak lower bounds are generated. Pardalos *et al.* (2004) formulated a quadratically constrained 0-1 programming problem to identify critical brain electrode sites that are relevant in the detection of precursors to temporal lobe seizures. They reported the first analysis of an automated online algorithm for detecting these precursors in ongoing EEG signals, effectively providing a prospective seizure warning system. Emet and Westerlund (2004) conducted a computational comparison of solving a cyclic chromatographic separation problem using MINLP methods, and reported that the extended cutting plane (ECP) method compares favourably against traditional outer approximation and branch-and-bound methods. A review of the recent advances on MINLP optimization of planning and design problems in the process industry was presented by Kallrath (2005).

Ghosh *et al.* (2005) studied mixed-integer formulations for the identifications of metabolic network fluxes via NMR data. They proposed a joint problem methodology that involves using the MILP solutions of the "front end" analysis to bound the desired solution of the data-to-fluxes problem. A branch-and-bound global optimization algorithm was used to solve the latter. Lin *et al.* (2005) revisited the nonconvex product portfolio problem introduced by Kallrath (2003), presented an improved formulation consisting of a concave objective function with linear constraints in the continuous and binary variables, proposed several techniques for tightening the model and accelerating its solution, developed a customized branch-and-bound approach which addresses the problem to global optimality, applied it to small and large instances, and demonstrated that global solutions can be obtained very efficiently in contrast to commercial MINLP solvers. Based on global equilibrium search techniques, Pardalos and Shylo (2006) proposed an algorithm to solve the job shop scheduling problem, and reported improved upper bounds for several well-known benchmark problems.

Sherali and Desai (2005a) formulated the hard clustering problem as a mixed-integer nonlinear program and solved it to global optimality with use of the RLT. The authors applied their method to a number of standard data sets found in the literature, as well as on larger problem instances which they syntesized, and reported robustness of the procedure and computational dominance over the k-means algorithm. They also developed a root-node heuristic procedure to reach a good quality solution early in the branch and bound process. The methodology was extended later to address also the fuzzy clustering problem Sherali and Desai (2005b).

Meyer and Floudas (2006) studied superstructures of pooling networks, which are important to the petrochemical, chemical, and wastewater treatment industries, and formulated this generalized pooling problem as a nonconvex MINLP that involves many bilinear terms in the constraint functions. They proposed a global optimization algorithm that is based on a novel piecewise linear reformulation-linearization technique (RLT) formulation that they developed. Using this approach, the authors verified the global solution of a combinatorially complex industrial problem that contained 156 bilinear terms and 55 binary variables. Karuppiah and Grossmann (2006) addressed the problem of optimal synthesis of an integrated water system, where water using processes and water treatment operations are jointly taken into account. The resulting MINLP was solved with a new deterministic spatial branch and contract algorithm, in which piecewise under- and overestimators are used for constructing the relaxations at each node. The extension of the model to include the uncertainty in various operating parameters was presented in a later publication Karuppiah and Grossmann (2008a). The extended MINLP is a deterministic equivalent of a two-stage stochastic programming model with recourse, where the uncertain parameters take on a finite number of realizations. This problem is addressed with a spatial branch-and-cut algorithm that uses Lagrangian decomposition Karuppiah and Grossmann (2008b). Karuppiah *et al.* (2008) presented an outer-approximation algorithm to globally solve a nonconvex MINLP formulation that corresponds to the continuous-time scheduling of refinery crude oil operations. The solution procedure relies on effective MILP relaxations that benefit from additional cuts derived after spatially decomposing the network.

Bergamini *et al.* (2007) formulated an MINLP model for the global optimization of heat exchanger networks, and presented a new solution methodology that is based on outer approximation and utilizes piecewise underestimation. Rigorous constraints obtained from physical insights are also included in the formulation, and the authors reported computationally efficient global solutions for problems with up to nine process streams. Foteinou *et al.* (2008) presented a mixed-integer optimization framework for the synthesis and analysis of regulatory networks. Their approach integrates prior biological knowledge regarding interactions between genes and corresponding transcription factors, in an effort to minimize the complexity of the problem. The original nonconvex formulation is appropriately linearized and the resulting MILP solved with use of standard commercial solvers. A case study involving gene expression and binding data from E.coli is presented.

Tan *et al.* (2007a) presented a novel clustering approach for use with DNA microarray data. The algorithm is based on the Global Optimum Search methodology Floudas *et al.* (1989); Floudas (1995) and includes a procedure to determine the optimal number of clusters to be used. The approach compared favourably with other methods, in terms of intra-cluster similarity and inter-cluster dissimilarity, when it was applied on a case study that involves data centered on the Ras signaling pathway of S.cerevisiae. Tan *et al.* (2007b) also studied the effects of different normalization and pre-clustering techniques on the clustering results produced with the previous approach. They reported that these can be significant and should be taken into consideration when attempting to cluster microarray data and, therefore, this work has significance in fine-tuning their novel clustering approach discussed above.

Rebennack *et al.* (2008) proposed a decomposition algorithm for the special class of MINLPs that have an assignment constraint, and formulated the column enumeration approach. The resulting master problem is a partitioning problem whose objective function coefficients are computed via easy-to-solve optimization subproblems. They applied the method on a problem of packing circles into several given rectangles. Exler *et al.* (2008) studied MINLP formulations with differential-algebraic constraints that arise in the integrated design of process and control systems. In order to globally solve this class of problems, they developed extensions of the metaheuristic Tabu Search. They applied their method on a wastewater treatment plant model Alex *et al.* (1999) and the Tennessee Eastman Process Downs and Vogel (1993).

3.3 Generalized disjunctive programming, GDP. Grossmann and Lee (2003) studied generalized disjunctive programming, GDP, problems which feature convex nonlinear inequalities in the disjunctions, proposed a convex nonlinear relaxation of the nonlinear GDP problem based on the convex hull representation of each of the disjunctions which was derived by variable disaggregation and reformulation, formulated the nonlinear GDP as an MINLP which was shown to produce improved bounds compared to big-M models, and presented comparative computational studies of the two formulations. Lee and Grossmann (2003) studied nonconvex GDP problems with bilinear equality constraints, derived convex underestimators and overestimators for the bilinear constraints using the reformulation/linearization approach, expressed the discrete choices as disjunctions which were subsequently relaxed by their convex hull representations, used their earlier two level global optimization approach Lee and Grossmann (2001), and presented computational studies for pooling problems, water usage problems, and wastewater network problems. Vecchietti *et al.* (2003) presented characterization and properties for various types of disjunctions, including an interesting result for improper disjunctions. They proposed also a cutting plane method that avoids the explicit generation of equations and variables of the convex hull, and used this method to solve a small process synthesis problem.

Montagna *et al.* (2004) formulated a generalized disjunctive programming model to address the problem of synthesis of a biotechnological process in which a set of biotechnological products must be elaborated. They solved the problem after transforming it into an MINLP with use of either a big-M or a convex hull reformulation, and they evaluted the computational performance of the two approaches. Stein *et al.* (2004) proposed different alternatives to address the combinatorial component of GDP and MINLP models and reformulate them into continuous optimization problems, investigated theoretical properties of the various reformulations with regard to their numerical solution, and demonstrated their results with process engineering applications that involve stationary as well as dynamic process models.

Lee and Grossmann (2005) presented a review of advances in the area of discrete/continuous mathematical programming. After a brief presentation of algebraic-based MILP and MINLP approaches, the authors focused on the alternative approach of logic-based GDP modeling. They reviewed various solution strategies, a hybrid model that integrates disjunctive with mixed-integer programming, as well as their two-level branch-and-bound procedure for the global optimization of non-convex GDP problems. Sawaya and Grossmann (2005) discussed issues regarding the computational implementation of the convex hull reformulation for nonlinear generalized disjunctive programming. In particular, they addressed the issue of division-by-zero avoidance in the inequalities respresenting the convex hulls, while preserving the convexity of the problem. They proposed to replace the original set of constraints with two approximating constraint sets, and they strengthened their approach by deriving theoretical conditions under which one of the sets is superior to the other. Sawaya and Grossmann (2007) also proposed a cutting plane method for solving linear generalized disjunctive programming problems. The method relies on cuts generated by an LP or QP separation problem that successively strengthen the equivalent big-M reformulation of the original GDP problem. The authors provided illustrative examples in the areas of strip-packing, retrofit planning, and zero-wait job-shop scheduling.

4 Differential-algebraic models, DAEs

Singer and Barton (2003), Singer and Barton (2004) studied the global optimization of integral objective functions subject to ordinary differential equations, derived convex relaxations for the integral based on a pointwise integrand scheme, developed a branch-and-bound global optimization approach on a Euclidean space which combines the integrand convex relaxations with differential inequalities, McCormick's composition approach, and outer approximation, and illustrated their approach with several small benchmark problems. Lee and Barton (2003) studied the global optimization of linear time varying hybrid systems which exhibit both discrete state and continuous state behavior, and extended their recently developed approach for the determination of the optimal mode sequence when the transition times are fixed Barton and Lee (2003), proposed a reformulation of the problem via binary variables while maintaining the linearity of the dynamical system, derived convex relaxations of Bolza-type functions using recent results for linear time varying continuous systems Lee *et al.* (2004), and applied it to benchmark problems and an isothermal plug flow reactor problem.

Adjiman and Papamichail (2003) developed further their deterministic spatial branch-and-bound approach for nonconvex models with ordinary differential equations Papamichail and Adjiman (2002), proposed three convex relaxations for the parameter estimation of the initial value problem, and presented computational results on several parameter estimation problems in kinetics. Chachuat and Latifi (2003) introduced a spatial branch-and-bound global optimization approach for problems with ordinary differential equations in the constraints, presented results on the first and second order derivatives for the initial value problem and the two point boundary value problem, compared the sensitivity and the adjoint approaches, developed convex underestimators using the αBB principles, and presented computational studies and comparisons of the sensitivity versus the adjoint approach for several problems. Banga *et al.* (2003a) studied integrated process design and operation, parameter estimation in bioprocess models, and focused on stochastic global optimization methods for dynamic systems, addressed handling of constraints in stochastic methods, presented hybrid approaches for dynamic optimization, and presented computational studies on the optimal control of bioreactors, the integrated design of a waste treatment plant (see also Moles *et al.* (2003)) where they provided comparisons for several algorithmic approaches, and discussed advances in the parameter estimation of bioprocesses. Banga *et al.* (2003b) reviewed and introduced optimization as a key technology for food processing and discussed stochastic global optimization methods and their potential applicability in food process engineering.

Papamichail and Adjiman (2004) presented a deterministic spatial branch and bound global optimization algorithm for problems with a set of first-order differential equations in the constraints. They also presented a proof of convergence in a separate publication Papamichail and Adjiman (2005). Barton and Lee (2004) presented a method for constructing convex relaxations of nonconvex NLPs with linear dynamic systems embedded. Chachuat *et al.* (2005) developed a decomposition approach for a general class of mixed-integer dynamic optimization problems that is capable of guarranteeing convergence to the global solution. They applied their method to the problem of optimal design and operation of a batch process consisting of a series reaction followed by a separation with no intermediate storage. Chachuat *et al.* (2006) presented an overview of global methods for dynamic programming, with emphasis on the control parameterization approach. Branch-and -bound algorithms for dynamic optimization are thoroughly discussed, and so is the outer-approximation algorithm for mixed-integer dynamic optimization problems. Future directions for research are proposed, and recently developed convex and concave relaxations for the solutions of nonlinear ODEs are also presented.

Barton *et al.* (2006) focused on the solution of optimization problems with continuous time automata embedded, and presented an overview of the control parameterization approach. They discussed the decomposition approach, restricted classes of subproblems such as the optimal mode sequences problem or the optimal transition times problem, and proposed future directions for research. Singer and Barton (2006) examined the global optimization of an integral objective function subject to nonlinear ordinary differential equations. They used McCormick's composition result to derive theory for the convex relaxation of an integral.

Lin and Stadtherr (2006, 2007) created a new interval solver for parametric ODEs, which is used to produce guarranteed bounds on the solution of dynamic systems with interval-valued parameters, and incorporated it within the framework of a global optimization solver for dynamic systems, including optimal control problems. They tested the new method on a collection of benchmark problems and reported more than an order of magnitude improvement in most cases. Long *et al.* (2006, 2007) applied a deterministic non-convex optimization method for nonlinear model predictive control of systems that exhibit nonlinear hybrid dynamics. They used discrete variables to model the multiple regimes of operation, and suggested some algorithm modifications to accelerate the solution of the resulting MINLP. Using a challenging wastewater treatment plant benchmark model, Egea *et al.* (2007b) conducted a computational comparison between various global optimization approaches, including surrogate-based methods. Egea *et al.* (2008) developed a new algorithm for the global optimization of costly nonlinear continuous problems. It is based on the scatter search metaheuristic Egea *et al.* (2007a), which was coupled with a kriging-based prediction method in order to discard the evaluation of solutions that are not likely to provide high quality function values. Frits *et al.* (2007) used an interval arithmetic based branch and bound framework to find the singular points and bifurcations in studying feasibility of batch extractive distillation. Angira and Santosh (2007) applied the trigonometric differential evolution algorithm on dynamic optimization problems encountered in chemical engineering. They reported better computational performance when compared to the standard differential evolution algorithm, particularly in problems involving a large number of control stages.

5 Semi-infinite programming

Zakovic and Rustem (2003) considered solving a minimization SIP problem through a two stage procedure that searches for global maximum violation of the constraints. They also considered a version that searches for any violation of the constraints, and performed a computational comparison between the two variants. They applied these algorithms on various minimax (with or without coupled constraints) problems, as well as a small process design problem. Geletu and Hoffmann (2004) outlined theoretical considerations for determining coarse approximations of a solution of a generalized semi-infinite programming problem via global optimization of an exact discontinuous penalty approach. They extended the theory of robust analysis of marginal functions so as to study the behavior of the marginal functional of a resulting auxiliary parametric SIP problem. Liu *et al.* (2004) presented a variant of the adaptive dual parameterization algorithm of Liu and Teo (2002) where, in each iteration, only a quadratic problem with a limited dimension and number of constraints is required to be solved.

Bhattacharjee *et al.* (2005a) used interval analysis methods to construct a finite reformulation of a semi-infinite program. This reformulation is generally overconstrained relatively to the original SIP and, therefore, its solution constitutes a feasible upper bound to the SIP solution. This bound can be further refined by a subdivision procedure that was shown to converge to the true solution within finite epsilon. Bhattacharjee *et al.* (2005b) presented a branch and bound based deterministic global optimization algorithm for semi-infinite programs. The upper bounds are generated by replacing the infinite number of real-valued constraints with a finite number of constrained inclusion bounds, while the lower ones are generated via convex relaxations of a discretized approximation to the original problem. Chang and Sahinidis (2005) studied the optimization of metabolic pathways under stability considerations. Their formulation is a bilevel optimization problem which they solve after transforming it to a semi-infinite optimization problem.

Floudas and Stein (2007) presented a novel adaptive convexification algorithm for the solution of semi-infinite optimization problems. The method adaptively constructs convex relaxations of the lower level problem based on αBB underestimators, replaces the relaxations with their equivalent KKT conditions, and solves the resulting mathematical programs with complementarity constraints. Liu (2007) presented a homotopy interior point method for globally solving semi-infinite programs. Mitsos *et al.* (2008a) considered nonconvex SIP problems and calculated relaxation-based bounds. They obtained lower bounds by solving a mathematical program with equilibrium constraints that resulted from discretization of the parameter set and KKT analysis of the lower-level maximization problem. Convex and linear relaxations of the latter were used for upper bounding.

6 Grey-Box and nonfactorable models

Sherali and Ganesan (2003) presented a pseudo-global optimization approach to address black-box problems. Their method utilizes a branch and bound framework where the lower bounds are calculated from response surface-generated polynomial approximations. The authors developed two alternative branching procedures and applied their methods in a conteinership design problem. Regis and Shoemaker (2005) proposed strategies that improve the performance of the radial basis function method of Gutmann (2001). They also presented a new approach for the constrained global optimization of expensive black box functions using response surface models Regis and Shoemaker (2007). Huang *et al.* (2006) proposed a new global optiimzation method for stochastic black-box systems that is based on sequential kriging meta-models. A kriging based method was also developed by Davis and Ierapetritou (2007, 2008) for the solution of NLPs and MINLPs containing black-box functions. Hu *et al.* (2007) presented a model reference adaptive search for global optimization.

7 Bilevel nonlinear optimization

Pistikopoulos *et al.* (2003) studied bilevel optimization models which are of linear-linear,
linear-quadratic, quadratic-linear, or quadratic-quadratic type, and introduced approaches from parametric programming to transform the bilevel problem into a family of single level optimization problems which can be solved to global optimality, and presented computational results on several small benchmark problems. De Saboia *et al.* (2004) conducted a computational study of two global algorithms for linear bilevel programming and proposed some modifications that improve their computational performance. Ryu *et al.* (2004) addressed a bilevel decision-making problem under uncertainty in the context of enterprise-wide supply chain optimization with one level corresponding to a plant planning problem, while the other to a

distribution network problem. The proposed strategy transforms the bilevel problem into a family of single parametric optimization problems that can be solved to global optimiality.

Gümüş and Floudas (2005) studied the global optimization of bilevel mixed-integer optimization problems, proposed an approach that is applicable to mixed-integer nonlinear outer problem and twice continuously differentiable nonlinear inner problem, introduced another approach based on the convex hull representation of the inner problem, which is applicable when the inner level problem features functions which are mixed integer nonlinear in the outer variables and linear, polynomial, or multilinear in the inner integer variables, and linear in inner continuous variables; and applied it to several challenging benchmark problems.

Campelo and Scheimberg (2005) studied the linear bilevel problem and derived the characterization of a strict local solution. Babahadda and Gadhi (2006) used the notion of convexificator to establish necessary optimality conditions for bilevel optimization problems. Solodov (2007) considered the bilevel problem of minimizing a nonsmooth convex function over the set of minimizers of another nonsmooth convex function, and developed a bundle-type algorithm for solving it. Tuy *et al.* (2007) extended their monotonic optimization approach to handle general bilevel programming problems. They showed how the latter can be transformed into monotonic programs that can be solved by either polyblock approximation or branch-reduce-and-bound methods using monotonicity cuts. Faisca *et al.* (2007) proposed a global optimization approach for bilevel programming that is based on reformulating the inner optimization problem as a multi-parametric one, where the parameters are the variables of the outer level problem. They used their approach to solve quadratic, linear, and mixed-integer linear bilevel problems, with or without the presence of uncertainty. Mitsos *et al.* (2008b) developed a global optimization algorithm for bilevel problems with nonconvexities in both the inner and outer programs, presented a KKT-based heuristic for tightening of the lower bounds, extended the method to include branching, and analyzed two branching heuristics.

8 Advances from Princeton University

In the remainder of this paper, we will present recent advances from Princeton University on (i) explicit facets for convex and concave envelopes for trilinear functions, (ii) convex underestimators for trigonometric functions, (iii) new convex underestimators based on a piecewise quadratic perturbation function, (iv) the generalized αBB convex underestimators, (v) tight convex underestimators for $\mathcal{C}^2$-continuous functions, (vi) functional forms of convex underestimators, and (vii) convexity of products of univariate functions

8.1 Explicit facets of convex and concave envelopes for trilinear monomials. Approximations of the convex envelope of nonconvex functions play a central role in deterministic global optimization algorithms and the efficiency of these algorithms is highly infuenced by the tightness of these approximations. Meyer and Floudas (2003, 2004) proposed explicit expressions defining the facets of the convex and concave envelopes for trilinear monomials, with mixed sign domains, as well as with positive or negative bounded domains for each variable. These advances are discussed in the sequel.

Facets of the convex envelope. The description of the nonvertical facets depends on the signs of the bounds of the three variables. For the shake of brevity, we present here only the case where all three variables are nonnegative. The complete set of cases, including the facets for the concave envelope, can be found in the papers by Meyer and Floudas (2003, 2004).

Let $\underline{x}_1 \geq 0$, $\underline{x}_2 \geq 0$, $\underline{x}_3 \geq 0$, and let symbols x, y, z denote a permutation of x_1, x_2, x_3. There always exists a mapping $\{x_1, x_2, x_3\} \to \{x, y, z\}$ such that the following relations apply:

$$\begin{array}{c} \overline{x}\underline{y}\underline{z} + \underline{x}\overline{y}\overline{z} \leq \underline{x}\overline{y}\underline{z} + \overline{x}\underline{y}\overline{z} \\ \overline{x}\underline{y}\underline{z} + \underline{x}\overline{y}\overline{z} \leq \overline{x}\overline{y}\underline{z} + \underline{x}\underline{y}\overline{z}. \end{array} \tag{8.1}$$

Using this mapping, the linear equalities defining the facets of the convex envelope of the trilinear monomial are:

$$\begin{aligned}
w &= \underline{y}\underline{z}x + \underline{x}\underline{z}y + \underline{x}\underline{y}z - 2\underline{x}\underline{y}\underline{z} \\
w &= \overline{y}\overline{z}x + \overline{x}\overline{z}y + \overline{x}\overline{y}z - 2\overline{x}\overline{y}\overline{z} \\
w &= \underline{y}\overline{z}x + \underline{x}\overline{z}y + \overline{x}\underline{y}z - \underline{x}\underline{y}\overline{z} - \overline{x}\underline{y}\overline{z} \\
w &= \overline{y}\underline{z}x + \overline{x}\underline{z}y + \underline{x}\overline{y}z - \overline{x}\overline{y}\underline{z} - \underline{x}\overline{y}\underline{z} \\
w &= \frac{\theta}{\overline{x} - \underline{x}}x + \overline{x}\underline{z}y + \overline{x}\underline{y}z + \left(-\frac{\theta\underline{x}}{\overline{x} - \underline{x}} - \overline{x}\overline{y}\underline{z} - \overline{x}\underline{y}\overline{z} + \underline{x}\overline{y}\overline{z}\right), \\
& \quad \text{where } \theta = \overline{x}\overline{y}\underline{z} - \underline{x}\overline{y}\overline{z} - \overline{x}\underline{y}\underline{z} + \overline{x}\underline{y}\overline{z} \\
w &= \frac{\theta}{\underline{x} - \overline{x}}x + \underline{x}\overline{z}y + \underline{x}\overline{y}z + \left(-\frac{\theta\overline{x}}{\underline{x} - \overline{x}} - \underline{x}\underline{y}\overline{z} - \underline{x}\overline{y}\underline{z} + \overline{x}\underline{y}\underline{z}\right), \\
& \quad \text{where } \theta = \underline{x}\underline{y}\overline{z} - \overline{x}\underline{y}\underline{z} - \underline{x}\overline{y}\overline{z} + \underline{x}\overline{y}\underline{z}.
\end{aligned} \tag{8.2}$$

Illustration. To construct the lower bounding facets of the convex envelope where $\mathbf{x} = [1, 2] \times [1, 2] \times [1, 2]$, we first observe that all bounds are positive (i.e., the case described above applies). As the bounds on all the variables are the same it makes no difference how we map $\{x_1, x_2, x_3\}$ onto $\{x, y, z\}$. After substitution, the facets become

$$\begin{aligned}
w &= 1x_1 + 1x_2 + 1x_3 - 2 \\
w &= 4x_1 + 4x_2 + 4x_3 - 16 \\
w &= 2x_1 + 2x_2 + 2x_3 - 6 \\
w &= 2x_1 + 2x_2 + 2x_3 - 6 \\
w &= 2x_1 + 2x_2 + 2x_3 - 6 \\
w &= 2x_1 + 2x_2 + 2x_3 - 6.
\end{aligned}$$

Note that the last four facets are identical and hence we only need three linear equalities to define the convex envelope.

Comparison with other bounding schemes. The recursive arithmetic interval (rAI) scheme for generating convex lower bounds for the multilinear monomial was compared with the arithmetic interval (AI) scheme and other bounding schemes studied by Ryoo and Sahinidis (2001). The separation distances between the trilinear monomial xyz and the lower bounding functions $f_{\mathrm{AI}}(x, y, z)$ and $f_{\mathrm{rAI}}(x, y, z)$ are defined as $d_{\mathrm{AI}}(x, y, z) = xyz - f_{\mathrm{AI}}(x, y, z)$ and $d_{\mathrm{rAI}}(x, y, z) = xyz - f_{\mathrm{rAI}}(x, y, z)$.

These separation distances are compared with $d_C(x, y, z)$, the separation distance between xyz and the convex envelope. Two graphs (Figure 1) are presented for each combination. In each graph, y and z are constant while the separation distances are plotted as a function of x. The AI and rAI systems are shown to generate poor bounds relative to the convex envelope.

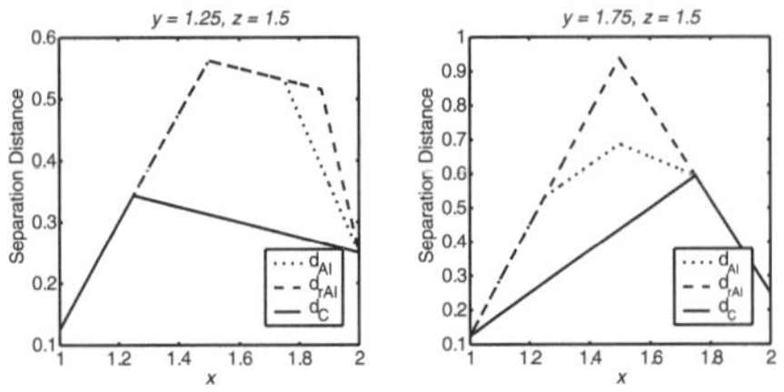

Figure 1 Comparison of lower bounding separation distances, $(x, y, z) \in [1, 2]^3$

8.2 Convex underestimators for trigonometric functions. Caratzoulas and Floudas (2005) proposed a convex underestimator for the function

$$f(x) = \alpha \sin(x + s), \quad x \in [x_L, x_U], \quad \alpha > 0\,. \tag{8.3}$$

The underestimation method can be applied to one-dimensional as well as multi-dimensional problems involving trigonometric polynomials, since the product of trigonometric functions can always be decomposed into the sum of sin and cos functions with arguments that are linear combinations of the problem variables. The general case $\sin(k\,x)$, $x \in [x_L, x_U]$, reduces to the above equation form by appropriate scaling of the independent variable.

Construction of underestimator. A summary of the method is presented here for the translated function $g(x) = f(x + x_L)$, $x \in [0, D]$, where $D = x_U - x_L$. The following three-parameter (a, b, x_s) trigonometric function is considered as the underestimator:

$$\phi(x) = -a \sin[k(x - x_s)] + b, \quad x \in [0, D], \quad a \geq 0\,, \tag{8.4}$$

where $k = 2\pi/L$, and L is the period of $\phi(x)$.

For $\phi(x)$ to be convex, the conditions $x_s \leq 0$ and $L \geq 2\,(D - x_s)$ must be satisfied. Caratzoulas and Floudas (2005) considered $L = 2\,(D - x_s) + M$, where $M \in \mathcal{R}^+$, an arbitrary positive, real number, that makes the period of $\phi(x)$ dependent on the phase shift x_s. They proved that any $M \geq 2\,D$ is sufficient, and by means of asymptotic analysis of the solution have further shown, rigorously, that the value of M does not affect how tight the underestimator will be.

Of the three equations necessary to uniquely determine the parameters a, b and x_s, two are obtained from the requirement that at the bounds of the domain $\phi(x)$ match $g(x)$, that is, $g(x_0) = \phi(x_0)$, $x_0 \in \{0, D\}$. If $g(x)$ is nonconvex and the domain includes at least one minimum, a third equation is obtained by setting $g(q) = t(q)$, where $t(x) = g(x_0) + g'(x_t)(x - x_0)$, $x_0 \in \{0, D\}$, is tangent to $g(x)$ at the point x_t and passes through the point x_0; $q \in (0, D)$ denotes the minimum of $g(x)$ nearest to x_0. From these equations, Caratzoulas and Floudas (2005) obtained

$$a = \frac{\Delta + \delta_{\Delta,0}\,(f_L - T)}{\sin(kx_s) + \sin\{k[D - (D - q)\,\delta_{\Delta,0} - x_s]\}}, \tag{8.5}$$

and

$$b = f_L - \frac{[\Delta + \delta_{\Delta,0}\,(f_L - T)]\sin(kx_s)}{\sin(kx_s) + \sin\{k[D - (D-q)\,\delta_{\Delta,0} - x_s]\}}\,, \tag{8.6}$$

where $f_L = f(x_L)$, $f_U = f(x_U)$, $\Delta = f_L - f_U = g(0) - g(D)$, $T = t(q)$, and $\delta_{\Delta,0} = 1$, if $\Delta = 0$, and zero otherwise.

For the phase shift, x_s, they obtained

$$\begin{array}{rll} \tan(kx_s) = & -\frac{\Delta\sin(kq) + (T-f_L)\sin(kD)}{\Delta(1-\cos(kq)) + (T-f_L)(1-\cos(kD))}, & \Delta \neq 0 \\ x_s = & -M/2 > 0, & \Delta = 0\,. \end{array} \tag{8.7}$$

This equation must be solved numerically (a few Newton iterations have been proven sufficient) and it was shown that it always has a solution, that is, for given q and L (i.e., $L = M$) there always exists a unique $x_s < 0$ satisfying it.

For $x_0 = 0$, and $q = q_l$ the minimum of $g(x)$ nearest to $x = 0$, one obtains the tangent line $t_l(x)$. For $x_0 = D$, and $q = q_u$ the minimum of $g(x)$ nearest to $x = D$, one obtains the tangent line $t_u(x)$. Thus, two sets of parameters are obtained, (a_l, b_l, x_{sl}) for $q = q_l$, and (a_u, b_u, x_{su}) for $q = q_u$, and the respective functions $\phi_l(x)$ and $\phi_u(x)$. If both $\phi_l(x)$ and $\phi_u(x)$ are underestimators, the tighter one is chosen — that is the one with the smaller amplitude parameter, a. Caratzoulas and Floudas (2005) proved the following:

Property 1: For $M \geq 2D$, the function $F(x) = \phi_l(x) - \phi_u(x)$ cannot have a single root in the interval $[q_l, q_u]$.

Property 2: If $t_l(q_l) > t_u(q_l)$ and $t_l(q_u) < t_u(q_u)$, a sufficient condition for $\phi_l(x)$ ($\phi_u(x)$) to be an underestimator is that the function $(\phi_l - t_u)(x)$ ($(\phi_u - t_l)(x)$) has a root in $[q_l, q_u]$.

Theorem: *At least one of the functions $\phi_l(x)$ and $\phi_u(x)$ constructed above is an underestimator.*

If $g(x)$ is non-convex and the domain does *not* include a minimum, in the rather trivial case where $g(0) > g(D)$ and the tangent to $g(x)$ passing through $x = 0$ does not exist, namely $x_t \notin [0, D]$, the underestimator is a line through the end points; similarly if $g(0) < g(D)$ and the tangent to $g(x)$ passing through $x = D$ does not exist. That would also be the case if $g(0) = g(D)$. If, however, either one of the two tangents exists, an underestimator of the same form as before is sought. By enforcing the same end-points matching conditions as before, one obtains the equations for the parameters a and b. However, in the absence of a minimum point, the condition $\phi(q) = t(q)$ cannot be employed. Instead, they set $(\mathrm{d}\phi/\mathrm{d}x)_{x=0} = 0$ if $g(0) < g(D)$, or $(\mathrm{d}\phi/\mathrm{d}x)_{x=D} = 0$ if $g(0) > g(D)$, to obtain

$$x_s = \begin{cases} -D - M/2, & \Delta < 0 \\ D - M/2, & \Delta > 0\,. \end{cases} \tag{8.8}$$

Caratzoulas and Floudas (2005) proved that the function $\phi(x)$ obtained in this manner is also an underestimator.

Maximum separation distance. Caratzoulas and Floudas (2005) investigated the behaviour of the solutions with respect to the parameter $M \geq 2\,D$ as that becomes very large. In all cases, they showed that the curvature, $a\,k^2$, of $\phi(x)$ approaches a finite value. Based on their asymptotic analysis, they also have investigated the $\max_{x\in[0,D]}\{\min_{x\in[0,D]}[g(x)] - \phi(x)\}$ and its dependence on the domain size, D, as a measure of high tight an underestimator $\phi(x)$ is. Specifically, they showed that, as $M \to \infty$,

$$\max_{x\in[0,D]}\left\{\min_{x\in[0,D]}[g(x)]-\phi(x)\right\}$$
$$\sim \begin{cases} \min_{x\in[0,D]}[g(x)]-f_L+\Delta/[4\,r\,D\,(1-r\,D)], & \Delta\neq 0 \\ \min_{x\in[0,D]}[g(x)]-f_L+(f_L-T)\,D^2/[4\,q\,(D-q)], & \Delta=0\,. \end{cases} \tag{8.9}$$

where $r\equiv[\Delta\,q+(T-f_L)\,D]/[\Delta\,q^2+(T-f_L)\,D^2]$, with $r\sim 1/D$ and $1-rD\sim\Delta\,q/[D\,(f_L-T)]$. As D increases, the quantity on the left-hand side grows linearly.

Illustration. As an example, let us consider the function $f(x)=\sin x+\sin\frac{10x}{3}+\ln x-0.84x$, with $x\in[1.5,12.484]$. This function has a unique minimum with an objective function value of -8.7429 located at $x=10.914$. Applying the proposed convex underestimation approach on this example to underestimate, individually, each of the first two terms in $f(x)$, the term $\ln x$, being concave, has been underestimated by a straight line connecting the end points of the domain. The first initial lower bound is -9.7818 at $x=9.656$. Using αBB with the theoretical value $\alpha=6.0007$, one obtains an initial lower bound of -185.2376 located at $x=6.992$. Figure 2 presents graphs of $f(x)$, its constituent trigonometric terms, their respective underestimators, and the overall underestimator.

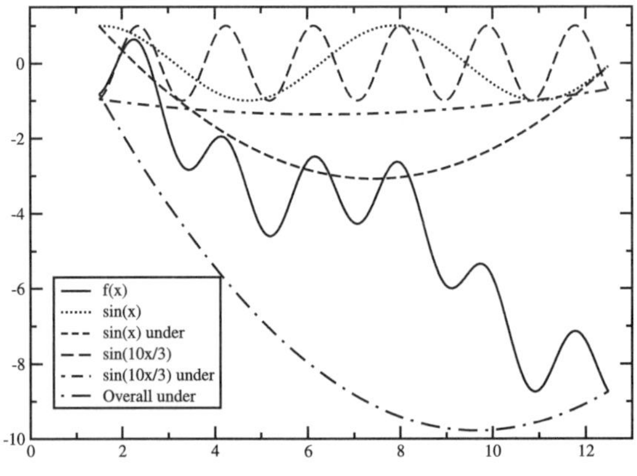

Figure 2 Functions $f(x)$, $\sin x$, $\sin(10x/3)$, and their underestimators

8.3 Convex underestimators by piecewise quadratic perturbation. In their effort to locate the global solution, deterministic global optimization algorithms, like the αBB Maranas and Floudas (1994); Androulakis *et al.* (1995); Adjiman *et al.* (1998a,b), employ a branch and bound framework. During this process, convex underestimation techniques are used to formulate relaxed convex problems that can be solved to optimality with the use of local solvers, thus providing valid lower bounds for the original problem. The tightness of the underestimators used is of fundamental importance for the computational performance of these algorithms, since a tighter relaxation can lead to faster fathoming and less nodes of the branch and bound tree to be visited Floudas (2000).

In the case of arbitrary nonconvex functions that do not exhibit an exploitable mathematical structure, the αBB general underestimator Androulakis *et al.* (1995);

Adjiman and Floudas (1996) can be used:

$$L(x) = f(x) - \sum_{v=1}^{V} \alpha_v (x_v - x_v^L)(x_v^U - x_v). \tag{8.10}$$

Originally introduced in the work of Maranas and Floudas (1994), this underestimator derives from the function by subtracting a positive quadratic ($\alpha_v \geq 0 \; \forall v$). Given sufficiently large values of the α_v parameters, all nonconvexities in the original function $f(x)$ can be overpowered, resulting into a convex underestimator $L(x)$ that is valid for the entire domain $[x^L, x^U]$. A number of rigorous methods have been devised in order to select appropriate values for these parameters Adjiman and Floudas (1996); Adjiman *et al.* (1998a); Hertz *et al.* (1999). Extensive computational testing of the algorithm Adjiman *et al.* (1998b) showed that the most efficient of those methods is the one based on the *scaled Gherschgorin* theorem, that is based on interval analysis of the Hessian matrix elements.

This section briefly discusses the work of Meyer and Floudas (2005b) who introduced a refinement of the classical method, via a smooth, piecewise quadratic, perturbation function q. The αBB convexification approach can be viewed as an approximate solution to a more general convexification problem, that of finding a convexifying perturbation function $q(x)$ which minimizes a measure, μ, of the separation between a nonconvex C^2-continuous function $f(x)$ and the convex underestimator $f(x) - q(x)$. The size of the considered domain affects the result of every step in the α calculation and strongly influences the tightness of the resulting convex underestimator. In particular, reducing the domain reduces the mismatch between the assumed quadratic functional form and the ideal form; it reduces the overestimation in the interval extension of the Hessian matrix; and the maximum separation distance has been shown to be a quadratic function of interval length Maranas and Floudas (1994). Constructing a convex underestimator using a number of different α vectors, each applying to a *subregion* of the full domain can lead to improved convex underestimators.

Form of new underestimators. Let $f(x)$ be a C^2-continuous function. For each variable x_i, let the interval $D_i = [\underline{x}_i, \overline{x}_i]$ be partitioned into N_i subintervals. The endpoints of these subintervals are denoted $x_i^0, x_i^1, \cdots, x_i^{N_i}$ where $\underline{x}_i = x_i^0 < x_i^1 < \cdots < x_i^k < \cdots < x_i^{N_i} = \overline{x}_i$. A smooth convex underestimator of $f(x)$ over a domain D is defined by

$$\phi(x) = f(x) - q(x), \quad x \in D, \tag{8.11}$$

where

$$q(x) = \sum_{i=1}^{n} q_i^k(x_i), \quad x_i \in D_i, \; \forall i \tag{8.12}$$

$$q_i^k(x_i) = \alpha_i^k (x_i - x_i^{k-1})(x_i^k - x_i) + \beta_i^k x_i + \gamma_i^k. \tag{8.13}$$

In each interval $[x_i^{k-1}, x_i^k]$, $\alpha_i^k \geq 0$ is chosen such that $\nabla^2 \phi(x)$, the Hessian matrix of $\phi(x)$, is positive semi-definite for all members of the set $\{x \in D : x_i \in [x_i^{k-1}, x_i^k]\}$. $q_i^k(x_i)$ is the quadratic function associated with variable i in the kth interval. The function $q(x)$ is a piecewise quadratic function constructed from the functions $q_i^k(x_i)$.

The continuity and smoothness properties of $q(x)$ are produced in a spline-like manner. For $q(x)$ to be smooth, the q_i^k functions and their gradients must match at the endpoints x_i^k. In addition, we require that $q(x) = 0$ at the vertices of the hyperrectangle D. To satisfy these requirements, the following conditions are imposed for all $i = 1, \ldots, n$:

$$\begin{aligned} q_i^1(x_i^0) &= 0 \\ q_i^k(x_i^k) &= q_i^{k+1}(x_i^k) \;\; \forall k = 1, \ldots, N_i - 1 \\ q_i^{N_i}(x_i^{N_i}) &= 0 \\ \left.\frac{dq_i^k}{dx_i}\right|_{x_i^k} &= \left.\frac{dq_i^{k+1}}{dx_i}\right|_{x_i^k} \quad \forall k = 1, \ldots, N_i - 1. \end{aligned} \tag{8.14}$$

Expanding and solving these equations, we obtain

$$\begin{aligned} \beta_i^1 &= \left(\sum_{k=1}^{N_i-1} s_i^k (x_i^k - x_i^{N_i})\right) / (x_i^{N_i} - x_i^0) \\ \beta_i^k &= \beta_i^1 + \sum_{j=1}^{k-1} s_i^j \;\; \forall k = 2, \ldots, N_i \\ \gamma_i^k &= -\beta_i^1 x_i^0 - \sum_{j=1}^{k-1} s_i^j x_i^j \;\; \forall k = 1, \ldots, N_i, \end{aligned} \tag{8.15}$$

where $s_i^k = -\alpha_i^k (x_i^k - x_i^{k-1}) - \alpha_i^{k+1} (x_i^{k+1} - x_i^k)$.

The following properties hold for function $\phi(x)$ Meyer and Floudas (2005b).

Property 1: $\phi(x) \; : \; D \; \rightarrow \; \mathbb{R}$ is a continously differentiable function ($\mathcal{C}^1$-continuity).

Property 2: $\phi(x)$ is a convex function, that is $\nabla^2 \phi(x) \succeq 0 \; \forall \; x \in D$.

Property 3: $\phi(x)$ is an underestimator of $f(x)$, that is $\phi(x) \leq f(x) \; \forall \; x \in D$.

Illustrative example. Consider the Lennard-Jones potential energy function, $f(x) = \frac{1}{x^{12}} - \frac{2}{x^6}$, in the interval $[\underline{x}, \overline{x}] = [0.85, 2.00]$. The first term of this function is a convex function and dominates when x is small, while the second term is a concave function which dominates when x is large. The minimum eigenvalue of this function in an interval $[\underline{x}, \overline{x}]$ can be calculated explicitly as follows:

$$\min_{xin[\underline{x},\overline{x}]} f''(x) = \begin{cases} \frac{156}{\overline{x}^{14}} - \frac{84}{\overline{x}^8}, & \text{if } \overline{x} \leq 1.21707 \\ -7.47810, & \text{if } [\underline{x}, \overline{x}] \ni 1.21707 \\ \frac{156}{\underline{x}^{14}} - \frac{84}{\underline{x}^8}, & \text{if } \underline{x} \geq 1.21707. \end{cases}$$

The classical αBB underestimator for this function and interval is $f(x) - \frac{7.47810}{2}(2 - x)(x - 0.85)$. Figure 3 depicts the potential energy function, the classical αBB underestimator, and two α spline underestimator (based on 2 and 16 subregions).

8.4 The generalized αBB global optimization approach. In this section, a new class of convex underestimators by Akrotirianakis and Floudas (2004a,b) is presented along with their key theoretical properties and an illustrative example which compares the quality of the new convex understimators.

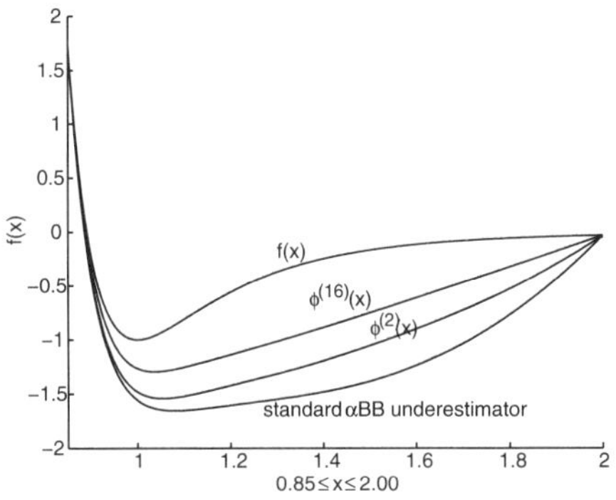

Figure 3 Lennard-Jones potential function and underestimators

Form of new underestimators. The new class of underestimating functions, $L_1(x;\gamma)$, of an arbitrary nonconvex function, $f(x)$ is

$$L_1(x;\gamma) = f(x) + \Phi(x;\gamma), x \in D, \tag{8.16}$$

where $\Phi(x;\gamma) = -\sum_{i=1}^{n}(1-e^{\gamma_i(x_i-x_i^L)})(1-e^{\gamma_i(x_i^U-x_i)})$, $\gamma = (\gamma_1,\gamma_2,\ldots,\gamma_n)^T$ is a vector of non-negative parameters, and D is a suitably defined box domain.

Akrotirianakis and Floudas (2004a) proved the following properties for the function $L_1(x;\gamma)$.

Property 1: $L_1(x;\gamma) \le f(x), \forall\, x \in D$, because $\Phi(x;\gamma) \le 0\ \forall\, x \in D$.

Property 2: $L_1(x^C;\gamma) = f(x^C)$, for every corner point x^C of D, because $\Phi(x^C;\gamma) = 0$ for all x^C of D.

Property 3: There exist certain values of the parameters γ_i so that $L_1(x;\gamma)$ is a convex function. This is due to the fact that the relaxation function $\Phi(x;\gamma)$ is convex for every $x \in D$ and $\gamma_i \ge 0, i = 1,2\ldots n$. Hence, if the parameters γ_i have large enough values then all the non-convexities in the original function $f(x)$ can be eliminated, thereby producing a convex function $L_1(x;\gamma)$.

Property 4: The maximum separation distance between the nonconvex function $f(x)$ and its underestimator $L_{G\alpha BB}(x;\gamma)$ is

$$\max_{x\in D}\{f(x) - L_1(x;\gamma)\} = \sum_{i=1}^{n}(1-e^{\frac{1}{2}\gamma_i(x_i^U-x_i^L)})^2. \tag{8.17}$$

Property 5: The underestimators constructed over supersets of the current set are always less tight than the underestimator constructed over the current box constraints.

Calculation of γ parameters. The values of the parameters $\gamma_i, i = 1,2,\ldots,n$ are determined by an iterative procedure that not only guarantees the convexity of the underestimator $L_1(x;\gamma)$ but also ensures that $L_1(x;\gamma)$ is tighter than the αBB underestimator

$$L_{\alpha BB}(x;\alpha) = f(x) - \sum_{i=1}^{n}\alpha_i(x_i - x_i^L)(x_i^U - x_i). \tag{8.18}$$

The initial values of the γ_i parameters are selected by solving the following system of non-linear equations:

$$\ell_i + \gamma_i^2 + \gamma_i^2 e^{\gamma(x_i^U - x_i^L)} = 0,\ i = 1, 2 \ldots, n, \tag{8.19}$$

where $\ell_i \leq 0, i = 1, 2, \ldots, n$. The parameters ℓ_i convey second order characteristics of the original nonconvex function into the construction process of the underestimator. Candidate values for these parameters can be provided by the scaled Gerschgorin method (Adjiman *et al.* (1998a)):

$$\alpha_i = \max\left\{0, -\frac{1}{2}(\underline{f}_{ii} - \sum_{j \neq i} \max\left\{|\underline{f}_{ij}|, |\overline{f}_{ij}|\right\} \frac{d_j}{d_i}\right\}, \tag{8.20}$$

where $\underline{f}_{ij}$ and $\overline{f}_{ij}$ are the lower and upper bounds of $\partial^2 f/\partial x_i x_j$ as calculated by interval analysis, and $d_i, i = 1, 2, \ldots, n$ are positive parameters. A common choice for those parameters is $d_i = x_i^U - x_i^L$, which reflects the fact that variables with a wider range have a larger effect on the quality of the underestimator than variables with a smaller range.

Akrotirianakis and Floudas (2004a) proved the following two important results regarding the relationship between the maximum separation distances between $f(x)$ and the two underestimators $L_1(x;\gamma)$ and $L_{\alpha BB}(x;\alpha)$.

Theorem 1: *Let* $\underline{\gamma} = (\underline{\gamma}_1, \underline{\gamma}_2, \ldots, \underline{\gamma}_n)^T$ *be the solution of the above system. Then, the two underestimators* $L_1(x;\underline{\gamma})$ *and* $L_{\alpha BB}(x;\underline{\alpha})$, *where*

$$\underline{\alpha} = (\frac{4(1 - e^{0.5\underline{\gamma}_1(x_1^U - x_1^L)})^2}{(x_1^U - x_1^L)^2}, \ldots, \frac{4(1 - e^{0.5\underline{\gamma}_n(x_n^U - x_n^L)})^2}{(x_n^U - x_n^L)^2})^T, \tag{8.21}$$

have the same maximum separation distance from $f(x)$.

Theorem 2: *Let* $\overline{\alpha} = (\overline{\alpha}_1, \overline{\alpha}_2, \ldots, \overline{\alpha}_n)^T$ *be the values of the* α *parameters as computed by (8.20). Then, the two underestimators* $L_1(x;\overline{\gamma})$ *and* $L_{\alpha BB}(x;\overline{\alpha})$, *where*

$$\overline{\gamma} = (\frac{2\log(1 + \sqrt{\overline{\alpha}_1}(x_1^U - x_1^L)/2)}{x_1^U - x_1^L}, \ldots, \frac{2\log(1 + \sqrt{\overline{\alpha}_n}(x_n^U - x_n^L)/2)}{x_n^U - x_n^L})^T, \tag{8.22}$$

have the same maximum separation distance from $f(x)$.

The above two theorems reveal that for any $\gamma \in [\underline{\gamma}, \overline{\gamma}]$ there exists an $\alpha \in [\underline{\alpha}, \overline{\alpha}]$, such that the underestimators $L_1(x;\gamma)$ and $L_{\alpha BB}(x;\alpha)$ have the same maximum separation distance from the nonconvex function $f(x)$. From all these pairs of underestimators, the only one that is known to be convex *a priori* is $L_{\alpha BB}(x;\overline{\alpha})$, since this is the one resulting from the classical αBB method. However, for most arbitrarily nonconvex functions the underestimators $L_{\alpha BB}(x;\alpha)$ and $L_1(x;\gamma)$ are convex within a large portion of the intervals $[\underline{\alpha}, \overline{\alpha}]$ and $[\underline{\gamma}, \overline{\gamma}]$ respectively. Based on the above observations, it is natural to search for a vector γ in the interval $[\underline{\gamma}, \overline{\gamma}]$ or for a vector α in the interval $[\underline{\alpha}, \overline{\alpha}]$, so that at least one of the underestimators $L_1(x;\gamma)$, $L_{\alpha BB}(x;\alpha)$ is convex.

Akrotirianakis and Floudas (2004a) proposed an approach that iteratively determines, using interval analysis, the minimum values of the γ or α parameters that result in an underestimator that is convex and tighter than the classical αBB method. They also developed the generalized αBB global optimization approach, denoted as GαBB, and perfomed extensive computational studies for box constrained global optimization problems (see Akrotirianakis and Floudas (2004b)).

Illustrative example. This example Floudas *et al.* (1999) consists of the global minimization of a potential function describing the pseudoethane molecule which takes the form:

$$\begin{aligned} f_1(x) \quad = \quad & \frac{588600}{(3r_0^2 - 4cos(\theta)r_0^2 - 2(sin^2(\theta)cos(x - \frac{2\pi}{3}) - cos^2(\theta))r_0^2)^6} \\ & - \frac{1079.1}{(3r_0^2 - 4cos(\theta)r_0^2 - 2(sin^2(\theta)cos(x - \frac{2\pi}{3}) - cos^2(\theta))r_0^2)^3} \\ & + \frac{600800}{(3r_0^2 - 4cos(\theta)r_0^2 - 2(sin^2(\theta)cos(x) - cos^2(\theta))r_0^2)^6} \\ & - \frac{1071.5}{(3r_0^2 - 4cos(\theta)r_0^2 - 2(sin^2(\theta)cos(x) - cos^2(\theta))r_0^2)^3} \\ & + \frac{481300}{(3r_0^2 - 4cos(\theta)r_0^2 - 2(sin^2(\theta + \frac{2\pi}{3})cos(x) - cos^2(\theta))r_0^2)^6} \\ & - \frac{1064.6}{(3r_0^2 - 4cos(\theta)r_0^2 - 2(sin^2(\theta + \frac{2\pi}{3})cos(x) - cos^2(\theta))r_0^2)^3}, \end{aligned}$$

where $x \in [0, 2\pi]$ is the dihedral angle, $r_0 = 1.54A$ is the covalent bond length, and $\theta = 109.5^o$ is the covalent bond angle.

The value of the α parameter computed by the classical αBB method using (8.20) is $\overline{\alpha} = 77.124$, and the corresponding value for the γ parameter is $\overline{\gamma} = 1.0673$. Solving for γ, we obtain $\underline{\gamma} = 0.8521$, and the corresponding value for the α parameter is $\underline{\alpha} = 18.579$. The convexity verification algorithm of Akrotirianakis and Floudas (2004a) checks whether there exist values of $\gamma \in [\underline{\gamma}, \overline{\gamma}]$ and $\alpha \in [\underline{\alpha}, \overline{\alpha}]$ such that the underestimator $L_{\alpha BB}(x; \alpha)$ is convex. After 16 iterations it concludes that if $\alpha = \underline{\alpha}$, then $L_{\alpha BB}(x; \alpha)$ is a convex underestimator of $f_1(x)$. The minima of the two underestimators $L_{\alpha BB}(x; \overline{\alpha})$ and $L_{\alpha BB}(x; \alpha)$ are -762.2377 and -184.4244, respectively. Figure 4 compares the two underestimators $L_{\alpha BB}(x; \overline{\alpha})$ and $L_{\alpha BB}(x; \alpha)$, and shows the improvement.

8.5 Tight convex underestimators for $\mathcal{C}^2$-continuous functions. Maranas and Floudas (1994) showed that the maximum separation distance between the original function $f(x)$ and the αBB underestimator $L(x)$ of equation (8.10) is a quadratic function of interval length. Because of this, as well as because of potentially less overestimation in the interval extension of the Hessian matrix elements, the underestimator would become tighter with shrinkage of the domain under consideration.

This section describes the work of Gounaris and Floudas (2008a,b) on the development of tight convex underestimators. The construction of these underestimators is based on a piecewise application of the αBB underestimator. We then identify those supporting line segments that have to be combined with convex parts of the original underestimators so as to form a $\mathcal{C}^1$-continuous convex underestimator that is valid for the overall domain under consideration. One can also consider only the lines defined by these linear segments, thus coming up with a piecewise linear underestimator that can easily be incorporated in the NLP relaxation as a set of linear constraints.

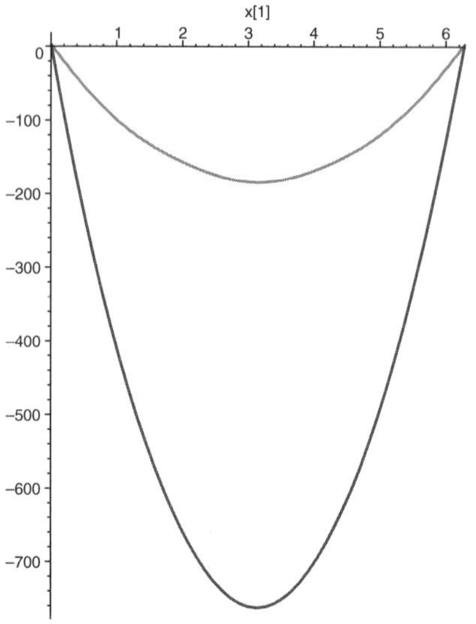

Figure 4 Comparison of the underestimators $L_{\alpha BB}(x;\overline{\alpha})$ and $L_{\alpha BB}(x;\alpha)$

The high quality results of the approach in the univariate case can be used so as to extend the method's applicability to functions with a higher number of variables. This is achieved by proper projections of the multivariate αBB underestimators into select two-dimensional planes. Furthermore, since the method utilizes projections into lower-dimensional spaces, they explored ways to recover some of the information lost in this process. In particular, they apply the method after having transformed the original problem in an orthonormal fashion. This leads to the construction of even tighter underestimators, through the accumulation of additional valid linear cuts in the relaxation.

Theoretical results for univariate functions. Let $f(x)$ be a univariate function that needs to be underestimated in $D = [x^L, x^U]$. We select an integer $N > 1$ and partition the complete domain in N segments of equal length. Thus, the ith subdomain would be defined as $D_i = [x^{i-1}, x^i]$, where $x^i = x^L + \frac{i}{N}(x^U - x^L)$, $i = 0, 1, \cdots, N$.

For every subdomain D_i, $i = 1, 2, \cdots, N$, we construct the corresponding αBB underestimator:

$$\begin{aligned} P_i(x) &= f(x) - \alpha^i (x - x^{i-1})(x^i - x) \\ \alpha^i &= \max\left\{0, -\tfrac{1}{2}\underline{f''}_{(D_i)}\right\} \end{aligned}, \tag{8.23}$$

where $\underline{f''}_{(D_i)}$ is a lower bound of the second derivative that is valid for the entire subdomain D_i. Note that although an underestimator $P_i(x)$ can be defined outside its respective subdomain, its convexity is only guaranteed for $x \in [x^{i-1}, x^i]$.

We define $P(x), x \in [x^L, x^U]$ to be the following branched function:

$$P(x) = P_i(x), \text{ if } x^{i-1} \le x \le x^i. \tag{8.24}$$

Function $P(x)$ is a *piecewise* convex valid underestimator of $f(x)$. Since it is not convex, a convexification technique has to be employed. The proposed technique involves the identification of those supporting line segments that are required for an overall understimator $U(x)$. The technique is based on two algorithms, called *"inner"* and *"outer"*, which are described in detail in Gounaris and Floudas (2008a). The underestimator $U(x)$ consists of the identified linear parts, as well as convex parts of the underestimators $P_i(x)$, therefore it is a $\mathcal{C}^1$-continuous branched function. This might pose some computational complications if the lower bounding (relaxation) problem is to be solved by local optimization solvers that require $\mathcal{C}^2$-continuity. In order to avoid this problem, one can take into account only the lines defined by the line segments. According to this alternative, we first identify the linear segments needed for the construction of underestimator $U(x)$, but we consider those as lines defined in $[x^L, x^U]$. Let there be K such lines denoted as $T_k(x), k = 1, 2, \cdots, K$ and arranged in order of ascending slope. If applicable, this set can be augmented with lines that are tangential to P_1 and P_N at the respective domain edges x^L and x^U.

Each of these lines T_k is a valid underestimator of function $f(x)$ across the whole domain. We define the function $V(x)$ to be the pointwise maximum of all these lines. $V(x)$ is convex, since it is the pointwise maximum of linear functions and it is obviously an underestimator, since it consists of pieces of other underestimators. At the expense of some tightness (in the regions where underestimator $U(x)$ consisted of convex parts), we now have a piecewise linear underestimator $V(x)$ that can be incorporated in the relaxation as a set of linear constraints. The whole lower bounding problem can now be formulated as a linear programming problem (LP).

Tightness of univariate underestimator. It is apparent that as the level of partitioning inreases, the underestimator $P(x)$ comes closer to the function, and therefore convex underestimators $U(x)$ and $V(x)$ approach the convex envelope of $f(x)$. Gounaris and Floudas (2008a) proved the following two theorems that are relevant with the tightness of the resulting underestimators in the univariate case.

Theorem 1: *There is some finite partitioning level N, for which the convex underestimator $U(x)$ is the convex envelope of function $f(x)$.*

Theorem 2: *There is some finite partitioning level N, for which underestimator $V(x)$ is ϵ-close to underestimator $U(x)$, that is*

$$\max_{x \in D} \{U(x) - V(x)\} < \epsilon, \tag{8.25}$$

where $\epsilon > 0$ is an arbitrarily small constant.

Extension to multivariate functions. Let $f(x)$ be a function of V variables that needs to be underestimated in a box domain $D = [x_1^L, x_1^U] \times \cdots \times [x_V^L, x_V^U]$. We choose integers $N_v > 1$, $v = 1, 2, \cdots, V$ and partition each range $[x_v^L, x_v^U]$ in N_v segments of equal length. Thus, the jth segment of the vth set would be defined as $[x_v^{j-1}, x_v^j]$, where $x_v^j = x_v^L + \frac{j}{N_v}(x_v^U - x_v^L)$, $j = 0, 1, \cdots, N_v$. The complete V-dimensional domain D has now been partitioned into $N = \prod_{v=1}^{V} N_v$ box subdomains of equal measures. Let D_i be such a V-dimensional subdomain. It is uniquely defined by a set of indices i_v, $1 \leq i_v \leq N_v$, $\forall v = 1, 2, \cdots, V$. Thus, the ith subdomain would be defined as $D_i = [x_1^{i_1 - 1}, x_1^{i_1}] \times \cdots \times [x_V^{i_V - 1}, x_V^{i_V}]$.

For every subdomain D_i, $i = 1, 2, \cdots, N$, we construct the corresponding αBB underestimator Maranas and Floudas (1994); Androulakis *et al.* (1995); Adjiman and Floudas (1996); Adjiman *et al.* (1998a,b):

$$P_i(x) = f(x) - \sum_{v=1}^{V} \alpha_v^i (x_v - x_v^{i_v-1})(x_v^{i_v} - x_v)$$
$$\alpha_v^i = \max\left\{0, -\frac{1}{2}\left(\underline{h_{vv}^{(i)}} - \sum_{\substack{u=1\\u\neq v}}^{V} \max\left\{|\underline{h_{vu}^{(i)}}|, |\overline{h_{vu}^{(i)}}|\right\} \frac{(x_u^{i_u} - x_u^{i_u-1})}{(x_v^{i_v} - x_v^{i_v-1})}\right)\right\}, \tag{8.26}$$

where $\underline{h_{vu}^{(i)}}$ and $\overline{h_{vu}^{(i)}}$ are respectively lower and upper bounds of $\partial^2 f/\partial x_v x_u$ that are valid for the entire subdomain D_i.

We select variable w, $1 \leq w \leq V$, which we designate to be the *active* variable, and enumerate all $M_w = N/N_w$ permutations of indices i_v, $v \neq w$. Every such permutation m, $1 \leq m \leq M_w$, corresponds to a subdomain

$$D_{wm} = [x_w^L, x_w^U] \times \prod_{\substack{v=1\\v\neq w}}^{V} [x_v^{i_v-1}, x_v^{i_v}],$$

which can be further divided into N_w subdomains

$$D_{wmj} = [x_w^{j-1}, x_w^j] \times \prod_{\substack{v=1\\v\neq w}}^{V} [x_v^{i_v-1}, x_v^{i_v}], \; j = 1, 2, \cdots, N_w.$$

These subdomains, belong to the set of the original subdomains D_i (for $i_w = j$) and therefore each one has an underestimator $P_{wmj}(x)$ associated with it, that is

$$P_{wmj}(x) = f(x) - \alpha_w^i (x_w - x_w^{j-1})(x_w^j - x_w) - \sum_{\substack{v=1\\v\neq w}}^{V} \alpha_v^i (x_v - x_v^{i_v-1})(x_v^{i_v} - x_v), \tag{8.27}$$

where index i satisfies $D_i = D_{wmj}$ and parameters α_v^i, $v = 1, 2, \cdots, V$ are calculated according to (8.26).

For every such subdomain D_{wmj}, $j = 1, 2, \cdots, N_w$, we define the following univariate function:

$$G_{wmj}(x_w) = \min_{\substack{x_v\\ \forall v\neq w}} P_{wmj}(x), \quad x_w^{j-1} \leq x_w \leq x_w^j. \tag{8.28}$$

Since they correspond to the minimum of a convex function over a subset of its variables, these functions are convex. Furthermore, each one is defined over a different segment of $[x_w^L, x_w^U]$. Therefore, each one can be considered as a convex *piece* of an overall *piecewise* convex underestimator. The latter is fully suitable for application of the convex underestimation method for univariate functions which was described in the previous section.

Let $V_{wm}(x_w)$ be the piecewise linear underestimator obtained by the univariate method, and let it be the pointwise maximum of K_{wm} associated lines, that is

$$V_{wm}(x_w) = \max\{T_{wmk}(x_w), \; \forall k = 1, 2, \cdots, K_{wm}\}, \quad x_w^L \leq x_w \leq x_w^U. \tag{8.29}$$

Without loss of generality, let us assume that the lines T_{wmk} are arranged in order of ascending slope, that is, $slope(T_{wm(k-1)}) < slope(T_{wmk}), k = 2, 3, \cdots, K_{wm}$, and that the set already includes the potential augmented tangents at the domain edges, designated earlier as T_0 and T_{K+1}.

Univariate underestimator $V_{wm}(x_w)$ could, in principle, be considered as a multivariate function that is dependent to only one variable, x_w, and defined over the whole multidimensional (dimension V) subdomain D_{wm}. That is

$$V_{wm}(x_w) \to V_{wm}(x), \quad x \in D_{wm}. \tag{8.30}$$

Function $V_{wm}(x)$ is piecewise affine and consists of segments of V-dimensional hyperplanes. Since these hyperplanes depend only on the wth variable, they are parallel to all standard basis vectors e_v with the exception of e_w (to which they are parallel only if the slope of the corresponding line T_{wmk} is zero). This function is a valid underestimator for the original function $f(x)$ across the whole subdomain D_{wm}.

Applying the aforementioned procedure for every permutation $m=1,2,\cdots,M_w$, we come up with a collection of such underestimating segments, each of which is a valid underestimator for the function $f(x)$ across a subset of its original domain D. In order to develop a convex underestimator that would be valid for the whole domain, we have to *combine* all these segments. Let $m = 0$ denote the combination of all permutations $m = 1, 2, \cdots, M_w$. This combination can be achieved back in the projection space, by computing the lower hull of the set of all underestimators $V_{wm}(x_w)$. In fact, one needs to consider only the vertex points of each underestimator $V_{wm}(x_k)$ (that is the points of intersection between two lines $T_{wm(k-1)}$ and T_{wmk}), as well as their end points $\left(x_w^L, T_{wm1}(x_w^L)\right)$ and $\left(x_w^U, T_{wm(K_{wm})}(x_w^U)\right)$. Any standard 2d convex hull algorithm (e.g., *Graham-Scan*) can be used for this purpose. The lower hull is a convex piecewise linear function $V_{w0}(x_w)$, and it is the pointwise maximum of K_{w0} lines, that is

$$V_{w0}(x_w) = \max\left\{T_{w0k}(x_w),\ \forall k = 1, 2, \cdots, K_{w0}\right\}, \quad x_w^L \leq x_w \leq x_w^U. \tag{8.31}$$

By construction, this function is a convex underestimator of all pieces $G_{wmj}(x_w)$ for all permutations, that is

$$V_{w0}(x_w) \leq G_{wmj}(x_w), \quad x_w \in [x_w^{j-1}, x_w^j],\ \forall j = 1, 2, \cdots, N_w,\ \forall m = 1, 2, \cdots, M_w. \tag{8.32}$$

Therefore, function $V_{w0}(x_w)$, if considered as $V_{w0}(x)$, is a valid underestimator for function $f(x)$ across its whole original domain D.

For any selection of the active variable w, the method will yield a convex (piecewise affine) underestimator which would be valid for the whole domain of interest, D. However, the method can be independently applied for every variable being active (one at a time), leading to a collection of valid underestimators. The pointwise maximum of all these is itself a valid convex underestimator, and is tighter (or equally tight) to the original function than any of its predescessors. Thus, the resulting underestimator is

$$V(x) = \max\left\{V_{w0}(x),\ \forall w = 1, 2, \cdots, V\right\}, \quad x \in D. \tag{8.33}$$

Note that the underestimator $V(x)$ is also piecewise hyperplanar, and can be represented in the problem relaxation as a set of linear constraints. Since we do not know explicitly which hyperplanes $T_{w0k}(x_w) \to T_{w0k}(x),\ k = 1, 2, \cdots, K_{w0},\ w = 1, 2, \cdots, V$ contribute some part of theirs to the overall underestimator $V(x)$, all of them should be included in this relaxation, despite the fact that some may end up being redundant.

Since our method produces piecewise affine underestimators $L \equiv V$, the resulting convex relaxation is just a linear programming problem (LP), which takes the form of (8.34).

$$\begin{aligned} \min_{\mu,x} \quad & \mu \\ s.t. \quad & \mu \geq T^{(0)}_{w0k}(x_w) \ , \left\{ \begin{array}{l} \forall k = 1,2,\cdots,K^{(0)}_{w0} \\ \forall w = 1,2,\cdots,V \end{array} \right\} \\ & T^{(q)}_{w0k}(x_w) \leq 0 \ , \left\{ \begin{array}{l} \forall k = 1,2,\cdots,K^{(q)}_{w0} \\ \forall w = 1,2,\cdots,V \\ \forall q = 1,2,\cdots,Q \end{array} \right\} \end{aligned} \tag{8.34}$$

Domain rotation. The methodology presented above involves the minimization of underestimators $P_{wmj}(x)$, over all their variables with the exception of one, variable x_w, which is designated as "active". Whenever such a projection into spaces of lower dimensionality is involved, there is the possibility that some useful information is lost. Some of this lost information will be recovered if we opt to apply the methodology for every variable being "active", one at a time, which basically calls for projecting into V different two-dimensional planes, each one being parallel to a different basis vector e_v, $v = 1,2,\cdots,V$. However, since there is a finite number of variables in our problem, there is a limited number of planes to which we can project. If we want to enhance further the collection of underestimators that we will eventually accumulate in the relaxation (thus improve our chances for better tightness/lower bound), we will have to project into additional planes, that do not correspond to some variable that is "natural" to the problem, rather than to some linear combination of theirs.

This can be achieved by applying an orthonormal transformation to the problem's variable space, that is

$$x \to x' = R \cdot x. \tag{8.35}$$

This transformation has to be orthogonal, which means that it should preserve the lengths of vectors and the angles between vectors. Furthermore, it should be an orientation-preserving transformation. A $V \times V$ matrix R that could provide such a transformation is called a *rotation matrix* and has to be a member of the special orthogonal group, that is

$$R \in SO(V) \Leftrightarrow \left\{ \begin{array}{l} R^{-1} = R^T \\ |R| = +1 \end{array} \right. . \tag{8.36}$$

In their work, Gounaris and Floudas (2008b) discuss the selection of a suitable such matrix. They rigorously address the issue of selecting a suitable "rotated" domain and some suitable level of partitioning, and they also present a method to calculate appropriate values for the α parameters in the transformed counterpart of the problem.

Examples. Figure 5 depicts the plots for the following univariate functions:

$$\begin{aligned} f_1(x) &= (3x - 1.4)\sin(18x) + 1.7, \\ f_2(x) &= x^2 - \cos(18x) \end{aligned}$$

and

$$f_3(x) = -\sum_{k=1}^{5} k \sin[(k+1)x + k].$$

Figure 6 depicts plots for two nonlinear bivariate functions.

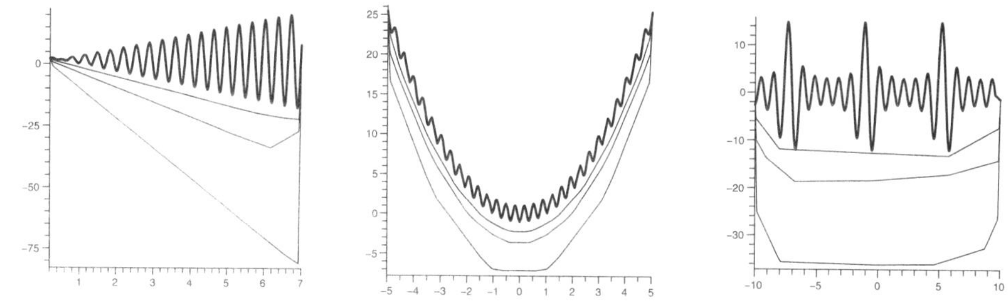

Figure 5 Univariate functions f_{1-3} with underestimators $V(x)$ for partitioning levels $N = 24$, 36 and 48 (increasing tightness)

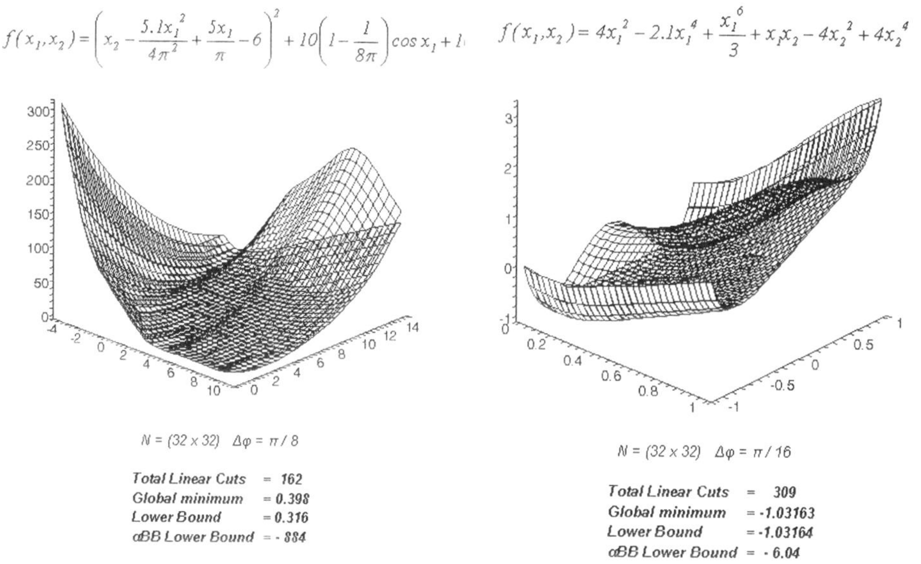

Figure 6 Piecewise planar underestimators of bivariate functions

8.6 Functional forms of convex underestimators. Global optimization methods are generally tunable and thus a set of choices need to made by the user. These choices may involve the selection of appropriate values for the method's parameters or/and the selection of an auxiliary function. The latter is often the case in global optimization techniques that use generic convex underestimators.

In this section, we present the work of Floudas and Kreinovich (2007a,b) on the functional forms of convex underestimators for twice continuously differentiable functions. It is shown that natural symmetry requirements enable one to analytically solve the problem of finding the optimal auxiliary function. It is further shown that the original Adjiman *et al.* (1998a,b) and the generalized Akrotirianakis and Floudas (2004a,b) αBB method are the only optimal schemes, since they involve the only functional forms that satisfy these requirements.

Selecting convex underestimators: The αBB method. A known efficient approach to designing a convex underestimator is the αBB global optimization algorithm Maranas and Floudas (1994); Androulakis *et al.* (1995); Adjiman *et al.* (1998a,b), in which we select an underestimator $L(x) = f(x) + \Phi(x)$, where

$$\Phi(x) = -\sum_{i=1}^{n} \alpha_i (x_i - x_i^L)(x_i^U - x_i). \tag{8.37}$$

Here, the parameters α_i are selected in such a way that the resulting function $L(x)$ is convex and still not too far away from the original objective function $f(x)$. For a thorough presentation of ways to select these parameters, see Adjiman and Floudas (1996); Adjiman *et al.* (1998a,b); Hertz *et al.* (1999).

In many optimization problems, the αBB techniques are very efficient, but in some non-convex optimization problems, it is desirable to improve their performance. One way to do that is to provide a more general class of methods, with more parameters to tune. In the αBB techniques, for each coordinate x_i, we have a single parameter α_i affecting this coordinate. Changing α_i is equivalent to a linear re-scaling of x_i. Indeed, if we change the unit for measuring x_i to a new unit which is λ_i times smaller, then all the numerical values become λ_i times larger: $x_i \to y_i = g_i(x_i)$, where $g_i(x_i) = \lambda_i x_i$. In principle, we can have two different re-scalings:

- $x_i \to y_i = g_i(x_i) = \lambda_i x_i$ on the interval $[x_i^L, x_i]$, and
- $x_i \to z_i = h_i(x_i) = \mu_i x_i$ on the interval $[x_i, x_i^U]$.

If we substitute the new values $y_i = g_i(x_i)$ and $z_i = h_i(x_i)$ into the formula (8.37), then we get the following expression

$$\Phi(x) = -\sum_{i=1}^{n} \alpha_i \cdot (g_i(x_i) - g_i(x_i^L)) \cdot (h_i(x_i^U) - h_i(x_i)). \tag{8.38}$$

For the above linear re-scalings, we get

$$\widetilde{\Phi}(x) = -\sum_{i=1}^{n} \widetilde{\alpha}_i (x_i - x_i^L)(x_i^U - x_i),$$

where $\widetilde{\alpha}_i = \alpha_i \lambda_i \mu_i$.

From this viewpoint, a natural generalization is to replace *linear* re-scalings $g_i(x_i)$ and $h_i(x_i)$ with *non-linear* ones, that is, to consider convex underestimators of the type $L(x) = f(x) + \Phi(x)$, where $\Phi(x)$ is described by the formula (8.38) with non-linear functions $g_i(x_i)$ and $h_i(x_i)$. Now, instead of selecting a number α_i for each coordinate i, we have an additional freedom of choosing arbitrary non-linear functions $g_i(x_i)$ and $h_i(x_i)$. The question of which are the best choices is naturally posed. In Akrotirianakis and Floudas (2004a,b), several different non-linear functions were tried, and it turned out that among the tested functions, the best results were achieved for the exponential functions $g_i(x_i) = \exp(\gamma_i x_i)$ and $h_i(x_i) = -\exp(-\gamma_i x_i)$. For these functions, the expression (8.38) can be somewhat simplified; indeed, $\alpha_i \cdot (g_i(x_i) - g_i(x_i^L)) \cdot (h_i(x_i^U) - h_i(x_i)) = \alpha_i (e^{\gamma_i x_i} - e^{\gamma_i x_i^L})(-e^{-\gamma_i x_i^U} + e^{-\gamma_i x_i}) = \widetilde{\alpha}_i (1 - e^{\gamma_i (x_i - x_i^L)})(1 - e^{\gamma_i (x_i^U - x_i)})$, where $\widetilde{\alpha}_i \stackrel{\text{def}}{=} \alpha_i e^{\gamma_i (x_i^U - x_i^L)}$.

Two related questions naturally arise and are addressed in the work of Floudas and Kreinovich (2007a,b):

- first, a *practical* question: an empirical choice is made by using only finitely many functions; is this choice indeed the best – or are there other, even better functions $g_i(x_i)$ and $h_i(x_i)$, which we did not discover because we did not try them?
- second, a *theoretical* question: how can we explain the above empricial fact?

Shift invariance. The starting point for measuring each coordinate x_i is often a matter of arbitrary choice. If a selection of the functions $g_i(x_i)$ and $h_i(x_i)$ is "optimal" (in some intuitive sense), then the results of using these optimal functions should not change if we simply change the starting point for measuring x_i, that is, replace each value x_i with a new value $x_i + s$ (where s is the shift in the starting point). Indeed, otherwise, if the "quality" of the resulting convex underestimators changes with shift, we could apply a shift and get better functions $g_i(x_i)$ and $h_i(x_i)$ – which contradicts the assumption that the selection of $g_i(x_i)$ and $h_i(x_i)$ is already optimal.

The "optimal" choices $g_i(x_i)$ and $g_i(x_i)$ can be determined from the requirement that each component $\alpha_i \cdot (g_i(x_i) - g_i(x_i^L)) \cdot (h_i(x_i^U) - h_i(x_i))$ in the sum (8.38) be invariant under the corresponding shift, that is, that they satisfy the following definition.

Definition 1: A pair of smooth functions $(g(x), h(x)))$ is *shift-invariant* if for every s and α, there exists $\widetilde{\alpha}(\alpha, s)$ such that for every x^L, x, and x^U, we have

$$\alpha \cdot (g(x) - g(x^L)) \cdot (h(x^U) - h(x)) = \widetilde{\alpha}(\alpha, s) \cdot (g(x+s) - g(x^L+s)) \cdot (h(x^U+s) - h(x+s)) \tag{8.39}$$

Floudas and Kreinovich (2007a,b) proved that if a pair of functions $(g(x), h(x))$ is shift-invariant, then each of the functions $g(x)$ and $h(x)$ has the form $A + C\exp(\gamma x)$ or $A + kx$.

Sign and scale invariance. In addition to shift, another natural symmetry is changing the sign. If we require that the expression (8.38) remain invariant under a replacement of x by $-x$, then we get the relation between $g(x)$ and $h(x)$: $h(x) = -g(-x)$. So, if a pair $(g(x), h(x))$ is shift-invariant and sign-invariant, then:

- either $g(x) = \exp(\gamma x)$ and $h(x) = -\exp(-\gamma x)$,
- or $g(x) = h(x) = x$.

In other words, the optimal generalized αBB scheme is either the original αBB Maranas and Floudas (1994); Androulakis *et al.* (1995); Adjiman *et al.* (1998a,b), or the scheme with exponential functions described in Akrotirianakis and Floudas (2004a,b).

Sign-invariance can be perceived as a special case of scale-invariance. The latter addresses a change in the unit for measuring x, that is, transformations $x \to \lambda x$.

It has already been shown that there are only two shift-invariant solutions: exponential and linear functions. Out of these two solutions, only the linear solution – corresponding to the original αBB – is scale-invariant. Thus, if we also require scale-invariance, we restrict ourselves only to original techniques and miss the (often better) exponential generalizations.

Although imposing both shift- and scale-invariance leads to restrictions, one could still choose to employ only the latter, formally expressed as follows:

Definition 2: A pair of smooth functions $(g(x), h(x))$ is *scale-invariant* if for every λ and α, there exists $\widetilde{\alpha}(\alpha, \lambda)$ such that for every x^L, x, and x^U, we have

$$\alpha\cdot(g(x)-g(x^L))\cdot(h(x^U)-h(x)) = \widetilde{\alpha}(\alpha,\lambda)\cdot(g(\lambda\cdot x)-g(\lambda\cdot x^L))\cdot(h(\lambda\cdot x^U)-h(\lambda\cdot x)) \tag{8.40}$$

It was also proven that if a pair of functions $(g(x), h(x))$ is scale-invariant, then each of the functions $g(x)$ and $h(x)$ has the form $A\cdot x^\gamma$ or $A+k\cdot\ln(x)$.

From the theoretical viewpoint, these functions may look as good as the exponential functions coming from shift invariance but, since they are not infinitely differentiable for all x, they do not work so well in practice.

Generalization of shift invariance. Instead of the expression (8.38), we can consider an even more general expression

$$\Phi(x) = -\sum_{i=1}^{n} \alpha_i\cdot a_i(a, x^L)\cdot b_i(x_i, x_i^U). \tag{8.41}$$

What can be concluded from shift-invariance in this more general case?

Definition 3: A pair of smooth functions $(a(x,x^L), b(x,x^U))$ is *shift-invariant* if for every s and α, there exists $\widetilde{\alpha}(\alpha, s)$ such that for every x^L, x, and x^U, we have

$$\alpha\cdot a(x,x^L)\cdot b(x,x^U) = \widetilde{\alpha}(\alpha,s)\cdot a(x+s, x^L+s)\cdot b(x+s, x^U+s). \tag{8.42}$$

Regarding such functions, Floudas and Kreinovich (2007a) proved the following proposition:

Proposition 1: *If a pair of fons $(a(x,x^L), b(x,x^U))$ is shift-invariant, then*

$$a(x,x^L)\cdot b(x,x^U) = A(x-x^L)\cdot B(x^U-x)\cdot e^{\gamma x^L}$$

for some functions $A(x)$ and $B(x)$ and for some real number γ.

Comment: If we additionally require that the expression $a(x,x^L)\cdot b(x,x^U)$ be invariant under $x\to -x$, then we conclude that $B(x)=A(x)$.

Another shift-invariance result comes from the following observation. Both the αBB expression $-(x-x^L)(x^U-x)$ and the generalized expression $-(1-e^{\gamma(x-x^L)})(1-e^{\gamma(x^U-x)})$ have the form $a(x-x^L)\cdot a(x^U-x)$ with $a(0)=0$. The differences $x-x^L$ and x^U-x come from the fact that we want these expressions to be shift-invariant. The product form makes sense, since we want the product to be 0 on each border $x=x^L$ and $x=x^U$ of the corresponding interval $[x^L, x^U]$.

On the other hand, it is well known that optimizing a product is more difficult than optimizing a sum; since we will be minimizing the expression $f(x)+\Phi(x)$, it is therefore desirable to be able to reformulate it in terms of the easier-to-minimize sum, e.g., as $b(x-x^L)+b(x^U-x)+c(x^U-x^L)$ for some functions b and c (for minimization purposes, c does not depend on x and is thus a constant). It is worth mentioning that both the αBB expression and its exponential generalization allow such representation. Note that $-(x-x^L)(x^U-x) = \frac{1}{2}(x-x^L)^2+\frac{1}{2}(x^U-x)^2-\frac{1}{2}(x^U-x^L)^2$; and $-(1-e^{\gamma(x-x^L)})(1-e^{\gamma(x^U-x)}) = -1+e^{\gamma(x-x^L)}+e^{\gamma(x^U-x)}-e^{\gamma(x^U-x^L)}$. Interestingly, the above two expressions are the only ones which have this easiness-to-compute property:

Definition 4: We say that a smooth function $a(x)$ from real numbers to real numbers describes an *easy-to-compute* underestimator if $a(0)=0$, $a'(0)\neq 0$, and there exist smooth functions $b(x)$ and $c(x)$ such that for every x, x^L, and x^U, we have

$$a(x - x^L) \cdot a(x^U - x) = b(x - x^L) + b(x^U - x) + c(x^U - x^L). \tag{8.43}$$

The condition $a'(0) \neq 0$ comes from the fact that otherwise, for small $\Delta x \stackrel{\text{def}}{=} x - x^L$ and $x^U - x$, each value $a(x - x^L)$ will be quadratic in $x - x^L$, the resulting product will be fourth order, and we will not be able to compensate for quadratic non-convex terms in the original objective function $f(x)$ – which defeats the purpose of using $f(x) + \Phi(x)$ as a *convex* underestimator.

Proposition 2: *The only functions which describe easy-to-compute underestimators are* $a(x) = kx$ *and* $a(x) = k(1 - e^{\gamma x})$.

The above result leads us to select linear and exponential functions as "the best" in some reasonable sense. To conclude, the work of Floudas and Kreinovich (2007a,b) has a much further-reaching effect than on the case of αBB-based convex underestimation mainly discussed here. A symmetry-based approach leads to optimal techniques also in the cases of optimal bisection (for selecting box-splitting strategies) and optimal selection of penalty and barrier functions. Other empirically optimal techniques can also be explained by symmetry-based arguments. These include the "epsilon-inflation" technique Kreinovich *et al.* (1997); Rump (1998), results in simulated annealing and genetic algorithms Nguyen and Kreinovich (1997), as well as in the optimal selection of probabilities in swarm optimization Kennedy *et al.* (2001); Iourinski *et al.* (2002).

8.7 Convexity of products of univariate functions. This section presents the work of Gounaris and Floudas (2008c) who investigated the characteristics that have to be possesed by a functional mapping $f:\mathbb{R} \mapsto \mathbb{R}$ so that it is suitable to be employed in a variable transformation of the type $x \to f(y)$ in the convexification of posynomials. The results provide rigorous reasoning on why transformations that have already appeared in the literature, like the exponential Maranas and Floudas (1995, 1997) or reciprocal Li *et al.* (2008), work properly in convexifying posynomial programs. At the same time, the results can be used to infer the convexity of multivariate functions that can be expressed as products of univariate factors, through conditions on these factors on an individual basis.

Convexity analysis for bivariate product. Let us consider the bivariate function $\mathcal{F}_2(y_1, y_2) = f_1(y_1)f_2(y_2)$, defined over the box domain $[y_1^L, y_1^U] \times [y_2^L, y_2^U]$.

The Hessian matrix of $\mathcal{F}_2$ is

$$\nabla^2 \mathcal{F}_2(y_1, y_2) = H_2(y_1, y_2) = \begin{bmatrix} f_1^{''}(y_1)f_2(y_1) & f_1^{'}(y_1)f_2^{'}(y_2) \\ f_1^{'}(y_1)f_2^{'}(y_1) & f_1(y_1)f_2^{''}(y_2) \end{bmatrix}, \tag{8.44}$$

where $f_i^{'}$ denotes the first and $f_i^{''}$ denotes the second derivative of the univariate function f_i with respect to the corresponding variable y_i.

The two eigenvalues of this matrix are

$$\lambda_{1,2}(y_1, y_2) = \frac{1}{2}\left\{\left(f_1 f_2^{''} + f_2 f_1^{''}\right) \pm \sqrt{\left(f_1 f_2^{''} - f_2 f_1^{''}\right)^2 + 4(f_1^{'})^2(f_2^{'})^2}\right\}. \tag{8.45}$$

For function $\mathcal{F}_2(y_1, y_2)$ to be convex, $H_2(y_1, y_2)$ needs to be positive semidefinite at every point (y_1, y_2), that is $\lambda_{1,2}(y_1, y_2) \geq 0 \;\; \forall (y_1, y_2) \in [y_1^L, y_1^U] \times [y_2^L, y_2^U]$.

This condition results equivalently into the following two conditions, both of which need to be satisfied $\forall(y_1, y_2)$.

$$f_1 f_2^{''} + f_2 f_1^{''} \geq 0 \tag{8.46}$$

$$f_1 f_1^{''} f_2 f_2^{''} \geq (f_1^{'})^2 (f_2^{'})^2 \tag{8.47}$$

Through a series of arguments, Gounaris and Floudas (2008c) showed that the above conditions are satisfied by a select set of functional forms, which they named *convexification transformation suitable functions* or, simply, *CTSF*. These functions possess the following properties in the given domain of interest:

Property i: Strict positivity

Property ii: Strict monotonicity

Property iii: $f(y)f^{''}(y) - (f^{'}(y))^2 \geq 0$

It can be shown that a CTSF is also strictly convex. Note that a strictly negative function that satisfies properties (ii) and (iii) (and therefore is concave) can also serve as a suitable function. However, without loss of generality, the definition is restricted to the strictly positive case, as one can always infer convexity working with the negative of the function under consideration. Gounaris and Floudas (2008c) also proved a number of related lemmas.

Lemma 1: *A linear combination of a CTSF $f(y)$, $h(y) = \alpha f(y) + \beta$, where $\alpha > 0$ and β sufficiently large, is also CTSF in the same domain.*

Lemma 2: *A monotone sum of two CTSF $f(y)$ and $g(y)$, $h(y) = f(y) + g(y)$, is also CTSF in the same domain.*

Lemma 3: *A monotone product of two CTSF $f(y)$ and $g(y)$, $h(y) = f(y)g(y)$, is also CTSF in the same domain.*

Lemma 4: *An exponentiated CTSF, $h(y) = (f(y))^{\kappa}$, where $\kappa > 0$, is also CTSF in the same domain.*

Properties (i)-(iii) are in agreement with the transformation functions that have been proposed in the literature for use in the convexification/relaxation of posynomial/signomial programs. In particular, both the exponential Maranas and Floudas (1995, 1997), $f(y) = e^y$ $(y \in \mathbb{R})$, and the reciprocal Li *et al.* (2008), $f(y) = 1/y$ $(y \in \mathbb{R}_+^*)$, are functions that possess them and thus are suitable for such transformations. This translates to the fact that bivariate functions such as $e^{(y_1+y_2)}$, $1/(y_1 y_2)$, or even e^{y_1}/y_2, are indeed convex (in $\mathbb{R}^2$,${\mathbb{R}_+^*}^2$ and $\mathbb{R} \times \mathbb{R}_+^*$, respectively). Of course, proving the convexity of $\mathcal{F}_2$, given functions f_1 and f_2, is typically not a difficult task, since the dimensionality of the Hessian matrix is small. The practical importance of the derived conditions will be more apparent when we generalize to higher dimensions.

Convexity analysis for multivariate product. The conditions derived previously, through the analysis of the bivariate product, are indeed sufficient conditions for the convexity of general products of univariate functions. The following theorem holds.

Theorem: *If univariate functions $f_i(y_i)$ are strictly positive and satisfy property (iii) in domain D_i, for every $i = 1, 2, \cdots, n$, then the n-variate function $\mathcal{F}_n(y_1, y_2, \cdots, y_n) = \prod_{i=1}^{n} f_i(y_i)$ is convex in domain $D = \times_{i=1}^{n} D_i$.*

The theorem can be proven by induction on the number of factors n. For the proof, see Gounaris and Floudas (2008c). Its usefulness in proving the convexity of multivariate products of univariate functions lies in the fact that each univariate factor can be considered individually, and if all are proven to satisfy the required properties, then the convexity of the multivariate product can be inferred. Note that this can offer a significant advantage especially when the number of factors and, thus, variables is large, because conventional Hessian eigenvalue analysis becomes non-tractable and one is usually restricted to examining convexity only within a local context.

Example. Let us present an example, where one can infer the convexity of a (relatively complex) multivariate function. We will show that

$$\mathcal{F}_4(y_1,y_2,y_3,y_4) = \left\{\frac{1.8}{y_2} + 1.2y_2 - \frac{3ln(1-y_1)}{y_2} - 2y_2 ln(1-y_1)\right\}\frac{e^{y_3-y_4}}{y_4^{1.2}}$$

is a convex function in $D = [\frac{1}{3},\frac{2}{3}]^4$.

Firstly, we observe that function $\mathcal{F}_4$ can be decomposed into a product of four univariate factors, as follows:

$$\begin{aligned}
\mathcal{F}_4(y_1,y_2,y_3,y_4) &= \left\{\frac{1.8}{y_2} + 1.2y_2 - \frac{3ln(1-y_1)}{y_2} - 2y_2 ln(1-y_1)\right\}\frac{e^{y_3-y_4}}{y_4^{1.2}} \\
&= \left\{\frac{3}{y_2}\left[0.6 - ln(1-y_1)\right] + 2y_2\left[0.6 - ln(1-y_1)\right]\right\}\frac{e^{y_3-y_4}}{y_4^{1.2}} \\
&= \{0.6 - ln(1-y_1)\}\left\{\frac{3}{y_2} + 2y_2\right\}\frac{e^{y_3}e^{-y_4}}{y_4^{1.2}} \\
&= \{0.6 - ln(1-y_1)\}\left\{\frac{2y_2^2+3}{y_2}\right\}\{e^{y_3}\}\left\{\frac{e^{-y_4}}{y_4^{1.2}}\right\}
\end{aligned}$$

Let us define $f_1(y_1) = 0.6 - ln(1-y_1)$,$f_2(y_2) = \frac{2y_2^2+3}{y_2}$,$f_3(y_3) = e^{y_3}$ and $f_4(y_4) = \frac{e^{-y_4}}{y_4^{1.2}}$. According to the theorem, function $\mathcal{F}_4$ will be convex, if all four factors $f_i(y_i)$ are CTSF (since they would satisfy properties i and iii).

(1) For function f_1, we firstly observe that function $h(y_1) = 1 - ln(1-y_1)$ is CTSF in domain $[\frac{1}{3},\frac{2}{3}]$. This is because (i) $\inf_{y\in[\frac{1}{3},\frac{2}{3}]} h(y) = h(\frac{1}{3}) = 1 - ln(\frac{2}{3}) > 0$, (ii) $\inf_{y\in[\frac{1}{3},\frac{2}{3}]} h'(y) = h'(\frac{1}{3}) = \frac{3}{2} > 0$, and (iii) $h(y)h''(y) - (h'(y))^2 = -ln(1-y)(\frac{1}{1-y})^2 \overset{y\in[\frac{1}{3},\frac{2}{3}]}{\geq} 0$. We further observe that $f_1(y_1) = h(y_1) - 0.4$. According to the relevant lemma, function h will maintain its suitability if we add to it a constant, as long as the latter is greater or equal than β^L, where:

$$\begin{aligned}
\beta^L &= \sup_{\substack{y_1\in[\frac{1}{3},\frac{2}{3}] \\ y_1:h''(y_1)>0}} \frac{-\left\{h(y_1)h''(y_1) - (h'(y_1))^2\right\}}{h''(y_1)} \\
&\overset{\text{note}^1}{=} \max_{y_1\in[\frac{1}{3},\frac{2}{3}]} ln(1-y_1) = ln(\frac{2}{3}) < -0.4
\end{aligned}$$

Therefore, function $f(y_1) = h(y_1) - 0.4$ is also CTSF in $[\frac{1}{3},\frac{2}{3}]$.

[1] The supremum has been replaced by maximum because the domain is closed and function h is continuous.

(2) For function $f_2 = \frac{2y_2^2+3}{y_2}$, we consider it as the product of two functions, namely $h_1(y_2) = \frac{1}{y_2}$ and $h_2(y_2) = 2y_2^2 + 3$. Function h_1 is the reciptrocal function, which is CTSF in every domain $D_y \subset \mathbb{R}_+^*$. Function h_2 is CTSF in $[\frac{1}{3}, \frac{2}{3}]$ since (i) $\inf_{y\in[\frac{1}{3},\frac{2}{3}]} h_2(y) = f(\frac{1}{3}) = \frac{10}{9} > 0$, (ii) $\inf_{y\in[\frac{1}{3},\frac{2}{3}]} h_2'(y) = h_2'(\frac{1}{3}) = \frac{2}{3} > 0$, and (iii) $h_2(y)h_2''(y) - (h_2'(y))^2 = 2 - 2y^2 \overset{y\in[\frac{1}{3},\frac{2}{3}]}{\geq} 0$. In addition, and despite the fact that h_1 is a decreasing and h_2 is an increasing function, their product is strictly monotone (decreasing), since $\max_{y_2\in[\frac{1}{3},\frac{2}{3}]} f_2'(y_2) = f_2'(\frac{2}{3}) = -\frac{19}{4} < 0$. Therefore, the relevant lemma dictates that f_2 will be CTSF in $[\frac{1}{3}, \frac{2}{3}]$.

(3) Function $f_3 = e^{y_3}$ is the exponential function, which is CTSF in every subset of $\mathbb{R}$.

(4) Function $f_4 = \frac{e^{-y_4}}{y_4^{1.2}}$ can be considered as the product of two functions, namely $h_1(y_4) = e^{-y_4}$ and $h_2(y_4) = \frac{1}{y_4^{1.2}}$. Both functions are strictly positive and strictly monotone. Regarding property (iii), we have $h_1h_1'' - (h_1')^2 = 0 \geq 0 \ \ \forall y_4 \in \mathbb{R}$ and $h_2h_2'' - (h_2')^2 = \frac{1.2}{y_4^{4.4}} \geq 0 \ \ \forall y_4 \in \mathbb{R}_+^*$. Therefore, both are CTSF in $[\frac{1}{3}, \frac{2}{3}]$. Since they both exhibit monotonicity of the same type (decreasing), their product, function f_4, will also be strictly monotone and, according to the relevant lemma, it will also be CTSF in $[\frac{1}{3}, \frac{2}{3}]$.

Therefore, function $\mathcal{F}_4$ is, indeed, convex.

9 Summary

This paper reviewed the advances in global optimization during the period 1998-2008. The focal point was novel theoretical, algorithmic, and applications oriented advances on deterministic global optimization methods for (i) general twice differentiable NLPs; (ii) mixed-integer nonlinear optimization problems MINLPs; (iii) models with differential-algebraic constraints; (iv) grey-box and nonfactorable models; (v) semi-infinite programming problems; and (vi) bilevel nonlinear optimization problems. Recent advances from Princeton University were also presented on explicit facets of convex and concave envelopes for trilinear monomials, convex underestimators for trigonometric functions, convex underestimators by piecewise quadratic perturbation, the generalized αBB global optimization approach, tight convex underestimators for $\mathcal{C}^2$-continuous functions, functional forms of convex underestimators, and convexity of products of univariate functions

Acknowledgements. Christodoulos A. Floudas gratefully acknowledges support from the National Science Foundation, the National Institutes of Health (R01 GM52032 and R24 GM069736), the Environmental Protection Agency (GAD R 832721-010), AspenTech Corporation, and BASF Corporation.

References

W.P. Adams and H.D. Sherali. A hierarchy of relaxations leading to the convex hull representation for general discrete optimization problems. *Annals of Operations Research*, 140(1):21–47, 2005.

C.S. Adjiman, I.P. Androulakis, and C.A. Floudas. A global optimization method, αBB, for general twice–differentiable NLPs – II. Implementation and computational results. *Computers & Chemical Engineering*, 22(9):1159–1179, 1998b.

C.S. Adjiman, S. Dallwig, C.A. Floudas, and A. Neumaier. A global optimization method, αBB, for general twice–differentiable NLPs – I. Theoretical advances. *Computers & Chemical Engineering*, 22(9):1137–1158, 1998a.

C.S. Adjiman and C.A. Floudas. Rigorous convex underestimators for general twice-differentiable problems. *Journal of Global Optimization*, 9:23–40, 1996.

C.S. Adjiman and I. Papamichail. A deterministic global optimization algorithm for problems with nonlinear dynamics. In C.A. Floudas and P.M. Pardalos, editors, *Frontiers in Global Optimization*, pages 1–24, Kluwer Academic Publishers, 2003.

I.G Akrotirianakis and C.A. Floudas. A new class of improved convex underestimators for twice continuously differentiable constrained NLPs. *Journal of Global Optimization*, 30(4):367–390, 2004a.

I.G. Akrotirianakis and C.A. Floudas. Computational experience with a new class of convex underestimators: Box-constrained NLP problems. *Journal of Global Optimization*, 29(3):249–264, 2004b.

J. Alex, R. Tschepetzki, U. Jumar, F. Obenaus, and K.H. Rosenwinkel. Analysis and design of suitable model structures for activated sludge tanks with circulating flow. *Water Science and Technology*, 39(4):55–60, 1999.

M.M. Ali, C. Khompatraporn, and Z.B. Zabinsky. A numerical evaluation of several stochastic algorithms on selected continuous global optimization test problems. *Journal of Global Optimization*, 31:635–672, 2005.

P. Amaral, J. Judice, and H.D. Sherali. A reformulation-linearization-convexification algorithm for optimal correction of an inconsistent system of linear constraints. *Computers & Operations Research*, 35(5):1494–1509, 2008.

I.P. Androulakis, C.D. Maranas, and C.A. Floudas. αBB: A global optimization method for general constrained nonconvex problems. *Journal of Global Optimization*, 7:337–363, 1995.

R. Angira and A. Santosh. Optimization of dynamic systems: A trigonometric differential evolution approach. *Computers & Chemical Engineering*, 31(9):1055–1063, 2007.

C. Audet, P. Hansen, and G. Savard. *Essays and Surveys in Global Optimization*. GERAD 25th Anniversary Series. Springer, 2005.

H. Babahadda and N. Gadhi. Necessary optimality conditions for bilevel optimization problems using convexificators. *Journal of Global Optimization*, 34(4):535–549, 2006.

J.R. Banga, E. Balsa-Canto, C.G. Moles, and A.A. Alonso. Improving food processing using modern optimization methods. *Trends in Food Science and Technology*, 14:131–144, 2003b.

J.R. Banga, C.G. Moles, and A.A. Alonso. Global optimization of bioprocesses using stochastic and hybrid methods. In C.A. Floudas and P.M. Pardalos, editors, *Frontiers in Global Optimization*, pages 45–70, Kluwer Academic Publishers, 2003a.

P.I. Barton and C.K. Lee. Global dynamic optimization of linear time varying hybrid systems. *Dynamics of Continuous Discrete and Impulsive Systems-Series B*, S:153–158, 2003.

P.I. Barton and C.K. Lee. Design of process operations using hybrid dynamic optimization. *Computers & Chemical Engineering*, 28(6-7):955–969, 2004.

P.I. Barton, C.K. Lee, and M. Yunt. Optimization of hybrid systems. *Computers & Chemical Engineering*, 30(10-12):1576–1589, 2006.

A. Beck, A. Ben-Tal, and M. Teboulle. Finding a global optimal solution for a quadratically constrained fractional quadratic problem with applications to the regularized total least squares. *SIAM Journal on Matrix Analysis and Applications*, 28(2):425–445, 2006.

H.P. Benson. On the global optimization of sums of nonlinear fractional functions over a convex set. *Journal of Optimization Theory and Applications*, 121(1):19–39, 2004.

H.P. Benson. Fractional programming with convex quadratic forms and functions. *European Journal of Operational Research*, 173(2):351–369, 2006.

H.P. Benson. Solving sum of ratios fractional programs via concave minimization. *Journal of Optimization Theory and Applications*, 135(1):1–17, 2007a.

H.P. Benson. A simplicial branch and bound duality-bounds algorithm for the linear sum-of-ratios problem. *European Journal of Operational Research*, 182(2):597–611, 2007b.

M.L. Bergamini, N.J. Scenna, and P.A. Aguirre. Global optimal structures of heat exchanger networks by piecewise relaxation. *Industrial & Engineering Chemistry Research*, 46(6):1752–1763, 2007.

K.J. Björk, P.O. Lindberg, and T. Westerlund. Some convexifications in global optimization of problems containing signomial terms. *Computers & Chemical Engineering*, 27:669–679, 2003.

B. Bhattacharjee, W.H. Green, and P.I. Barton. Interval methods for semi-infinite programming. *Computational Optimization and Applications*, 30(1):63–93, 2005a.

B. Bhattacharjee, P. Lemonidis, W.H. Green, and P.I. Barton. Global solution of semi-infinite programs. *Mathematical Programming*, 103(2):283–307, 2005b.

E.G. Birgin, C.A. Floudas, and J.M. Martinez. Global optimization using an augmented Lagrangian method with variable lower-level constraints. *Mathematical Programming*, submitted for publication, 2008.

E. Bringas, R. Karuppiah, M.F.S. Roman, I. Ortiz, and I.E. Grossmann. Optimal groundwater remediation network design using selective membranes. *Industrial & Engineering Chemistry Research*, 46(17):5555–5569, 2007.

G. Borradaile and P. Van Hentenryck. Safe and tight linear estimators for global optimization. *Mathematical Programming*, 102:495–517, 2005.

M. Campelo and S. Scheimberg. A study of local solutions in linear bilevel programming. *Journal of Optimization Theory and Applications*, 125(1):63–84, 2005.

S. Caratzoulas and C.A. Floudas. A trigonometric convex underestimator for the base functions in Fourier space. *Journal of Optimization Theory and Applications*, 124(2):339–362, 2005.

L.G. Casado, J.A. Martinez, I. Garcia, and Y.D. Sergeyev. New interval analysis support functions using gradient information in a global minimization algorithm. *Journal of Global Optimization*, 25:345–362, 2003.

B. Chachuat and M.A. Latifi. A new approach in deterministic global optimization of problems with ordinary differential equations. In C.A. Floudas and P.M. Pardalos, editors, *Frontiers in Global Optimization*, pages 83–108, Kluwer Academic Publishers, 2003.

B. Chachuat, A.B. Singer, and P.I. Barton. Global mixed-integer dynamic optimization. *AIChE Journal*, 51(8):2235–2253, 2005.

B. Chachuat, A.B. Singer, and P.I. Barton. Global methods for dynamic optimization and mixed-integer dynamic optimization. *Industrial & Engineering Chemistry Research*, 45(25):8373–8392, 2006.

Y.J. Chang and N.V. Sahinidis. Optimization of metabolic pathways under stability considerations. *Computers & Chemical Engineering*, 29(3):467–479, 2005.

W. Chaovalitwongse, P.M. Pardalos, and O.A. Prokopyev. A new linearization technique for multi-quadratic 0-1 programming problems. *Operations Research Letters*, 32(6):517–522, 2004.

A. Chinchuluun, P.M. Pardalos, and R. Enkhbat. Global minimization algorithms for concave quadratic programming problems. *Optimization*, 54(6):627–639, 2005.

M. D'Apuzzo, M. Marino, A. Migdalas, P.M. Pardalos, and G. Toraldo. Parallel computing in global optimization. In E.J. Kontoghiorghes, editor, *Handbook of Parallel Computing and Statistics*, pages 259–282, Chapman & Hall/CRC, 2006.

E. Davis and M. Ierapetritou. A kriging method for the solution of nonlinear programs with black-box functions. *AIChE Journal*, 53(8):2001–2012, 2007.

E. Davis and M. Ierapetritou. A kriging based method for the solution of mixed-integer nonlinear programs containing black-box functions. *Journal of Global Optimization*, in press, 2008.

C.H.M De Saboia, M. Campelo, and S. Scheimberg. A computational study of global algorithms for linear bilevel programming. *Numerical Algorithms*, 35(2-4):155–173, 2004.

X.S. Ding and F. Al-Khayyal. Accelerating convergence of cutting plane algorithms for disjoint bilinear programming. *Journal of Global Optimization*, 38(3):421–436, 2007.

J.P.K. Doye, R.H. Leary, M. Locatelli, and F. Schoen. Global optimization of Morse clusters by potential energy transformations. *INFORMS Journal on Computing*, 16(4):371–379, 2004.

J.J. Downs and E.F. Vogel. A plant-wide industrial-process control problem. *Computers & Chemical Engineering*, 17(3):245–255, 1993.

V. Dua, K.P. Papalexandri, and E.N. Pistikopoulos. Global optimization issues in multiparametric continuous and mixed-integer optimization problems. *Journal of Global Optimization*, 30(1):59–89, 2004.

M. Dur, R. Horst, and M. Locatelli. Necessary and sufficient global optimality conditions for convex minimization revisited. *Journal of Mathematical Analysis & Applications*, 217:637–639, 1998.

J.A. Egea, M. Rodriguez-Fernandez, J.R. Banga, and R. Marti. Scatter search for chemical and bio-process optimization. *Journal of Global Optimization*, 37(3):481–503, 2007a.

J.A. Egea, D. Vries, A.A. Alonso, and J.R. Banga. Global optimization for integrated design and control of computationally expensive process models. *Industrial & Engineering Chemistry*

Research, 46(26):9148–9157, 2007b.

J.A. Egea, E. Vazquez, J.R. Banga, and R. Marti. Improved scatter search for the global optimization of computationally expensive dynamic models. *Journal of Global Optimization*, in press, 2008.

S. Emet and T. Westerlund. Comparisons of solving a chromatographic separation problem using MINLP methods. *Computers & Chemical Engineering*, 28(5):673–682, 2004.

O. Exler, L.T. Antelo, J.A. Egea, A.A. Alonso, and J.R. Banga. A tabu search-based algorithm for mixed-integer nonlinear problems and its application to integrated process and control system design. *Computers & Chemical Engineering*, 32(8):1877–1891, 2008.

N.P. Faisca, V. Dua, B. Rustem, P.M. Saraiva, and E.N. Pistikopoulos. Parametric global optimization for bilevel programming. *Journal of Global Optimization*, 38(4):609–623, 2007.

S.C. Fang, D.Y. Gao, R.L Sheu, and S.Y. Wu. Canonical dual approach to solving 0-1 quadratic programming problems. *Journal of Industry and Management Optimization*, 4(1):125–142, 2008.

T.A. Feo and M.G.C. Resende. A probabilistic heuristic for a computationally difficult set covering problem. *Operations Research Letters*, 8(2):67–71, 1989.

T.A. Feo and M.G.C. Resende. Greedy randomized adaptive search procedures. *Journal of Global Optimization*, 6(2):109–133, 1995.

C.A. Floudas. *Nonlinear and Mixed-Integer Optimization: Fundamentals and Applications*. Oxford University Press, 1995.

C.A. Floudas. *Deterministic Global Optimization : Theory, Methods and Applications*. Nonconvex Optimization and its Applications. Kluwer Academic Publishers, 2000.

C.A. Floudas. Research challenges, opportunities and synergism in systems engineering and computational biology. *AIChE Journal*, 51(7):1872–1884, 2005.

C.A. Floudas. Computational methods in protein structure prediction. *Biotechnology and Bioengineering*, 97(2):207–213, 2007.

C.A. Floudas, A. Aggarwal, and A.R. Ciric. Global optimum search for non convex NLP and MINLP problems. *Computers & Chemical Engineering*, 13(10):1117–1132, 1989.

C.A. Floudas, I.G. Akrotirianakis, S. Caratzoulas, C.A. Meyer, and J. Kallrath. Global optimization in the 21st century: Advances and Challenges. *Computers & Chemical Engineering*, 29(6):1185–1202, 2005.

C.A. Floudas, H.K. Fung, S.R. McAllister, M. Monningmann, and R. Rajgaria. Advances in protein structure prediction and de novo protein design: A review. *Chemical Engineering Science*, 61(3):966–988, 2006.

C.A. Floudas and H.T. Jongen. Global optimization: Local minima and transition points. *Journal of Global Optimization*, 32:409–415, 2005.

C.A. Floudas and V. Kreinovich. Towards optimal techniques for solving global optimization problems: Symmetry-based approach. In A. Törn and J. Zilinskas, editors, *Models and Algorithms for Global Optimization*, pages 21–42, Springer, 2007a.

C.A. Floudas and V. Kreinovich. On the functional form of convex underestimators for twice continuously differentiable functions. *Optimization Letters*, 1:187–192, 2007b.

C.A. Floudas and P.M. Pardalos. *Frontiers in Global Optimization*. Nonconvex Optimization and its Applications. Kluwer Academic Publishers, 2003.

C.A. Floudas and P.M. Pardalos, Eds. Encyclopedia of Optimization, 2nd Edition. Kluwer Academic Publishers, 2008.

C.A. Floudas, P.M. Pardalos, C.S. Adjiman, W.R. Esposito, Z.H. Gümüş, S.T. Harding, J.L. Klepeis, C.Meyer, and C.A. Schweiger. *Handbook of Test Problems in Local and Global Optimization*. Kluwer Academic Publishers, 1999.

C.A. Floudas and O. Stein. The adaptive convexification algorithm: A feasible point method for semi-infinite programming. *SIAM Journal on Optimization*, 18(4):1187–1208, 2007.

C.A. Floudas and V. Visweswaran. A global optimization algorithm (GOP) for certain classes of nonconvex NLPs .1. Theory. *Computers & Chemical Engineering*, 14(12):1397–1417, 1990.

P.T. Foteinou, E. Yang, G.K. Saharidis, M.G. Ierapetritou, and I.P. Androulakis. A mixed-integer optimization framework for the synthesis and analysis of regulatory networks. *Journal of Global Optimization*, in press, 2008.

L. Freitas, G.M. Platt, and N. Henderson. Novel approach for the calculation of critical points in binary mixtures using global optimization. *Fluid Phase Equilibria*, 225(1-2):29–37, 2004.

E.R. Frits, M.C. Markot, Z. Lelkes, Z. Fonyo, T. Csendes, and E. Rev. Use of an interval global optimization tool for exploring feasibility of batch extractive distillation. *Journal of Global Optimization*, 38(2):297–313, 2007.

D.Y. Gao. Perfect duality theory and complete solutions to a class of global optimization problems. *Optimization*, 52(4-5):467–493, 2003.

D.Y. Gao. Canonical duality theory and solutions to constrained nonconvex quadratic programming - Dedicated to Professor Ivar Ekeland on the occasion of his 60th birthday. *Journal of Global Optimization*, 29(4):377–399, 2004.

D.Y. Gao. Sufficient conditions and perfect duality in nonconvex minimization with inequality constraints. *Journal of Industry and Management Optimization*, 1(1):53–63, 2005.

D.Y. Gao. Solutions and optimality criteria to box constraint nonconvex minimization problems. *Journal of Industry and Management Optimization*, 3(2):293–304, 2007.

R.R. Gattupalli and A. Lucia. Molecular conformation of n-alkanes using terrain/funneling methods. *Journal of Global Optimization*, in press, 2008.

A. Geletu and A. Hoffmann. A conceptual method for solving generalized semi-infinite programming problems via global optimization by exact discontinuous penalization. *European Journal of Operational Research*, 157(1):3–15, 2004.

S. Ghosh, T. Zhu, I.E. Grossmann, M.M. Ataai, M.M. Domach. Closing the loop between feasible flux scenario identification for construct evaluation and resolution of realized fluxes via NMR. *Computers & Chemical Engineering*, 29(3):459–466, 2005.

C.E. Gounaris and C.A. Floudas. Tight convex underestimators for C^2-continuous problems: I. Univariate functions. *Journal of Global Optimization*, 42(1):51–67, 2008a.

C.E. Gounaris and C.A. Floudas. Tight convex underestimators for C^2-continuous problems: II. Multivariate functions. *Journal of Global Optimization*, 42(1):69–89, 2008b.

C.E. Gounaris and C.A. Floudas. Convexity of products of univariate functions and convexification transformations for geometric programming. *Journal of Optimization Theory and Applications*, 138(3):407–427, 2008c.

V.Goyal and M.G. Ierapetritou. Framework for evaluating the feasibility/operability of nonconvex processes. *AIChE Journal*, 49(5):1233–1240, 2003a.

V.Goyal and M.G. Ierapetritou. MINLP optimization using simplicial approximation method for classes of non-convex problems. In C.A. Floudas and P.M. Pardalos, editors, *Frontiers in Global Optimization*, pages 165–196, Kluwer Academic Publishers, 2003.

I.E. Grossmann and S. Lee. Generalized convex disjunctive programming: Nonlinear convex hull relaxation. *Computational Optimization and Applications*, 26(1):83–100, 2003.

Z.H. Gümüş and C.A. Floudas. Global optimization of mixed-integer bilevel programming problems. *Computational Management Science*, 2:181–212, 2005.

H.M. Gutmann. A radial basis function method for global optimization. *Journal of Global Optimization*, 19(3):201–227, 2001.

E. Hansen and G.W. Walster. *Global Optimization Using Interval Analysis.* Pure and Applied Mathematics. Marcel Dekker, 2004.

N. Henderson, U. Freitas, and G.M. Platt. Prediction of critical points: A new methodology using global optimization. *AIChE Journal*, 50(6):1300–1314, 2004.

D. Hertz, C.S. Adjiman, and C.A. Floudas. Two results on bounding the roots of interval polynomials. *Computers & Chemical Engineering*, 23:1333, 1999.

J.B. Hiriart-Urruty and J.S. Ledyav. A note in the characterization of the global maxima of a convex function over a convex set. *Journal of Convex Analysis*, 3:55–61, 1996.

M.J. Hirsch, C.N. Meneses, P.M. Pardalos, and M.G.C. Resende. Global optimization by continuous grasp. *Optimization Letters*, 1(2):201–212, 2007.

R. Horst and H. Tuy. *Global Optimization: Deterministic Approaches.* Springer, 2003.

J.Q. Hu, M.C. Fu, and S.I. Marcus. A model reference adaptive search method for global optimization. *Operations Research*, 55(3):549–568, 2007.

D. Huang, T.T. Allen, W.I. Notz, and N. Zeng. Global optimization of stochastic black-box systems via sequential kriging meta-models. *Journal of Global Optimization*, 34(3):441–466, 2006.

D. Iourinski, S.A. Starks, V. Kreinovich and S.F. Smith. Swarm intelligence: Theoretical proof that empirical techniques are optimal. In: *Proceedings of the 2002 World Automation Congress WAC 2002, Orlando, FL*, 107–112, 2002.

H.W. Jiao, Y.R. Guo, and P.P. Shen. Global optimization of generalized linear fractional programming with nonlinear constraints. *Applied Mathematics and Computation*, 183(2):717–728, 2006.

J. Kallrath. Exact computation of global minima of a noncovex portfolio optimization problem. In C.A. Floudas and P.M. Pardalos, editors, *Frontiers in Global Optimization*, pages 237–254, Kluwer Academic Publishers, 2003.

J. Kallrath. Solving planning and design problems in the process industry using mixed integer and global optimization. *Annals of Operations Research*, 140(1):339–373, 2005.

J. Kallrath. Cutting circles and polygons from area-minimizing rectangles. *Journal of Global Optimization*, in press, 2008.

R. Karuppiah, K.C. Furman, and I.E. Grossmann. Global optimization for scheduling refinery crude oil operations. *Computers & Chemical Engineering*, 32(11):2745–2766, 2008.

R. Karuppiah and I.E. Grossmann. Global optimization for the synthesis of integrated water systems in chemical processes. *Computers & Chemical Engineering*, 30(4):650–673, 2006.

R. Karuppiah and I.E. Grossmann. Global optimization of multiscenario mixed integer nonlinear programming models arising in the synthesis of integrated water networks under uncertainty. *Computers & Chemical Engineering*, 32:145–160, 2008a.

R. Karuppiah and I.E. Grossmann. A Lagrangean based branch-and-cut algorithm for global optimization of nonconvex mixed-integer nonlinear programs with decomposable structures. *Journal of Global Optimization*, 41(2):163–186, 2008b.

J. Kennedy, R. Eberhart and Y. Shi. *Swarm Intelligence.* Morgan Kaufmann, 2001.

P. Kesavan, R.L. Allgor, E.P. Gadzke, and P.I. Barton. Outer approximation algorithms for separable nonconvex mixed-integer nonlinear problems. *Mathematical Programming*, 100(3):517–535, 2004.

J.L. Klepeis and C.A. Floudas. Ab initio tertiary structure prediction of proteins. *Journal of Global Optimization*, 25:113, 2003a.

J.L. Klepeis and C.A. Floudas. ASTRO-FOLD: A combinatorial and global optimization framework for ab initio prediction of three-dimensional structures of proteins from the amino-acid sequence. *Biophysical Journal*, 85:2119, 2003b.

J.L. Klepeis, M. Pieja, and C.A. Floudas. A new class of hybrid global optimization algorithms for peptide structure prediction: Integrated hybrids. *Computer Physics Communications*, 151:121, 2003a.

J.L. Klepeis, M.Pieja, and C.A. Floudas. A new class of hybrid global optimization algorithms for peptide structure prediction: Alternating hybrids and application to Met-Enkephalin and Melittin. *Biophysical Journal*, 84:869, 2003b.

V. Kreinovich, S.A. Starks and G. Meyer. On a theoretical justification of the choice of epsilon-inflation in PASCAL-XSC. *Reliable Computing*, 3:437–452, 1997.

C.K. Lee and P.I. Barton. Global dynamic optimization of linear hybrid systems. In C.A. Floudas and P.M. Pardalos, editors, *Frontiers in Global Optimization*, pages 289–312, Kluwer Academic Publishers, 2003.

A. Lee and I.E. Grossmann. A global optimization algorithm for nonconvex generalized disjunctive programming and applications to process systems. *Computers & Chemical Engineering*, 25:1675–1697, 2001.

S. Lee and I.E. Grossmann. Global optimization of nonlinear generalized disjunctive programming with bilinear equality constraints: Applications to process networks. *Computers & Chemical Engineering*, 27(11):1557–1575, 2003.

S. Lee and I.E. Grossmann. Logic-based modeling and solution of ninlinear discrete/continuous optimization problems. *Annals of Operations Research*, 139(1):267–288, 2005.

C.K. Lee, A.B. Singer, and P.I. Barton. Global optimization of linear hybrid systems with explicit transitions. *Systems & Control Letters*, 51(5):363–375, 2004.

H.L. Li and J.F. Tsai. Treating free variables in generalized geometric global optimization programs. *Journal of Global Optimization*, 33(1):1–13, 2005.

H.L. Li, J.F. Tsai, and C.A. Floudas. Convex underestimation for posynomial functions of positive variables. *Optimization Letters*, 2(3):333–340, 2008.

T. Li, Y.J. Wang, Z. Liang, and P.M. Pardalos. Local saddle point and a class of convexification methods for nonconvex optimization problems. *Journal of Global Optimization*, 38(3):405–419, 2007.

D. Li, Z.Y. Wu, H.W.J. Lee, X.M. Wang, and L.S. Zhang. Hidden convex minimization. *Journal of Global Optimization*, 31(2):211–233, 2005.

L. Liberti. Linearity embedded in nonconvex programs. *Journal of Global Optimization*, 33(2):157–196, 2005.

L. Liberti and N. Maculan. *Global Optimization: From Theory to Implementation*. Nonconvex Optimization and its Applications. Kluwer Academic Publishers, 2006.

L. Liberti and C.C. Pantelides. Convex envelopes of monomials of odd degree. *Journal of Global Optimization*, 25:157–168, 2003.

X.X. Lin, C.A. Floudas, and J. Kallrath. Global solution approach for a nonconvex MINLP problem in product portfolio optimization. *Journal of Global Optimization*, 32(3):417–431, 2005.

Y.D. Lin and M.A. Stadtherr. LP Strategy for the interval-Newton method in deterministic global optimization. *Industrial & Engineering Chemistry Research*, 43(14):3741–3749, 2004a.

Y.D. Lin and M.A. Stadtherr. Advances in interval methods for deterministic global optimization in chemical engineering. *Journal of Global Optimization*, 29(3):281–296, 2004b.

Y.D. Lin and M.A. Stadtherr. Deterministic global optimization for parameter estimation of dynamic systems. *Industrial & Engineering Chemistry Research*, 45(25):8438–8448, 2006.

Y.D. Lin and M.A. Stadtherr. Deterministic global optimization of nonlinear dynamic systems. *AIChE Journal*, 53(4):866–875, 2007.

J. Linderoth. A simplicial branch-and-bound algorithm for solving quadratically constrained quadratic programs. *Mathematical Programming*, 103(2):251–282, 2005.

G.X. Liu. A homotopy interior point method for semi-infinite programming problems. *Journal of Global Optimization*, 37(4):631–646, 2007.

W.B. Liu and C.A. Floudas. A remark on the GOP algorithm for global optimization. *Journal of Global Optimization*, 3(4):519–521, 1993.

Y. Liu and K.L. Teo. An adaptive dual parametrization algorithm for quadratic semi-infinite programming problems. *Journal of Global Optimization*, 24(2):205–217, 2002.

Y. Liu, K.L. Teo, and S.Y. Wu. A new quadratic semi-infinite programming algorithm based on dual parametrization. *Journal of Global Optimization*, 29(4):401–413, 2004.

C.E. Long, P.K. Polisetty, and E.P. Gatzke. Nonlinear model predictive control using deterministic global optimization. *Journal of Process Control*, 16(6):635–643, 2006.

C.E. Long, P.K. Polisetty, and E.P. Gatzke. Deterministic global optimization for nonlinear model predictive control of hybrid dynamic systems. *International Journal of Robust and Nonlinear Control*, 17(13):1232–1250, 2007.

A. Lucia, P.A. DiMaggio, M.L. Bellows, and L.M. Octavio. The phase behavior of n-alkane systems. *Computers & Chemical Engineering*, 29(11-12):2363–2379, 2005.

A. Lucia and Y. Feng. Global terrain methods. *Computers & Chemical Engineering*, 26:529–546, 2002.

A. Lucia and Y. Feng. Multivariable terrain methods. *AIChE Journal*, 49:2553, 2003.

A. Lundell, J. Westerlund, and T. Westerlund. Some transformation techniques with applications in global optimization. *Journal of Global Optimization*, in press, 2008.

Y.Q. Luo, X.G. Yuan, and Y.J. Liu. An improved PSO algorithm for solving non-convex NLP/MINLP problems with equality constraints. *Computers & Chemical Engineering*, 31(3):153–162, 2007.

C.D. Maranas and C.A. Floudas. Global minimum potential energy conformations of small molecules. *Journal of Global Optimization*, 4:135–170, 1994.

C.D. Maranas and C.A. Floudas. Finding all solutions of nonlinearly constrained systems of equations. *Journal of Global Optimization*, 7:143–182, 1995.

C.D. Maranas and C.A. Floudas. Global optimization in generalized geometric programming. *Computers & Chemical Engineering*, 21:351–370, 1997.

M.G. Marcovecchio, M.L. Bergamini, and P. Aguirre. On saddle points of augmented Lagrangians for constrained nonconvex optimization. *Journal of Global Optimization*, 34(3):339–368, 2006.

D. Maringer and P. Parpas. Global optimization of higher order moments in portfolio selection. *Journal of Global Optimization*, in press, 2008.

J.A. Martinez, L.G. Casado, I. Garcia, Y.D. Sergeyev, and B. Toth. On an efficient use of gradient information for accelerating interval global optimization algorithms. *Numerical Algorithms*, 37(1-4):61–69, 2004.

C.A. Meyer and C.A. Floudas. Trilinear monomials with positive or negative domains: Facets of convex and concave envelopes. In C.A. Floudas and P.M. Pardalos, editors, *Frontiers in Global Optimization*, pages 327–352, Kluwer Academic Publishers, 2003.

C.A. Meyer and C.A. Floudas. Convex hull of trilinear monomials with mixed-sign domains. *Journal of Global Optimization*, 29:125–155, 2004.

C.A. Meyer and C.A. Floudas. Convex envelopes for edge-concave functions. *Mathematical Programming*, 103(2):207–224, 2005a.

C.A. Meyer and C.A. Floudas. Convex underestimation of twice continuously differentiable functions by piecewise quadratic perturbation: Spline αBB underestimators. *Journal of Global Optimization*, 32:221–258, 2005b.

C.A. Meyer and C.A. Floudas. Global optimization of a combinatorially complex generalized pooling problem. *AIChE Journal*, 52:1027–1037, 2006.

A. Mitsos and P.I. Barton. A dual extremum principle in thermodynamics. *AIChE Journal*, 53(8):2131–2147, 2007.

A. Mitsos, P. Lemonidis, and P.I. Barton. Global solution of bilevel programs with a nonconvex inner program. *Journal of Global Optimization*, in press, 2008b.

A. Mitsos, P. Lemonidis, C.K. Lee, and P.I. Barton. Global solution of bilevel programs with a nonconvex inner program. *SIAM Journal of Optimization*, 19(1):77–113, 2008a.

C.G. Moles, G. Gutierrez, A.A. Alonso, and J.R. Banga. Integrated process design and control via global optimization. *Industrial & Engineering Chemistry Research*, 81:507–517, 2003.

N.P. Moloi and M.M. Ali. An iterative global optimization algorithm for potential energy minimization. *Computational Optimization and Applications*, 30(2):119–132, 2005.

J.M. Montagna, O.A. Iribarren, and A.R. Vecchietti. Synthesis of biotechnological processes using generalized disjunctive programming. *Industrial & Engineering Chemistry Research*, 43(15):4220–4232, 2004.

S.A. Munawar and R.D. Gudi. A nonlinear transformation based hybrid evolutionary method for MINLP solution. *Chemical Engineering Research & Design*, 83(A10):1218–1236, 2005.

A. Nahapetyan and P.M. Pardalos. A bilinear relaxation based algorithm for concave piecewise linear network flow problems. *Journal of Industry and Management Optimization*, 3(1):71–85, 2007.

A. Nahapetyan and P.M. Pardalos. A bilinear reduction based algorithm for solving capacitated multi-item dynamic pricing problems. *Computers & Operations Research*, 35(5):1601–1612, 2008.

A. Neumaier. Complete search in continuous global optimization and constraint satisfaction. In A. Iserles, editor, *Acta Numerica*, pages 271–369. Cambridge University Press, 2004.

A. Neumaier, O. Shcherbina, W. Huyer, and T. Vinko. A comparison of complete global optimization solvers. *Mathematical Programming*, 103(2):335–356, 2005.

H.T. Nguyen and V. Kreinovich. *Applications of Continuous Mathematics to Computer Science*. Kluwer Academic Publishers, 1997.

D.V. Nichita, S. Gomez, and E. Luna-Ortiz. Multiphase equilibria calculation by direct minimization of Gibbs free energy using tunneling global optimization method. *Journal of Canadian Petroleum Technology*, 43(5):13–16, 2004.

D.V. Nichita, C.D.D. Valencia, and S. Gomez. Volume-based thermodynamics global phase stability analysis. *Chemical Engineering Communications*, 193(10):1194–1216, 2006.

J. Nie, J. Demmel, and M. Gu. Global minimization of rational functions and the nearest GCDs. *Journal of Global Optimization*, 40(4):697–718, 2008.

G.M. Ostrovsky, L.E.K. Achenie, and M. Sinha. A reduced dimension branch-and-bound algorithm for molecular design. *Computers & Chemical Engineering*, 27:551–567, 2003.

I. Papamichail and C.S. Adjiman. A rigorous global optimization algorithm for problems with ordinary differential equations. *Journal of Global Optimization*, 24:1–33, 2002.

I. Papamichail and C.S. Adjiman. Global optimization of dynamic systems. *Computers & Chemical Engineering*, 28(3):403–415, 2004.

I. Papamichail and C.S. Adjiman. Proof of convergence for a global optimization algorithm for problems with ordinary differential equations. *Journal of Global Optimization*, 33(1):83–107, 2005.

P.M. Pardalos, W. Chaovalitwongse, L.D. Iasemidis, J.C. Sackellares, D.S. Shiau, P.R. Carney, O.A. Prokopyev, and V.A. Yatsenko. Seizure warning algorithm based on optimization and nonlinear dynamics. *Mathematical Programming*, 101(2):365–385, 2004.

P.M. Pardalos, O.A. Prokopyev, O.V. Shylo, and V.P. Shylo. Global equilibrium search applied to the unconstrained binary quadratic optimization problem. *Optimization Methods & Software*, 23(1):129–140, 2008.

P.M. Pardalos and O.V. Shylo. An algorithm for the job shop scheduling problem based on global equilibrium search techniques. *Computational Management Science*, 3(4):331–348, 2006.

P. Parpas, B. Rustem, and E.N. Pistikopoulos. Linearly constrained global optimization and stochastic differential equations. *Journal of Global Optimization*, 36(2):191–217, 2006.

P. Parpas, B. Rustem, and E.N. Pistikopoulos. Global optimization of robust chance constrained problems. *Journal of Global Optimization*, in press, 2008.

E.N. Pistikopoulos, V. Dua, and J. Ryu. Global optimization of bilevel programming problems via parametric programming. In C.A. Floudas and P.M. Pardalos, editors, *Frontiers in Global Optimization*, pages 457–476, Kluwer Academic Publishers, 2003.

K.V. Price, R.M. Storn, and J.A. Lampinen. *Differential Evolution: A Practical Approach to Global Optimization*. Natural Computing Series. Springer, 2005.

O.A. Prokopyev, H.X. Huang, and P.M. Pardalos. On complexity of unconstrained hyperbolic 0-1 programming problems. *Operations Research Letters*, 33(3):312–318, 2005a.

O.A. Prokopyev, C.N. Meneses, C.A.S. Oliveira, and P.M. Pardalos. On multiple-ratio hyperbolic 0-1 programming problems. *Pacific Journal of Optimization*, 1(2):327–345, 2005b.

S. Rebennack, J. Kallrath, and P.M. Pardalos. Column enumeration based decomposition techniques for a class of non-convex MINLP problems. *Journal of Global Optimization*, in press, 2008.

R.G. Regis and C.A. Shoemaker. COnstrained global optimization of expensive black box functions using radial basis functions. *Journal of Global Optimization*, 31(1):153–171, 2005.

R.G. Regis and C.A. Shoemaker. Improved strategies for radial basis function methods for global optimization. *Journal of Global Optimization*, 37(1):113–135, 2007.

S.M. Rump. A note on epsilon-inflation. *Reliable Computing*, 4:371–375, 1998.

H.S. Ryoo and N.V. Sahinidis. Analysis of bounds for multilinear functions. *Journal of Global Optimization*, 19:403–424, 2001.

H.S. Ryoo and N.V. Sahinidis. Global optimization of multiplicative programs. *Journal of Global Optimization*, 26:387–418, 2003.

J.H. Ryu, V. Dua, and E.N. Pistikopoulos. A bilevel programming framework for enterprise-wide process networks under uncertainty. *Computers & Chemical Engineering*, 28(6-7):1121–1129, 2004.

N.V. Sahinidis, M. Tawarmalani, and M. Yu. Design of alternative refrigerants via global optimization. *AIChE Journal*, 49(7):1761, 2003.

N.W. Sawaya and I.E. Grossmann. A cutting plane method for solving linear generalized disjunctive programming problems. *Computers & Chemical Engineering*, 29(9):1891–1913, 2005.

N.W. Sawaya and I.E. Grossmann. Computational implementation of non-linear convex hull reformulation. *Computers & Chemical Engineering*, 31(7):856–866, 2007.

H.D. Schafroth and C.A. Floudas. Predicting peptide binding to MHC pockets via molecular modeling, implicit solvation, and global optimization. *Proteins: Structure, Function, and Bioinformatics*, 54:534–556, 2004.

H. Schichl. Global optimization in the COCONUT project. In: *Numerical Software with Results Verification*, Lecture Notes in Computer Science 2991, Springer, 243–249, 2004.

A.M. Scurto, G. Xu, J.F. Brennecke, and M.A. Stadtherr. Phase behavior and reliable computation of high-pressure solid-fluid equilibrium with cosolvents. *Industrial & Engineering Chemistry Research*, 42(25):6464–6475, 2003.

O. Shcherbina, A. Neumaier, D. Sam-Haroud, X.H. Vu, and T.V Nguyen. Benchmarking global optimization and constraint satisfaction codes. In: *Global Optimization and Constraint Satisfaction*, Lecture Notes in Computer Science 2861, Springer, 211–222, 2003.

P.P. Shen. Linearization method of global optimization for generalized geometric programming. *Applied Mathematics and Computation*, 162:353–370, 2005.

P.P. Shen and G.X. Yuan. Global optimization for the sum of generalized polynomial fractional functions. *Mathematical Methods of Operations Research*, 65(3):445–459, 2007.

H.D. Sherali and J. Desai. A global optimization RLT-based approach for solving the hard clustering problem. *Journal of Global Optimization*, 32(2):281–306, 2005a.

H.D. Sherali and J. Desai. A global optimization RLT-based approach for solving the fuzzy clustering problem. *Journal of Global Optimization*, 33(4):597–615, 2005b.

H.D. Sherali and V. Ganesan. A pseudo-global optimization approach with application to the design of conteinerships. *Journal of Global Optimization*, 26(4):335–360, 2003.

H.D. Sherali, Y. Lee, and Y. Kim. Partial convexification cuts for 0-1 mixed-integer programs. *European Journal of Operational Research*, 165(3):625–648, 2005.

A.B. Singer and P.I. Barton. Global solution of optimization problems with dynamic systems embedded. In C.A. Floudas and P.M. Pardalos, editors, *Frontiers in Global Optimization*, pages 477–498, Kluwer Academic Publishers, 2003.

A.B. Singer and P.I. Barton. Global solution of optimization problems with parameter-embedded linear dynamic systems. *Journal of Optimization Theory and Applications*, 121(3):613–646, 2004.

A.B. Singer and P.I. Barton. Global optimization with nonlinear ordinary differential equations. *Journal of Global Optimization*, 34(2):159–190, 2006.

A.B. Singer, J.W. Taylor, P.I. Barton, and W.H. Green. Global dynamic optimization for parameter estimation in chemical kinetics. *Journal of Physical Chemistry A*, 110(3):971–976, 2006.

M. Sinha, L.E.K. Achenie, and R. Gani. Blanket wash solvent blent design using interval analysis. *Industrial & Engineering Chemistry Research*, 42:516–527, 2003.

M.V. Solodov. A bundle method for a class of bilevel nonsmooth convex minimization problems. *SIAM Journal on Optimization*, 18(1):242–259, 2007.

M. Srinivas and G.P. Rangaiah. Implementation and evaluation of random tunneling algorithm for chemical engineering applications. *Computers & Chemical Engineering*, 30(9):1400–1415, 2006.

M. Srinivas and G.P. Rangaiah. Differential evolution with tabu list for global optimization and its application to phase equilibrium and parameter estimation problems. *Industrial & Engineering Chemistry Research*, 46(10):3410–3421, 2007.

O. Stein, J. Oldenburg, and W. Marquardt. Continuous reformulations of discrete-continuous optimization problems. *Computers & Chemical Engineering*, 28(10):1951–1966, 2004.

R. Storn and K. Price. Differential evolution - A simple and efficient heuristic for global optimization over continuous spaces. *Journal of Global Optimization*, 11(4):341–359, 1997.

X.L. Sun, D. Li, and K.I.M. McKinnon. On saddle points of augmented Lagrangians for constrained nonconvex optimization. *SIAM Journal on Optimization*, 15(4):1128–1146, 2005.

M.P. Tan, J.R. Broach, and C.A. Floudas. A novel clustering approach and prediction of optimal number of clusters: Global optimum search with enhanced positioning. *Journal of Global Optimization*, 39:323–346, 2007a.

M.P. Tan, J.R. Broach, and C.A. Floudas. Evaluation of normalization and pre-clustering issues in a novel clustering approach: Global optimum search with enhanced positioning. *Journal of Bioinformatics and Computational Biology*, 5(4):875–893, 2007b.

F. Tardella. On the existence of polyhedral convex envelopes. In C.A. Floudas and P.M. Pardalos, editors, *Frontiers in Global Optimization*, pages 563–573, Kluwer Academic Publishers, 2003.

F. Tardella. Existence and sum decomposition of vertex polyhedral envelopes. *Optimization Letters*, 2(3):363–375, 2008.

M. Tawarmalani and N.V. Sahinidis. Global optimization of mixed-integer nonlinear programs: A theoretical and computational study. *Mathematical Programming*, 99(3):563–591, 2004.

M. Tawarmalani and N.V. Sahinidis. A polyhedral branch-and-cut approach to global optimization. *Mathematical Programming*, 103(2):225–249, 2005.

A. Törn and J. Zilinskas. *Models and Algorithms for Global Optimization.* Optimization and its Applications. Springer, 2007.

J.F. Tsai. Global optimization of nonlinear fractional programming problems in engineering design. *Engineering Optimization*, 37(4):399–409, 2005.

J.F. Tsai and M.H. Lin. An optimization approach for solving signomial discrete programming problems with free variables. *Computers & Chemical Engineering*, 30(8):1256–1263, 2006.

J.F. Tsai and M.H. Lin. Finding all solutions of systems of nonlinear equations with free variables. *Engineering Optimization*, 39(6):649–659, 2007.

J.F. Tsai, M.H. Lin, and Y.C. Hu. On generalized geometric programming problems with non-positive variables. *European Journal of Operational Research*, 178(1):10–19, 2007.

H. Tuy, A. Migdalas, and N.T. Hoai-Phuong. A novel approach to bilevel nonlinear programming. *Journal of Global Optimization*, 38(4):527–554, 2007.

H. Tuy, P.T. Trach, and H. Konno. Optimization of polynomial fractional functions. *Journal of Global Optimization*, 29(1):19–44, 2004.

S. Ulas, U.M. Diwekar, and M.A. Stadtherr. Uncertainties in parameter estimation and optimal control in batch distillation. *Computers & Chemical Engineering*, 29(8):1805–1814, 2005.

A. Vaia and N.V. Sahinidis. Simultaneous parameter estimation and model structure determination in FTIR spectroscopy by global MINLP optimization. *Computers & Chemical Engineering*, 27:763–779, 2003.

A. Vecchietti, S. Lee, and I.E. Grossmann. Modeling of discrete/continuous optimization problems: Characterization and formulation of disjunctions and their relaxations. *Computers & Chemical Engineering*, 27(3):433–448, 2003.

Y.J. Wang and Z. Liang. A deterministic global optimization algorithm for generalized geometric programming. *Applied Mathematics and Computation*, 168:722–737, 2005.

Y.J. Wang, P.P. Shen, and Z. Liang. A branch-and-bound algorithm to globally solve the sum of several linear ratios. *Applied Mathematics and Computation*, 168(1):89–101, 2005.

T. Westerlund. Some transformation techniques in global optimization. In L. Liberti and N. Maculan, editors, *Global Optimization: From Theory to Implementation*, pages 45–74, Springer, 2006.

Z.Y. Wu, F.S. Bai, and L.S. Zhang. Convexification and concavification for a general class of global optimization problems. *Journal of Global Optimization*, 31(1):45–60, 2005a.

Z.Y. Wu, H.W.J. Lee, and X.M. Yang. A class of convexification and concavification methods for non-monotone optimization problems. *Optimization*, 54(6):605–625, 2005b.

Z.Y. Wu, D. Li, L.S. Zhang, and X.M. Wang. Peeling off a nonconvex cover of an actual convex problem: Hidden convexity. *SIAM Journal on Optimization*, 18(2):507–536, 2007.

R. Yamamoto and H. Konno. An efficient algorithm for solving convex-convex quadratic fractional programs. *Journal of Optimization Theory and Applications*, 133(2):241–255, 2007.

L.X. Yan, K. Shen, and S.H. Hu. Solving mixed integer nonlinear programming problems with line-up competition algorithm. *Computers & Chemical Engineering*, 28(12):2647–2657, 2004.

C.T. Young, Y. Zheng, C.W. Yeh, and S.S. Jang. Information-guided genetic algorithm approach to the solution of MINLP problems. *Industrial & Engineering Chemistry Research*, 46(5):1527–1537, 2007.

Z.B. Zabinsky. *Stochastic Adaptive Search for Global Optimization.* Nonconvex Optimization and its Applications. Kluwer Academic Publishers, 2003.

S. Zakovic and B. Rustem. Semi-infinite programming and applications to minimax problems. *Annals of Operations Research*, 124(1-4):81–110, 2003.

A. Zhigljavsky and A. Zilinskas. *Stochastic Global Optimization.* Optimization and its Applications. Springer, 2007.

W.X. Zhu and Q.X. Fu. A sequential convexification method (SCM) for continuous global optimization. *Journal of Global Optimization*, 26:167–182, 2003.

Y. Zhu and T. Kuno. Global optimization of nonconvex MINLP by a hybrid branch-and-bound and revised generalized benders decomposition approach. *Industrial & Engineering Chemistry Research*, 42:528–539, 2003.

Y. Zhu and T. Kuno. A disjunctive cutting-plane-based branch-and-cut algorithm for 0-1 mixed-integer convex nonlinear programs. *Industrial & Engineering Chemistry Research*, 45(1):187–196, 2006.

J. Zilinskas and I.D.L. Bogle. Evaluation ranges of functions using balanced random interval arithmetic. *Informatica Lithuan*, 14(3):403–416, 2003.

S. Zlobec. On the Liu-FLoudas convexification of smooth programs. *Journal of Global Optimization*, 32(3):401–407, 2005.

S. Zlobec. Characterization of convexifiable functions. *Optimization*, 55(3):251–261, 2006.

Fields Institute Communications
Volume **55**, 2009

Optimization in Biomedical Research

O. Erhun Kundakcioglu
Department of Industrial and Systems Engineering
University of Florida
Gainesville, FL 32611 USA
erhun@ufl.edu

Panos M. Pardalos
Department of Industrial and Systems Engineering
University of Florida
Gainesville, FL 32611 USA
pardalos@ufl.edu

Abstract. This survey concerns a rapidly emerging interdisciplinary research area. In recent years, numerous biomedical applications have benefited from optimization techniques. We present current tools in biomedical research that utilize optimization theory and refer to studies that identify directions for future research.

1 Introduction

Medical researchers and practitioners need to obtain information from data sets. This is a difficult task to perform without appropriate tools. The computational requirements for massive data sets have also created many challenging problems for the mathematical community as well as computer scientists. Optimization comes into the picture with most state-of-the-art machine learning tools. We discuss these tools with applications in biomedicine and employed optimization techniques. Furthermore, a survey on optimization applications in treatment planning and other biomedical problems is presented.

The remainder of the paper is organized as follows. Section 2 explains the role of optimization in the most commonly used machine learning techniques in biomedical research. We investigate unsupervised learning, classification, regression techniques, and kernel applications. In Section 3, we explore treatment planning applications and other optimization frameworks in biomedical research. Finally, Section 4 concludes this paper.

2000 *Mathematics Subject Classification.* Primary 92Bxx; Secondary 90Bxx.
Key words and phrases. Mathematical programming, optimization, data mining, machine learning, biomedicine, neuroscience, computational biology.

2 Machine learning applications

We first explore the unsupervised learning techniques and proceed with supervised learning techniques (i.e., classification and regression). Kernel methods conclude this section.

Unsupervised learning is the case where objects are not labeled with any class information. Density estimation is performed or clusters of objects are constructed based on similarities between their features.

On the other hand, *supervised learning* refers to the capability of a system to learn from a set of input/output pairs. The input is usually a *vector of features* for an object, and the output is the *label* for the class this object belongs to. The set of objects with a feature vector and a class label is called a *training set*. Based on this information, a function is derived and applied on a *test set*. The output of the *regression* function is a continuous value which is useful to forecast a label that can take any value. In *classification*, on the other hand, the output is a discrete class label that is used for categorical discrimination. The term *supervised* originates from the fact that the labels for the objects in the training set are provided as input, and therefore are determined by an outside source, which can be considered as the *supervisor*.

2.1 Unsupervised learning. Unsupervised learning is a machine learning technique where labels of inputs are not available.

Clustering is the partitioning of a data set into subsets (clusters), so that the data in each subset (ideally) share some common trait - often proximity according to some defined distance measure. Data clustering is a common technique for statistical data analysis, which is used in many fields, including data mining, pattern recognition, image analysis and bioinformatics. The computational task of classifying data set into k clusters is often referred to as k-clustering.

One of the most widely used clustering techniques is k-means clustering. Underlying optimization problem for k-means clustering for a data set of $\{\mathbf{x}_1, \ldots, \mathbf{x}_N\}$ can be formulated as

$$\min \quad J = \sum_{n=1}^{N} \sum_{k=1}^{K} r_{nk} \|\mathbf{x}_n - \mu_k\|^2 \tag{2.1a}$$

$$\text{subject to} \quad r_{nk} \in \{0, 1\} \qquad n = 1, \ldots, N; k = 1, \ldots, K \tag{2.1b}$$

where μ_k is a prototype associated with the kth cluster representing the centers of clusters. r_{nk} is the binary indicator variable that is 1 if and only if data point $\mathbf{x}_n$ is assigned to cluster k. This is also known as the 1-of-K coding scheme. The objective function is sometimes called a *distortion measure*.

Formulation (2.1) is usually solved with *k-means algorithm* which essentially is an expectation-maximization (EM) applied to mixtures of Gaussians. The most common form of the algorithm uses an iterative refinement heuristic known as Lloyd's algorithm [64]. EM algorithm consists of successive optimizations with respect to r_{nk} and μ_k. First some initial values of μ_k are chosen and J is minimized with respect to r_{nk} in the E (expectation) step. In the second phase, which is the M (maximization) step, J is minimized with respect to μ_k keeping r_{nk} fixed. Since each phase reduces the value of the objective function J, the convergence of the algorithm is assured.

K-means algorithm has been effectively applied in molecular biology for gene expression data analysis. Functionally related genes are identified by partitioning them into groups based on the similarity between their expression profiles. There are also several extensions of k-means that have been successfully applied to other biological problems [68]. See [61] for details on clustering applications in genomics. Next, we explore biclustering, another clustering method that utilize optimization techniques.

Biclustering is a data mining technique allowing clustering of data samples and data attributes (features) at the same time. It is sometimes referred to as *co-clustering*, *two-mode* or *two way clustering*. The technique is performed in such a way that each class of features created within the biclustering is related to a class of data samples by a certain property distinguishing it from other classes.

A data set (e.g., from microarray experiments) is normally given as a rectangular $m \times n$ matrix A, where each column represents a data sample (e.g., patient) and each row represents a feature (e.g., gene)

$$A = (a_{ij})_{m \times n}$$

where a_{ij} is the expression of ith feature in j^{th} sample.

Biclustering is applied by simultaneous partitioning of the samples and features (i.e., columns and rows of matrix A, respectively) into k classes. Let $S_1, S_2, \ldots, S_k$ denote the classes of the samples (columns) and $F_1, F_2, \ldots, F_k$ denote the classes of features (rows). Biclustering can be formally defined as a collection of pairs of sample and feature subsets $\mathcal{B} = \{(S_1, F_1), (S_2, F_2), \ldots, (S_k, F_k)\}$ such that

$$\begin{aligned}
S_1, S_2, \ldots, S_k &\subseteq \{a_{\cdot j}\}_{j=1,\ldots,n}, \\
\bigcup_{r=1}^{k} S_r &= \{a^j\}_{j=1,\ldots,n}, \\
S_v \bigcap S_u &= \emptyset \Leftrightarrow v \neq u, \\
F_1, F_2, \ldots, F_k &\subseteq \{a_{i\cdot}\}_{i=1,\ldots,m}, \\
\bigcup_{r=1}^{k} F_r &= \{a_i\}_{i=1,\ldots,m}, \\
F_v \bigcap F_u &= \emptyset \Leftrightarrow v \neq u,
\end{aligned}$$

where $\{a_{\cdot j}\}_{j=1,\ldots,n}$ and $\{a_{i\cdot}\}_{i=1,\ldots,m}$ denote the set of columns and rows of the matrix A, respectively.

A pair (S_k, F_k) is called a *bicluster*. The ultimate goal in a biclustering problem is to find a partitioning for which samples from the same class have *similar* values for that class' characteristic features. The visualization of a reasonable biclustering should reveal a block-diagonal or "checkerboard" pattern as in Fig. 1. A detailed survey on biclustering techniques can be found in [20] and [69].

The criteria used to relate clusters of samples and features may have different properties. Most commonly, it is required that the submatrix corresponding to a bicluster is either overexpressed (i.e., mostly includes values above average), or has a lower variance than the whole data set. However, biclustering in general may rely on any kind of common patterns among elements of a bicluster. One exact solution approach for the consistent biclustering framework with feature selection is based

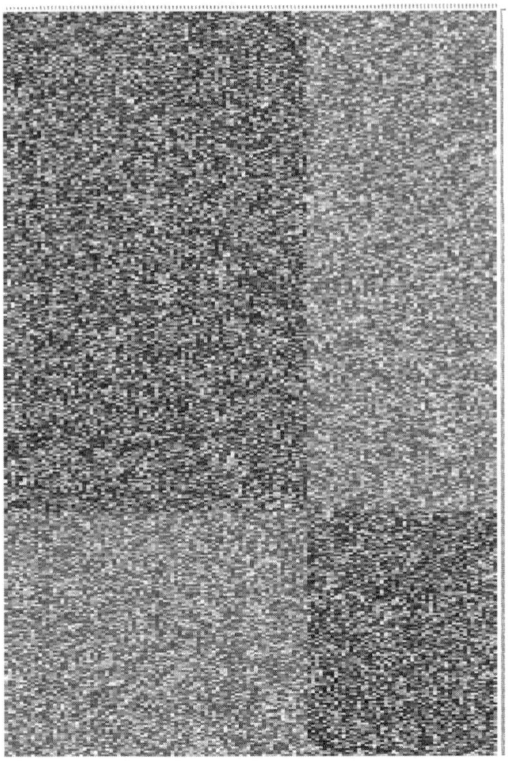

Figure 1 An example of checkerboard pattern after biclustering

on 0–1 fractional programming [18]. We explore consistent biclustering in Section 2.4.

In [29], Divina and Aguilar-Ruiz address the biclustering of gene expression data with evolutionary computation. Their approach is based on evolutionary algorithms and searches for biclusters following a sequential covering strategy. To avoid overlapping among biclusters, a weight is assigned to each element of the expression matrix. Weights are adjusted each time a bicluster is found. This is different from other methods that substitute covered elements with random values. Experimental results confirm the quality of the proposed method to avoid overlapping among biclusters.

One of the main goals in the analysis of large and heterogeneous gene expression data sets is to identify groups of genes that are co-expressed in subsets of experimental conditions. Identification of these local structures plays a key role in understanding the biological events associated with different physiological states as well as to identify gene expression signatures. Classical one-way clustering techniques, especially hierarchical clustering, have been commonly applied to cluster genes and samples separately in order to identify these types of local patterns. In the last few years, many authors have proposed the use of biclustering algorithms to identify gene-experiment relationships.

bioNMF estimates biclusters using a novel method based on a modified variant of the Non-negative Matrix Factorization algorithm [84]. This algorithm produces a suitable decomposition as product of three matrices that are constrained to have non-negative elements.

Biclustering is used to analyze one or several of six expression matrices collected from yeast [104, 93] and to analyze one or more of eleven different expression matrices with human gene expression levels [104, 47, 17]. Almost all these data sets contain expression data related to the study of cancer. Some contain data from cancerous tissues at different stages of the disease; others from individuals suffering from different types of cancer; and the remaining data sets contain data collected from individuals with a particular cancer or healthy people. These data sets are used to test the applicability of biclustering approaches in three major tasks: Identification of coregulated genes, gene functional annotation, and sample classification.

Biclustering techniques is also applied to the problem of identification of coregulated genes [93, 8]. More specifically, the objective is to identify sets of genes that, under specific conditions, exhibit coherent activations that indicate coregulation. These results are used to simply identify sets of coregulated genes or, more ambitiously, to identify specific regulation processes. A less obvious application is to use the biclustering results directly to perform automatic gene functional annotation [104, 93]. The idea underlying this approach is to use biclusters where a large majority of genes belong to a specific class in the gene ontology to guess the class of nonannotated genes.

Another significant area of application is related with sample and/or tissue classification [17, 104, 48]. An example is the diagnosis of leukemia where the goal is to identify different responses to treatment and group of genes to be used as the most effective probe [99].

The applications of biclustering mentioned above analyze data from gene expression matrices. However, biclustering can be used in the analysis of other biological data. Liu and Wang [49] apply biclustering to a drug activity data set. The goal in this study is to find groups of chemical compounds with similar behaviors when subsets of compound descriptors are taken into account. Lazzeroni and Owen [55] analyze nutritional data to identify subsets of foods with similar properties on a subset of food attributes.

In [36], it is shown how several problems in different areas of data mining and knowledge discovery can be viewed as finding the optimal covering of a finite set. Many such problems arise in biomedical and bioinformatics research. For example, protein functional annotation based on sequence information is an ubiquitous bioinformatics problem. It consists of finding a set of homolog (high similarity) sequences of known function to a given amino acid sequence of unknown function from the various annotated sequence databases. These can then be used as clues in suggesting further experimental analysis of the new protein. In [36] it is shown that these optimization problems can be stated as maximization of submodular functions on the set of candidate subsets. This generalization may be especially useful when conclusions from data mining need to be interpreted by human experts as in diagnostic hypothesis generation, logical methods of data analysis, conceptual clustering, and proteins functional annotations.

In [37], a novel electroencephalogram (EEG)-based, brain-state identification method is presented which could form the basis for forecasting a generalized epileptic seizure. 25 rats are exposed to hyperbaric oxygen until the appearance of a generalized EEG seizure. EEG segments from the preexposure, early exposure, and the period up to and including the seizure are processed by the fast wavelet transform. Features extracted from the wavelet coefficients are input to the unsupervised optimal fuzzy clustering (UOFC) algorithm. The UOFC is useful for classifying similar discontinuous temporal patterns in the semistationary EEG to a set of clusters which may represent brain-states. The unsupervised selection of the number of clusters overcomes the a priori unknown and variable number of states. The usually vague brain state transitions are naturally treated by assigning each temporal pattern to one or more fuzzy clusters.

Next, we start the supervised learning techniques with classification framework.

2.2 Linear classification. Given a training data set comprising the features and the corresponding labels, the goal in classification is to take an input vector $\mathbf{x}$ and assign to one of K discrete classes $C_1, C_2, \ldots, C_K$. The input space is divided into decision regions whose boundaries are called the *decision boundaries*. This regions are formed using the *discriminant functions*.

In a typical *binary classification* problem, each pattern vector $\mathbf{x}_i \in \mathbb{R}^n$, $i = 1, \ldots, l$ belongs to one of two classes C^+ and C^-. A vector is given the label $y_i = 1$ if $\mathbf{x_i} \in C^+$ or $y_i = -1$ if $\mathbf{x}_i \in C^-$. Using the information coming from this *training set*, the classification problem consists of determining which class new pattern vectors from the *test set* belong to.

In this section, we discuss infamous Support Vector Machine classifiers. Developed by Vladimir Vapnik [111], Support Vector Machines (SVMs) are the state-of-the-art supervised machine learning methods. SVM classifiers classify pattern vectors which are assumed to belong to two linearly separable sets from two different classes. Although there are infinitely many hyperplanes that separate the two classes, SVM classifier finds the hyperplane (ψ, b) that maximizes the distance from the convex hulls of both classes by solving a quadratic convex optimization problem. The success and robustness of SVM classifiers rely on strong fundamentals from the statistical learning theory, from which generalization bounds are derived.

A hyperplane can be represented by $\langle\psi\cdot\mathbf{x}\rangle + b = 0$, where ψ is the normal vector and b is the offset parameter. There is an inherent degree of freedom in specifying a hyperplane as $(\lambda\psi, \lambda b)$. A *canonical hyperplane* is the hyperplane with a functional margin of 1 with the closest pattern vector, i.e., $\min_{i=1,\ldots,l} |\langle\psi\cdot\mathbf{x}_i\rangle + b| = 1$.

The geometric margin is maximized as follows:

$$\max_{\psi} \frac{1}{2}\Big(\min_{\mathbf{x}_i\in C^+} \langle\mathbf{x}_i \cdot \frac{\psi}{\|\psi\|}\rangle - \max_{\mathbf{x}_i\in C^-} \langle\mathbf{x}_i \cdot \frac{\psi}{\|\psi\|}\rangle\Big) = \max_{\psi} \frac{2}{2\|\psi\|} = \min_{\psi} \|\psi\|. \tag{2.2}$$

Maximizing the geometric interclass margin while satisfying the canonical separating hyperplane condition for the pattern vectors boils down to the following optimization problem:

$$\min \quad \frac{1}{2}\|\psi\|^2 \tag{2.3a}$$

$$\text{subject to} \quad y_i(\langle\psi\cdot\mathbf{x}_i\rangle + b) \geq 1 \qquad i = 1, \ldots, l. \tag{2.3b}$$

Note that minimizing $\|\psi\|^2/2$ will give the same optimal solution with minimizing $\|\psi\|$, hence formulations (2.2) and (2.3) are equivalent. From the solution to (2.3), a new pattern vector $\mathbf{x}^*$ can be classified as positive if $\langle\psi\cdot\mathbf{x}^*\rangle + b > 0$, and negative otherwise.

This optimization problem for the maximal margin classifier is only feasible if the two classes of pattern vectors are linearly separable. However, most of the real life classification problems contain nonseparable data. In this case, the goal is to construct the hyperplane that also minimizes the misclassification errors. In order to keep track of the errors, *slack variables* ξ_i are introduced for each pattern vector $\mathbf{x}_i$ in the training set. The slack variables allow misclassification for pattern vectors with a penalty of C and the resulting formulation is called *soft margin formulation.*

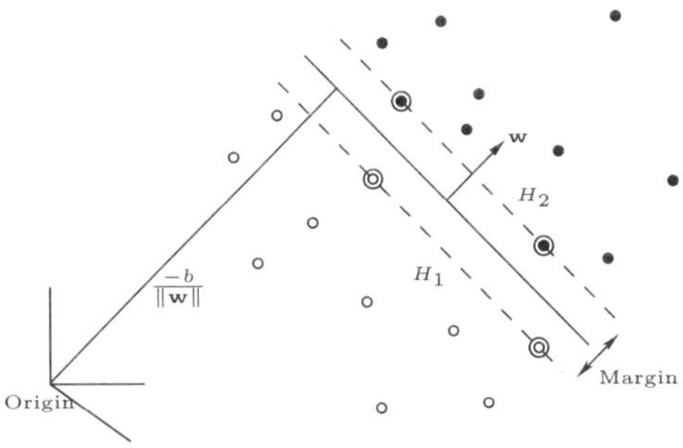

Figure 2 Maximal margin classifier

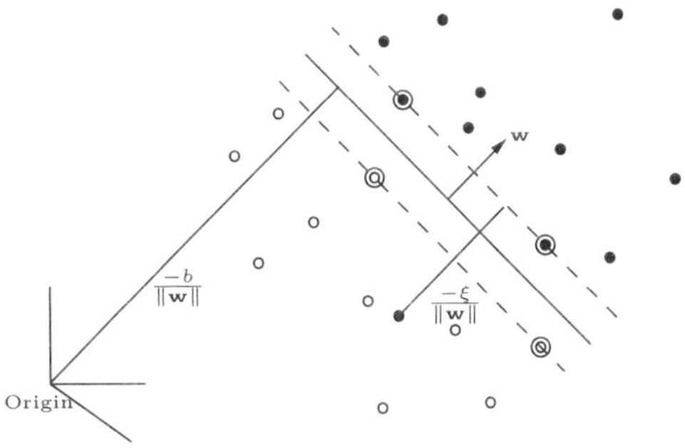

Figure 3 Soft margin classifier

In Fig. 3, *soft margin classifier* is demonstrated that incurs penalty for misclassified pattern vectors. Hard margin formulation can be generalized to soft margin formulation as follows.

$$\min \quad \frac{1}{2}\|\psi\|^2 + C\sum_{i=1}^{l}\xi_i^2 \tag{2.4a}$$

$$\text{subject to} \quad y_i(\langle\psi\cdot\mathbf{x}_i\rangle + b) \geq 1 - \xi_i \qquad i = 1, \ldots, l. \tag{2.4b}$$

In (2.4), it is unnecessary to enforce nonnegativity of the slack variables explicitly since a solution cannot be optimal when $\xi_i < 0$ for any pattern vector. It should be noted that the 2-norm of the slack variables are penalized in the objective of (2.4). An alternative formulation involves penalization of the 1-norm of slack variables. In this case, nonnegativity constraints on the slack variables are necessary as follows:

$$\begin{aligned} \min \quad & \frac{1}{2}\|\psi\|^2 + C\sum_{i=1}^{l}\xi_i && && (2.5a)\\ \text{subject to} \quad & y_i(\langle\psi\cdot\mathbf{x}_i\rangle + b) \geq 1-\xi_i && i=1,\ldots,l && (2.5b)\\ & \xi_i \geq 0 && i=1,\ldots,l. && (2.5c) \end{aligned}$$

It should also be noted that (2.4) and (2.5) is essentially the minimization of a convex function with linear inequality constraints. This problem can be solved efficiently by numerous methods [10]. See [26] for further details on formulation and implementation details of SVMs.

Sequential Minimal Optimization (SMO) solves SVM problems by iteratively selecting subsets only of size 2 and optimizing the target function with respect to them [86]. It has good convergence properties and it is easily implemented. The key point is that for a working set of 2, the optimization subproblem can be solved analytically without explicitly invoking a quadratic optimizer.

See [95] for a detailed survey on applications of SVM in neuroscience and [61] for classification applications in genomics. Next, we discuss some generalizations of the linear classification problem.

The multiple instance (MI) learning is introduced in [28] in the context of drug activity prediction and developed in [5, 66]. The problem consists of classifying positive and negative *bags* of points in the n-dimensional real space $\mathbb{R}^n$ where each bag contains a number of points. Patterns $\mathbf{x}_1,\ldots\mathbf{x}_l$ are grouped into bags $X_1,\ldots X_m$ with $X_j = \{\mathbf{x}_i : i \in I_j\}$, $I_j \subseteq \{1,\ldots,n\}$, and $\bigcup_j I_j = \{1,\ldots,n\}$. Each bag X_j is associated with a label $y_j \in \{1,-1\}$. Classification is performed such that at least one point for each positive bag is classified as positive, and all the points for all negative bags are classified as negative. In [28], a hypothesis class of axis-parallel rectangles are assumed, and algorithms are developed to deal with the drug activity prediction problem. An efficient algorithm is described in [66] for learning axis-aligned rectangles with respect to product distributions from MI examples in the PAC model. Auer gives a more efficient algorithm in [5]. Blum and Kalai [14] show that learning from multiple-instance examples is reducible to PAC-learning with two sided noise and to the statistical query model. Integer programming, expectation maximization, and kernel formulations are also proposed for MI classification problem [112, 118, 35, 4, 73]. Ray and Craven provide a benchmark of several multiple instance classification algorithms and their non-multiple-instance counterparts in [89]. Kundakcioglu and Pardalos [51] formulate MI classification problem as the following mixed 0–1 quadratic programming problem:

$$\begin{aligned} \min_{\psi,b,\xi,\eta} \quad & \frac{1}{2}\|\psi\|^2 + C\sum_{i=1}^{l}\xi_i^2 && && (2.6a)\\ \text{subject to} \quad & \langle\psi\cdot\mathbf{x}_i\rangle + b \geq 1-\xi_i - M(1-\eta_i) && \forall i : i\in I_j \wedge y_j = 1 && (2.6b)\\ & -\langle\psi\cdot\mathbf{x}_i\rangle - b \geq 1-\xi_i && \forall i : i\in I_j \wedge y_j = -1 && (2.6c)\\ & \sum_{i\in I_j}\eta_i \geq 1 && \forall j : y_j = 1 && (2.6d)\\ & \eta_i \in \{0,1\} && \forall i : i\in I_j \wedge y_j = 1 && (2.6e) \end{aligned}$$

In this formulation M is a sufficiently large number that ensures that the constraint is active if and only if $\eta_i = 1$. η_i is a binary variable that is 1 if i^{th} instance is one of the actual positive examples. In [51], a branch and bound algorithm is proposed for this $\mathcal{NP}$-hard problem that outperforms a commercial solver for large scale problems. MI classification has very successful implementations in application areas such as drug design [43, 28] and protein family modeling [106].

Seref *et al.* [96] introduce a generalized support vector classification framework, called the *Selective Support Vector Machine*: Let S_i, $i = 1, \dots, l$ be mutually exclusive sets of pattern vectors such that all pattern vectors $\mathbf{x}_{i,k}$, $k = 1, \dots, |S_i|$ have the same class label y_i. Select only one pattern vector $\mathbf{x}_{i,k^0}$ from each set S_i such that the margin between the set of selected positive and negative pattern vectors are maximized. This problem is formulated as a quadratic mixed integer programming problem, which is a generalization of the standard support vector classifiers and multiple instance classifiers.

$$\min \quad \frac{1}{2}\|\psi\|^2 + C\sum_{i=1}^{l}\sum_{k=1}^{|S_i|}\xi_{i,k}^2 \tag{2.7a}$$

$$\text{subject to} \quad y_i(\langle\psi\cdot\mathbf{x}_{i,k}\rangle + b) \geq 1 - \xi_{i,k} - M(1 - v_{i,k}) \quad i = 1\dots,l; k = 1,\dots,|S_i| \tag{2.7b}$$

$$\sum_{k=1}^{|S_i|} v_{i,k} = 1 \qquad i = 1,\dots,l \tag{2.7c}$$

$$v_{i,k} \in \{0,1\} \qquad i = 1,\dots,l; k = 1,\dots,|S_i|. \tag{2.7d}$$

This quadratic mixed 0–1 programming problem is shown to be $\mathcal{NP}$-hard [97]. An alternative approach is proposed with the free slack concept as follows:

$$\min \quad \frac{1}{2}\|\psi\|^2 + \frac{C}{2}\sum_{i=1}^{n}\xi_{i,k}^2 \tag{2.8a}$$

$$\text{subject to} \quad y_i(\langle\psi\cdot\mathbf{x}_{i,k}\rangle + b) \geq 1 - \xi_{i,k} - v_{i,k} \quad i = 1,\dots,l; k = 1,\dots,|S_i| \tag{2.8b}$$

$$\sum_{i=1}^{l}\sum_{k=1}^{|S_i|} v_{i,k} \leq V \tag{2.8c}$$

$$v_{i,k} \geq 0 \qquad i = 1,\dots,l; k = 1,\dots,|S_i|. \tag{2.8d}$$

Dual formulation for (2.8) is derived for nonlinear classification. The necessity of dual formulation for nonlinear classification is explained in Section 2.5. Formulations with free slack provide flexibility to the separating hyperplane to identify the pattern vectors with larger interclass margin. Iterative elimination and direct selection methods are developed to select such pattern vectors using the alternative formulations. These methods are compared with a naïve method on simulated data.

The iterative elimination method for selective SVM is also applied to neural data from a visuomotor categorical discrimination task to classify highly cognitive brain activities [94]. Standard and novel kernel based nonlinear classification methods are applied on a neural data recorded during a visuomotor task performed by a macaque monkey. The stages of the visuomotor task are the initial response of the visual cortex, the categorical discrimination of the visual stimuli and the appropriate response for the visual stimuli. A standard SVM classifier and an SVM based adaptive scaling method are used for feature selection in order to detect relevant time intervals and their spatial mapping on the brain. The first and the third stages of the visuomotor task are detectable with the standard SVM classifier. However, for the second stage, SVM classifier performs poorly. Dynamic Time Warping (DTW) is also applied in order to reduce the temporal variances. Motivated by the improvement in the first stage after DTW, selective SVM is applied. It is shown that the results obtained after selective SVM are exceptionally better compared to DTW for both classification accuracy and feature selection. The results clearly point to the involvement of the superior temporal gyrus and the inferior parietal lobe in the categorical discrimination stage of the visuomotor task.

There are a number of biomedical studies which introduce different methods for image *segmentation*. Segmentation is the process of partitioning a digital image into different sections in order to change the representation of the image. This new representation may involve certain characteristics in the image such as curves, edges, color, intensity or texture. Segmented images are usually used to determine brain abnormalities using image classification, and it is shown that SVMs perform very well. Lee et al. [56] use SVMs and a new SVM based method called *support vector random fields* for segmenting brain tumors from MR images. Rinaldi et al. [90] classify brain inflammation in *multiple sclerosis* patients based on the peripheral immune abnormalities from MR images using nonlinear SVMs. They determine that brain inflammation in patients with multiple sclerosis is associated with changes in subsets of peripheral lymphocytes. Thus, SVM classification helps detect a potential biomarker candidate for the prognosis of patients in the early stages of multiple sclerosis. Quddus et al. [87] combine SVMs and *boosting*, another machine learning method, to perform segmentation (via nonlinear classification) on white matter lesions in the MRI scans. Their composite classification method is shown to be faster and just as reliable as manual selection. In another study by Martinez-Ramon et al. [75] a similar approach is used to create segments of the brain with respect to their functions. Later, these segments are aggregated using boosting, which is used for multi-class SVM classification of an fMRI group study with interleaved motor, visual, auditory, and cognitive task design.

Automated tumor detection algorithms are a sought-after tool for assisting physicians to make more accurate and rapid detection of tumors. Support vector machine classifiers have contributed significantly in this area. Menze et. al. utilize SVM classification of MRI images to serve as an automated diagnostic tool for the detection of recurrent brain tumors [79]. They report that SVM among other methods is able to rule out lipid and lactate signals as being too unreliable, and that choline and N-acetylaspartate are the main sources of information (most important features).

Diabetes mellitus (DM) is a common disease in the industrialized countries and it is a prominent risk factor for ischemic cerebrovascular accidents. Diabetes alone is responsible for 7% of deaths in stroke patients. Diabetes mellitus often results in brain micro-blood flow disorders that may cause cerebral infarction. However, assessing the function of cerebral micro-vessels is difficult, since they are located within the bony skull. Kalatzis et. al. [45] apply SVMs to distinguish between blood flow data in patients with diabetes versus control subjects using *SPECT* images from cerebral abnormalities. Li et al. [63] use SVMs with *floating search method* to find relevant features for assessing the degree of malignancy in brain glioma from MRI findings and clinical data prior to operations. Li et al. [62] develop a novel algorithm that combines bagging of SVMs with embedded feature selection for the individual observations and compare the new algorithm using publicly available data sets. Next, we continue discussing Support Vector method within the regression framework.

2.3 Linear regression. Given a training data set comprising the features and the corresponding target values, the goal in regression is to take an input vector $\mathbf{x}$ and predict the value of one or more continuous *target* variables. In a typical *regression* problem, each pattern vector in the training set $\mathbf{x}_i \in \mathbb{R}^n$, $i = 1, \ldots, l$ is associated with a label (target value) $y_i \in \mathbb{R}$. A regression function is obtained using the training set and that function predicts the target values for the test set.

The Support Vector method can also be applied to the case of regression, maintaining all the main features that characterize the maximal margin algorithm. This method is called the Support Vector Regression (SVR). As in the classification case the learning algorithm minimizes a convex functional.

As with the classification, we motivate the approach by seeking to optimize the generalization bounds given for regression. These rely on defining a loss function that ignores errors within a certain distance of the true value. This type of function is referred to as an ϵ-insensitive loss function. With many reasonable choices of loss function, the solution is characterized as the minimum of a convex functional. Another motivation for considering the ϵ-insensitive loss function is that it will ensure sparseness of the dual variables similar to the classification case. The idea of representing the solution by means of a small subset of training points has enormous computational advantages. ϵ-insensitive loss function has that sparseness advantage, while still ensuring existence of a global minimum and the optimization of a reliable generalization bound.

In this section we first describe the ϵ-insensitive loss and then derive two approaches from the bounds relating to the 1-norm or 2-norm of the loss vector.

The linear ϵ-insensitive loss function $L_1^\epsilon(\mathbf{x}, y, f)$ is defined as

$$L_1^\epsilon(\mathbf{x}, y, f) = |y - f(\mathbf{x})|_\epsilon = \max(0, |y - f(\mathbf{x})| - \epsilon)$$

and the quadratic ϵ-insensitive loss function $L_2^\epsilon(\mathbf{x}, y, f)$ is defined as

$$L_2^\epsilon(\mathbf{x}, y, f) = |y - f(\mathbf{x})|_\epsilon^2 = \max(0, |y - f(\mathbf{x})|^2 - \epsilon).$$

Similar to the classification case, a parameter C is introduced to measure the trade-off between complexity and losses. The SVR problem for the quadratic ϵ-insensitive loss function is

$$\min \quad \frac{1}{2}\|\psi\|^2 + C\sum_{i=1}^{l}(\xi_i^2 + \hat{\xi}_i^2) \tag{2.9a}$$

$$\text{subject to} \quad ((\langle\psi \cdot \mathbf{x}_i\rangle + b) - y_i \leq \epsilon + \xi_i \qquad i = 1, \ldots, l \tag{2.9b}$$

$$y_i - ((\langle\psi \cdot \mathbf{x}_i\rangle + b) \leq \epsilon + \hat{\xi}_i \qquad i = 1, \ldots, l. \tag{2.9c}$$

The SVR problem for the linear ϵ-insensitive loss function is

$$\min \quad \frac{1}{2}\|\psi\|^2 + C\sum_{i=1}^{l}(\xi_i + \hat{\xi}_i) \tag{2.10a}$$

$$\text{subject to} \quad ((\langle\psi \cdot \mathbf{x}_i\rangle + b) - y_i \leq \epsilon + \xi_i \qquad i = 1, \ldots, l \tag{2.10b}$$

$$y_i - ((\langle\psi \cdot \mathbf{x}_i\rangle + b) \leq \epsilon + \hat{\xi}_i \qquad i = 1, \ldots, l \tag{2.10c}$$

$$\xi_i, \hat{\xi}_i \geq 0 \qquad i = 1, \ldots, l. \tag{2.10d}$$

Linear regression is used in identification of a directly proportional relationship between two physicochemical properties and drug activity prediction [44]. Breast cancer prognosis is studied extensively using linear programming and a regression framework in [101] and [72].

2.4 Other supervised learning methods. In this section we explore supervised learning methods that do not fall under the category of linear classification. We particularly present "consistent biclustering" and its biomedical applications.

Given a classification of the samples, S_r, let $S = (s_{jr})_{n\times k}$ denote a 0–1 matrix where $s_{jr} = 1$ if sample j is classified as a member of the class r (i.e., $a^j \in S_r$), and $s_{jr} = 0$ otherwise. Similarly, given a classification of the features, F_r, let $F = (f_{ir})_{m\times k}$ denote a 0–1 matrix where $f_{ir} = 1$ if feature i belongs to class r (i.e., $a_i \in F_r$), and $f_{ir} = 0$ otherwise. Construct corresponding *centroids* for the samples and features using these matrices as follows

$$C_S = AS(S^TS)^{-1} = (c^S_{i\xi})_{m\times r}, \tag{2.11}$$

$$C_F = A^TF(F^TF)^{-1} = (c^F_{j\xi})_{n\times r}. \tag{2.12}$$

The elements of the matrices, $c^S_{i\xi}$ and $c^F_{j\xi}$, represent average expression of the corresponding sample and feature in class ξ, respectively. In particular,

$$c^S_{i\xi} = \frac{\sum_{j=1}^n a_{ij}s_{j\xi}}{\sum_{j=1}^n s_{j\xi}} = \frac{\sum_{j|a^j\in S_\xi} a_{ij}}{|S_\xi|},$$

and

$$c^F_{j\xi} = \frac{\sum_{i=1}^m a_{ij}f_{i\xi}}{\sum_{i=1}^m f_{i\xi}} = \frac{\sum_{i|a_i\in F_\xi} a_{ij}}{|F_\xi|}.$$

Using the elements of matrix C_s, one can assign a feature to a class where it is over-expressed. Therefore feature i is assigned to class $\hat{r}$ if $c^S_{i\hat{r}} = \max_\xi\{c^S_{i\xi}\}$, i.e.,

$$a_i \in \hat{F}_{\hat{r}} \Longrightarrow c^S_{i\hat{r}} > c^S_{i\xi}, \qquad \forall\xi, \xi \neq \hat{r}. \tag{2.13}$$

Note that the constructed classification of the features, $\hat{F}_r$, is not necessarily the same as classification F_r. Similarly, one can use the elements of matrix C_F to classify the samples. Sample j is assigned to class $\hat{r}$ if $c^F_{j\hat{r}} = \max_\xi\{c^F_{j\xi}\}$, i.e.,

$$a^j \in \hat{S}_{\hat{r}} \Longrightarrow c^F_{j\hat{r}} > c^F_{j\xi}, \qquad \forall\xi, \xi \neq \hat{r}. \tag{2.14}$$

As before, the obtained classification $\hat{S}_r$ does not necessarily coincide with classification S_r.

Biclustering $\mathcal{B}$ is referred to as a *consistent biclustering* if relations (2.13) and (2.14) hold for all elements of the corresponding classes, where matrices C_S and C_F are defined according to (2.11) and (2.12), respectively.

A data set is *biclustering-admitting* if some consistent biclustering for that data exists. Furthermore, the data set is called *conditionally biclustering-admitting* with respect to a given (partial) classification of some samples and/or features if there exists a consistent biclustering preserving the given (partial) classification.

Let $\mathcal{B}$ be a consistent biclustering. Then there exist convex cones $P_1, P_2, \ldots, P_k \subseteq \mathbb{R}^m$ such that only samples from S_r belong to the corresponding cone P_r, $r = 1, \ldots, k$. Similarly, there exist convex cones $Q_1, Q_2, \ldots, Q_k \subseteq \mathbb{R}^n$ such that only features from class F_r belong to the corresponding cone Q_r, $r = 1, \ldots, k$.

It also follows from the conic separability that convex hulls of classes do not intersect [18].

By definition, a biclustering is consistent if $F_r = \hat{F}_r$ and $S_r = \hat{S}_r$. However, a given data set might not have these properties. The features and/or samples in the data set might not clearly belong to any of the classes and hence a consistent biclustering might not be constructed. In such cases, one can remove a set of features and/or samples from the data set so that there is a consistent biclustering for the truncated data. Selection of a representative set of features that satisfies certain properties is a widely used technique in data mining applications. This feature selection process may incorporate various objective functions depending on the desirable properties of the selected features, but one general choice is to select the maximal possible number of features in order to lose minimal amount of information provided by the training set.

Given a set of training data, construct matrix S and compute the values of $c^S_{i\xi}$ using (2.11). Classify the features according to the following rule: feature i belongs to class $\hat{r}$ (i.e., $a_i \in F_{\hat{r}}$), if $c^S_{i\hat{r}} > c^S_{i\xi}$, $\forall \xi \neq \hat{r}$. Finally, construct matrix F using the obtained classification. Let x_i denote a binary variable, which is one if feature i is included in the computations and zero otherwise. Consistent biclustering problem is formulated as follows.

$$\max_x \quad \sum_{i=1}^{m} x_i \tag{2.15a}$$

$$\text{subject to} \quad \frac{\sum_{i=1}^m a_{ij} f_{i\hat{r}} x_i}{\sum_{i=1}^m f_{i\hat{r}} x_i} > \frac{\sum_{i=1}^m a_{ij} f_{i\xi} x_i}{\sum_{i=1}^m f_{i\xi} x_i}, \quad \forall \hat{r}, \xi = 1, \ldots, k, \hat{r} \neq \xi, j \in S_{\hat{r}} \tag{2.15b}$$

$$x_i \in \{0, 1\}, \qquad i = 1, \ldots, m. \tag{2.15c}$$

The optimization problem (2.15) is proved to be $\mathcal{NP}$-hard [52] and is a specific type of *fractional 0–1 programming problem*, which can be solved using the approach described in [18]. When it has a feasible solution, the corresponding feature selection makes the data set conditionally biclustering-admitting with respect to the given classification of samples.

Two generalizations of (2.15) and an improved heuristic procedure is proposed in [81]. In this method, a linear programming problem with continuous variables is solved at each iteration. Numerical experiments on the data, which consists of samples from patients diagnosed with *acute lymphoblastic leukemia (ALL)* or *acute myeloid leukemia (AML)* diseases (see [38, 7, 9, 114, 116]), confirm that the algorithm outperforms the previous results in the quality of solution as well as computation time.

Busygin *et al.* [16] apply biclustering to analyze the electroencephalogram (EEG) data. Some biomedical applications of biclustering are DNA microarray analysis and drug design [20, 69, 105]. However, biclustering is shown to be also useful for feature selection which is the major concern of many biomedical studies [16]. Revealing subsets of channels whose Lyapunov exponents consistently change with switching the VNS stimulation ON and OFF is claimed to be very much in line with discovering upregulated and downregulated genes in a microarray data set. Therefore, each EEG channel is represented as a feature and data samples taken within the stimulation periods versus samples taken outside of these periods are analyzed. It is shown that the method of biclustering is able to perform successful feature selection. Another study where epilepsy treatment with vagus nerve stimulation is by Uthman *et al.* [110]. See [23, 91] for other applications of optimization to epileptic brain disorders.

Recently, Bertsimas and Shioda introduce mixed-integer optimization methods to the classical statistical problems of classification and regression and construct a software package called CRIO (classification and regression via integer optimization) [11]. CRIO separates data points into different polyhedral regions. In classification, each region is assigned a class, while in regression each region has its own distinct regression coefficients. Computational experimentations with generated and real data sets show that CRIO is comparable to and often outperforms the current leading methods in classification and regression. These results illustrate the potential for significant impact of integer optimization methods on computational statistics and data mining.

Logical Analysis of Data (LAD) is a technique that is used for risk prediction in medical applications [3]. This method is based on combinatorial optimization and boolean logic. The goal is essentially classifying groups of patients at low and high mortality risk and LAD is shown to outperform standard methods used by cardiologists.

Another supervised learning method is by Mammadov *et al.* [70] where a multi-label classifier is considered. See [58, 57] for surveys on classification and disease prediction methods that use mathematical programming techniques.

2.5 Kernel methods. In this section, we show how kernel method is used to extend linear learning techniques to their nonlinear counterparts. The equivalent dual formulations of linear parametric models can be used where predictions are based on linear combinations of *kernel function* evaluated at the training points. The kernel function for fixed nonlinear feature space mapping $\phi(\mathbf{x})$ is given by

$$K(\mathbf{x}, \mathbf{x}') = \phi(\mathbf{x})^T \phi(\mathbf{x}').$$

Introduced by Aizerman *et al.* [2], this concept facilitates the introduction of interesting extensions. The kernel formulated as an inner product in the feature space can be replaced with some other choice of kernel provided that the input vector $\mathbf{x}$ appears only in the form of inner products. This kernel substitution is called the *kernel trick* and is widely used for nonlinear variants of SVM [98], Fisher discriminant [80], PCA [92], and other machine learning techniques.

We will not give detailed biomedical applications in this section since the application areas for the previously mentioned methods are also valid for their nonlinear variants. We will only focus on support vector formulations and kernel substitutions.

Dual formulations for both 1-norm and 2-norm SVM classifiers can be obtained using the optimization theory. The significance of the dual formulations is that they allow the *kernel trick* to be introduced for nonlinear classification. The standard method to obtain the dual formulation of the SVM classification problem consists of two parts. First the *Lagrangian function* of the primal problem if derived. This function provides a lower bound for the solution of the primal problem. Next, stationarity is imposed on the Lagrangian function. Equivalent expressions for each primal variable are substituted back in the Lagrangian function and added as constraints where appropriate. The dual problem is obtained by maximizing the resulting function with the new constraints. The dual problem is a concave maximization problem, which can also be solved efficiently. The Lagrangian function for the 2-norm SVM classification problem is given as

$$L(\psi, b, \xi, \alpha) = \frac{1}{2}\|\psi\|^2 + C\sum_{i=1}^{l}\xi_i^2 - \sum_{i=1}^{l}\alpha_i\left[y_i(\langle\psi\cdot\mathbf{x}_i\rangle + b) - 1 + \xi_i\right].$$

Differentiating L with respect to the primal variables ψ, b, and ξ, and using stationarity, we obtain

$$\frac{\partial L}{\partial \psi} = \psi - \sum_{i=1}^{l} y_i\alpha_i\mathbf{x}_i = 0$$

$$\frac{\partial L}{\partial b} = \sum_{i=1}^{l} y_i\alpha_i = 0$$

$$\frac{\partial L}{\partial \xi_i} = 2C\xi_i - \alpha_i = 0.$$

The expressions are substituted back in the Lagrangian function and the following dual formulation is obtained which realizes the optimal hyperplane $\psi_0 = \sum_{i=1}^{l} y_i\alpha_i^0\mathbf{x}_i$.

$$\max \quad \sum_{i=1}^{l}\alpha_i - \frac{1}{2}\sum_{i=1}^{l}\sum_{j=1}^{l} y_iy_j\alpha_i\alpha_j\langle\mathbf{x}_i\cdot\mathbf{x}_j\rangle - \frac{1}{4C}\sum_{i=1}^{l}\alpha_i^2 \tag{2.16a}$$

$$\text{subject to} \quad \sum_{i=1}^{l} y_i\alpha_i = 0 \tag{2.16b}$$

$$\alpha_i \geq 0 \qquad i = 1, \ldots, l. \tag{2.16c}$$

Kernel trick can be applied and the 2-norm formulation becomes,

$$\max \quad \sum_{i=1}^{l}\alpha_i - \frac{1}{2}\sum_{i=1}^{l}\sum_{j=1}^{l} y_iy_j\alpha_i\alpha_jK(\mathbf{x}_i, \mathbf{x}_j) - \frac{1}{4C}\sum_{i=1}^{l}\alpha_i^2 \tag{2.17a}$$

$$\text{subject to} \quad \sum_{i=1}^{l} y_i\alpha_i = 0 \tag{2.17b}$$

$$\alpha_i \geq 0 \qquad i = 1, \ldots, l. \tag{2.17c}$$

Note that from Karush-Kuhn-Tucker complementarity conditions, the constraints in the primal problem are binding for those with the corresponding dual variable $\alpha_i^0 > 0$. Therefore b_0 can be calculated using

$$y_i[\langle\psi_0\cdot\mathbf{x}_i\rangle + b_0] = 1 - \xi_i^0 \qquad i : \alpha_i^0 > 0$$

$$\langle\psi_0\cdot\mathbf{x}_i\rangle + b_0 = y_i\left(1 - \frac{\alpha_i^0}{2C}\right) \qquad i : \alpha_i^0 > 0$$

$$b_0 = y_i\left(1 - \frac{\alpha_i^0}{2C}\right) - \sum_{j=1}^{l} y_j\alpha_j^0K(\mathbf{x}_j, \mathbf{x}_i) \qquad i : \alpha_i^0 > 0. \tag{2.18}$$

The derivation for the 1-norm dual formulation is very similar to that of 2-norm. The Lagrangian function for the 1-norm SVM classification problem is given as follows.

$$L(\psi, b, \xi, \alpha, \mathbf{r}) = \frac{1}{2}\|\psi\|^2 + C\sum_{i=1}^{l} \xi_i - \sum_{i=1}^{l} \alpha_i \left[y_i(\langle \psi \cdot \mathbf{x}_i \rangle + b) - 1 + \xi_i\right] - \sum_{i=1}^{l} r_i \xi_i$$

Differentiating L with respect to the primal variables and using stationarity, we obtain

$$\frac{\partial L}{\partial \psi} = \psi - \sum_{i=1}^{l} y_i \alpha_i \mathbf{x}_i = 0$$
$$\frac{\partial L}{\partial b} = \sum_{i=1}^{l} y_i \alpha_i = 0$$
$$\frac{\partial L}{\partial \xi_i} = C - \alpha_i - r_i = 0.$$

The expressions are substituted back in the Lagrangian function and the following dual formulation is obtained.

$$\max \quad \sum_{i=1}^{l} \alpha_i - \frac{1}{2}\sum_{i=1}^{l}\sum_{j=1}^{l} y_i y_j \alpha_i \alpha_j \langle \mathbf{x}_i \cdot \mathbf{x}_j \rangle \tag{2.19a}$$
$$\text{subject to} \quad \sum_{i=1}^{l} y_i \alpha_i = 0 \tag{2.19b}$$
$$0 \le \alpha_i \le C \qquad i = 1, \dots, l. \tag{2.19c}$$

Kernel trick can be applied and the 1-norm formulation becomes,

$$\max \quad \sum_{i=1}^{l} \alpha_i - \frac{1}{2}\sum_{i=1}^{l}\sum_{j=1}^{l} y_i y_j \alpha_i \alpha_j K(\mathbf{x}_i, \mathbf{x}_j) \tag{2.20a}$$
$$\text{subject to} \quad \sum_{i=1}^{l} y_i \alpha_i = 0 \tag{2.20b}$$
$$0 \le \alpha_i \le C \qquad i = 1, \dots, l. \tag{2.20c}$$

This problem is equivalent to the maximal margin hyperplane, with the additional constraint that all the α_i are upper bounded by C. This gives rise to the *box constraints* that is frequently used to refer to this formulation, since the vector α is constrained to lie inside the box with side length C in the positive orthant. The trade-off parameter between accuracy and regularization directly controls the size of the α_i. That is, the box constraints limit the influence of outliers, which would otherwise have large Lagrange multipliers. The constraint also ensures that the feasible region is bounded and hence the primal always has a non-empty feasible region.

Note that Karush-Kuhn-Tucker complementarity conditions can be used to obtain b_0 similar to the 2-norm case. However, in 1-norm case we look for both constraints to be binding, i.e., $\alpha_i^0 > 0, r_i^0 > 0$.

$$\begin{aligned} y_i(\langle \psi_0 \cdot \mathbf{x}_i \rangle + b_0) &= 1 & i : C > \alpha_i^0 > 0 \\ \langle \psi_0 \cdot \mathbf{x}_i \rangle + b_0 &= y_i & i : \alpha_i^0 > 0 \\ b_0 &= y_i - \sum_{j=1}^{l} y_j \alpha_j^0 K(\mathbf{x}_j, \cdot \mathbf{x}_i) & i : \alpha_i^0 > 0. \end{aligned} \tag{2.21}$$

One of the most commonly used kernels both theoretically and practically is the *Gaussian kernel* which is represented as,

$$K(\mathbf{x}_i, \mathbf{x}_j) = e^{-\frac{\|\mathbf{x}_i - \mathbf{x}_j\|^2}{\sigma}} \tag{2.22}$$

where, σ is referred to as the bandwidth. The smaller bandwidths are better in classifying intricate patterns, but worse in generalization.

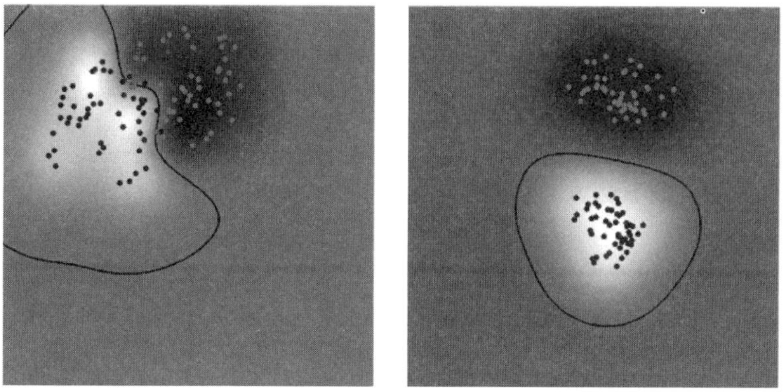

Figure 4 Examples of nonlinear classification using SVM with Gaussian Kernel

In [96], different relaxations for the selective SVMs are proposed. One of the relaxations gives pooled free slack for the data points instead of using binary variables.

$$\min \quad \frac{1}{2}\|\psi\|^2 + \frac{C}{2}\sum_{i=1}^{l} \xi_i^2 \tag{2.23a}$$

$$\text{subject to} \quad y_i(\langle \psi \cdot \mathbf{x}_{i,k} \rangle + b) \geq 1 - \xi_i - v_{i,k} \qquad i = 1, \ldots, l, k = 1, \ldots, |S_i| \tag{2.23b}$$

$$\sum_{i=1}^{l} \sum_{k=1}^{|S_i|} v_{i,k} \leq l \tag{2.23c}$$

$$v_{i,k} \geq 0 \qquad i = 1, \ldots, l, k = 1, \ldots, |S_i|. \tag{2.23d}$$

The Lagrangian function for (2.23) is given as follows,

$$
\begin{aligned}
L(\psi, b, \xi, \mathbf{v}, \alpha, \beta, \mathbf{r}) = & \frac{1}{2}\|\psi\|^2 + \frac{C}{2}\sum_{i=1}^{l}\xi_i^2 \\
& - \sum_{i=1}^{l}\sum_{k=1}^{|S_i|}\alpha_{i,k}\left[y_i(\langle\psi \cdot \mathbf{x}_{i,k}\rangle + b) - 1 + \xi_i + v_{i,k}\right] \\
& - \beta(l - \sum_{i=1}^{l}\sum_{k=1}^{|S_i|} v_{i,k}) \\
& - \sum_{i=1}^{l}\sum_{k=1}^{|S_i|} r_{i,k}v_{i,k}.
\end{aligned} \tag{2.24}
$$

In this equation, α, β and $\mathbf{r}$ are Lagrangian multipliers. Taking partial derivatives with respect to the primal variables ψ, b, and each ξ_i and $v_{i,k}$, and imposing stationarity, we get the following expressions:

$$
\begin{aligned}
\partial L/\partial\psi = \psi - \sum_{i=1}^{l}\sum_{k=1}^{|S_i|} y_i\alpha_{i,k}\mathbf{x}_{i,k} = 0; \quad & \partial L/\partial b = \sum_{i=1}^{l}\sum_{k=1}^{|S_i|} y_i\alpha_{i,k} = 0 \\
\partial L/\partial\xi_{i,k} = C\xi_{i,k} - \sum_{k=1}^{|S_i|}\alpha_{i,k} = 0; \quad & \partial L/\partial v_{i,k} = \beta - \alpha_{i,k} - r_{i,k} = 0.
\end{aligned} \tag{2.25}
$$

Substituting the partial derivatives (2.25) in (2.24), we can write the following dual formulation:

$$
\begin{aligned}
\max \quad & \sum_{i=1}^{l}\sum_{k=1}^{|S_i|}\alpha_{i,k} - \frac{1}{2}\sum_{i=1}^{l}\sum_{k=1}^{|S_i|}\sum_{j=1}^{l}\sum_{m=1}^{|S_j|} y_iy_j\alpha_{i,k}\alpha_{j,m}\langle\mathbf{x}_{i,k}\cdot\mathbf{x}_{j,m}\rangle \\
& - \frac{1}{2C}\sum_{i=1}^{l}\left(\sum_{k=1}^{|S_i|}\alpha_{i,k}\right)^2 - \beta l && \text{(2.26a)} \\
\text{subject to} \quad & \sum_{i=1}^{l}\sum_{k=1}^{|S_i|} y_i\alpha_{i,k} = 0 && \text{(2.26b)} \\
& 0 \le \alpha_{i,k} \le \beta \qquad i = 1, \ldots, l, k = 1, \ldots, |S_i|. && \text{(2.26c)}
\end{aligned}
$$

Kernel induced maps to nonlinear feature spaces can be used by replacing the linear dot product $\langle \mathbf{x}_{i,k} \cdot \mathbf{x}_{j,m} \rangle$ in (2.26) with a kernel and we obtain

$$\max \quad \sum_{i=1}^{l}\sum_{k=1}^{|S_i|} \alpha_{i,k} - \frac{1}{2}\sum_{i=1}^{l}\sum_{k=1}^{|S_i|}\sum_{j=1}^{l}\sum_{m=1}^{|S_j|} y_i y_j \alpha_{i,k}\alpha_{j,m} K(\mathbf{x}_{i,k}, \mathbf{x}_{j,m}) - \frac{1}{2C}\sum_{i=1}^{l}\left(\sum_{k=1}^{|S_i|}\alpha_{i,k}\right)^2 - \beta l \tag{2.27a}$$

$$\text{subject to} \quad \sum_{i=1}^{l}\sum_{k=1}^{|S_i|} y_i \alpha_{i,k} = 0 \tag{2.27b}$$

$$0 \leq \alpha_{i,k} \leq \beta \qquad i = 1, \ldots, l, k = 1, \ldots, |S_i|. \tag{2.27c}$$

Different relaxations for selective SVM classifiers with different special properties are derived in [96, 97].

For the Support Vector Regression, derivation of the dual and primal variable calculations via complementary slackness conditions are similar to that of SVM classifiers. For the sake of completeness, we only present the dual for SVR.

2-norm dual for SVR is as follows:

$$\max \quad \sum_{i=1}^{l} y_i\alpha_i - \epsilon\sum_{i=1}^{l}|\alpha_i| - \frac{1}{2}\sum_{i=1}^{l}\sum_{j=1}^{l}\alpha_i\alpha_j\left(K(\mathbf{x}_i, \mathbf{x}_j) + \frac{1}{C}\delta_{ij}\right) \tag{2.28a}$$

$$\text{subject to} \quad \sum_{i=1}^{l}\alpha_i = 0. \tag{2.28b}$$

b_0 is calculated using $\sum_{j=1}^{l}\alpha_j^0 K(\mathbf{x}_j, \mathbf{x}) + b_0 - y_i = -\epsilon - \frac{\alpha_j^0}{C}$ for any $\alpha_j^0 > 0$. 1-norm dual for SVR is as follows:

$$\max \quad \sum_{i=1}^{l} y_i\alpha_i - \epsilon\sum_{i=1}^{l}|\alpha_i| - \frac{1}{2}\sum_{i=1}^{l}\sum_{j=1}^{l}\alpha_i\alpha_j K(\mathbf{x}_i, \mathbf{x}_j) \tag{2.29a}$$

$$\text{subject to} \quad \sum_{i=1}^{l}\alpha_i = 0 \tag{2.29b}$$

$$-C \leq \alpha_i \leq C \qquad i = 1, \ldots, l. \tag{2.29c}$$

b_0 is calculated using $\sum_{j=1}^{l}\alpha_j^0 K(\mathbf{x}_j, \mathbf{x}) + b_0 - y_i = -\epsilon$ for any $0 < \alpha_j^0 < C$.

3 Biomedical treatment and other applications

In this section, we first discuss multi-quadratic 0–1 programming with applications in biomedical research on epilepsy. Epilepsy consists of more than 40 clinical syndromes affecting 50 million people worldwide. At least 30% of patients with epilepsy continue to have seizures despite treatment with antiepileptic drugs. Epileptic seizure occurrences seem to be random and unpredictable. However, recent studies in epileptic patients suggest that seizures are deterministic rather than random. Subsequently, studies of the spatiotemporal dynamics in electroencephalograms (EEGs), from patients with temporal lobe epilepsy, demonstrate a preictal transition of approximately 30 minutes to 1 hour duration before the ictal onset. The enormous number of neurons and dynamic nature of connections between them makes the analysis of brain function especially challenging. In order to perform a quantitative analysis of brain, one can treat certain groups of neurons (functional units of the brain) as vertices of a graph and investigate the connections between these functional units. *Lyapunov exponent* is an important measure that characterizes chaotic behavior of nonlinear system. Global Lyapunov exponent highlights how fast nearby orbits of the system converge or diverge in infinitely large time interval. On contrary, local Lyapunov exponent characterize local predictability around a point in phase space. Lyapunov exponent has proven its efficiency in EEG analysis for predicting epileptic seizures.

Let a system be set by

$$\dot{X}(t) = F(X), \text{ where } X : \mathbb{R} \to \mathbb{R}^n, F : \mathbb{R}^n \to \mathbb{R}^n.$$

The maximal Lyapunov exponent λ can be defined as

$$\lambda = \lim_{t\to\infty} \lim_{\delta X(0)\to 0} \frac{1}{t} \log_2 \frac{\delta X(t)}{\delta X(0)}. \tag{3.1}$$

For short term maximal Lyapunov exponent ($STL_{\max}$) we can take reasonable t instead of external limit. In real life we often deal with one dimensional time series of noisy data (such as EEG signal) instead of explicit system of equations. Wolf *et al.* [115] suggest an algorithm for Lyapunov Exponent calculation from time series. Pardalos *et al.* [83] and Chaovalitwongse *et al.* [24] use modification of Wolfs algorithm described in [42] for $STL_{\max}$ calculation that handles noisy non-stationary data.

$$STL_{\max} = \frac{1}{t_M - t_0} \sum_{k=1}^{M} \log_2 \frac{L'(t_k)}{L(t_{k-1})}. \tag{3.2}$$

Since the brain is a nonstationary system, algorithms used to estimate measures of the brain dynamics should be capable of automatically identifying and appropriately weighing existing transients in the data. In a chaotic system, orbits originating from similar initial conditions (nearby points in the state space) diverge exponentially (expansion process). Rate of divergence is an important aspect of the system dynamics and is reflected in the value of Lyapunov exponents. During the last decade, advances in studying brain are associated with extensive use of EEG which can be treated as the quantitative representation of the brain function. EEG data essentially represent time series recorded from the electrodes located in different functional units of brain. We utilize the concept of T-index to measure entrainment of two brain sites at a time moment. T-index at time t between electrode sites i and j is defined as

$$T_{i,j}(t) = \frac{\sqrt{N}\,|E\{STL_{\max,i} - STL_{\max,j}\}|}{\sigma_{i,j}(t)}$$

where $E\{\}$ is the sample average difference for the $STL_{\max,i} - STL_{\max,j}$ estimated over a moving window $w_t(\lambda)$. At the moment of a seizure some brain sites exhibit the convergence of their EEG signals, which is characterized by drop of the corresponding T-index below T_{critical}. A natural graph representing the brain is where each vertex corresponds to a functional unit/electrode, and there is a edge between two of them if T-index is below T_{critical}. The number of edges in this graph dramatically increases at seizure points, and it decreases immediately after seizures.

One aspect of the analysis of the epileptic brain is finding a maximum clique in this graph. It provides us with the largest set of critical electrode sites most entrained during the seizure. If the number of critical sites is set equal to k, we can formulate the problem of selecting the optimal group of critical site as a multi-quadratic 0–1 programming as follows.

Let $x_i \in \{0, 1\}$ denote if site i is selected. a_{ij} is the T-index between sites i and j during the seizure point. b_{ij} is the T-index between sites i and j 10 minutes after the onset of seizure.

$$\min \quad x^T A x \tag{3.3a}$$

$$\text{subject to} \quad \sum_{i=1}^{n} x_i = k \tag{3.3b}$$

$$x^T B x \geq T_{\text{critical}} k(k-1) \tag{3.3c}$$

$$x \in \{0, 1\}^n. \tag{3.3d}$$

Pardalos *et al.* [83] develop a novel linearization technique to reformulate a quadratically constrained quadratic 0–1 programming problem as an equivalent mixed integer programming (MIP) problem. The practical importance of this reformulation is that number of 0–1 variables remains the same and number of additional continuous variables is $O(n)$, where n is the number of 0–1 variables.

Next, we discuss *radiation therapy*, which is arguably the most popular domain in treatment planning, where optimization techniques are used. Radiation therapy is a method used to treat cancer with high-energy radiation that destroys the ability of cancerous cells to reproduce. There are two types of radiation therapy. The first one is *external beam radiotherapy*, or teletherapy, where high-energy rays are aimed to the cancerous tissues. Multileaf collimator shapes the beam by blocking out some parts of the beam. To precisely shape the beam, multileaf collimators consist of a small array of metal leaves for each beam. Each beam is specified by a set of evenly spaced strips and the treatment plan is defined by a collection of beams with the amount of radiation to be delivered along each strip within each beam. The second radiation therapy method is *brachytherapy*, also know as sealed source radiotherapy or endocurietherapy, where a radioactive source is placed inside or next to the area requiring treatment. Brachytherapy is commonly used to treat localized prostate cancer and cancers of the head and neck. Both types of therapy need to be planned to localize the radiation area so that the destroyed tissue is minimized.

Many optimization algorithms are developed for the treatment planning in radiation therapy which employ techniques such as multiobjective optimization [53, 54] (e.g. investigating tradeoffs between tumor coverage and critical organ sparing [25]), linear programming [65], mixed-integer programming [60, 59], nonlinear programming [13, 33], simulated annealing [113], and inverse planning with a genetic algorithm-based framework [12].

Recently, Men *et al.* consider the problem of intensity-modulated radiation therapy (IMRT) treatment planning using direct aperture optimization [77]. In contrast to the heuristic approaches, an exact approach is used that explicitly formulates the fluence map optimization (FMO) problem as a convex optimization problem in terms of all multileaf collimator (MLC) deliverable apertures and their associated intensities. However, the number of deliverable apertures, and therefore the number of decision variables and constraints in the new problem formulation, is typically enormous. To overcome this, an iterative approach is used that employs a subproblem whose optimal solution either provides a suitable aperture to add to a given pool of allowable apertures or concludes that the current solution is optimal. This approach has the additional advantage that it can explicitly account for transmission of dose through the part of an aperture that is blocked by the MLC system, yielding a more precise assessment of the treatment plan than what is possible using a traditional beamlet-based FMO problem. Tests on clinical head-and-neck cancer cases show the efficacy of this approach, yielding treatment plans comparable in quality to plans obtained by the traditional method with a reduction of more than 75% in the number of apertures and a reduction of more than 50% in beam-on time, with only a modest increase in computational effort. The results also show that delivery efficiency is very insensitive to the addition of traditional MLC constraints; however, jaws-only treatment requires about a doubling in beam-on time and number of apertures used.

For a review on optimization methods in radiation therapy, the reader is referred to [100, 32]. Acosta *et al.* study the influence of dose grid resolution on beam selection strategies in radiotherapy treatment design [1]. Censor *et al.* [22] study a unified model for handling dose constraints and radiation source constraints in a single mathematical framework based on the split feasibility problem. See [15] for description of other treatment problems. Future research directions for radiation therapy are discussed in [59].

Another problem that has been extensively studied is the non-unique probe selection. This problem consists of selecting oligonucleotide probes for use in hybridization experiments in which target viruses or bacteria are to be identified in biological samples. The presence or absence of these targets is determined by observing whether selected probes bind to their corresponding sequences. The goal is to select a probe set that is able to uniquely identify targets while containing a minimal number of probes. Ragle *et al.* [88] present the first exact method for finding optimal solutions to the non-unique probe selection problem within practical computational limits, without the a priori elimination of candidate probes. Previous methods have employed heuristics to find approximate solutions that are not provably optimal, and as a result, no knowledge has been obtained regarding the quality of those solutions relative to optimality. The computational results show that the method can find the optimal solution within 10 minutes, and is capable of reducing the number of probes required over state-of-the-art heuristic techniques by as much as 20%.

Using d-disjunct matrix, Thai *et al.* [108] present two $(1 + (d + 1)\log n)$-approximation algorithms to identify at most d targets for the non-unique probe selection problem. Based on their selected non-unique probes, the decoding algorithms with linear time complexity are also presented. The proposed algorithms with fault tolerant solutions can identify at most d targets in the presence of experimental errors.

Other optimization based studies in biomedicine are in DNA microarray experiments [109, 50, 19], intensity modulated proton therapy [85], ultrasound-mediated DNA transfection [117], protein design and gene networks [78, 6, 34, 102, 76, 30, 107], human motion analysis [27], imaging [31, 21, 67], ultrasound surgery [41], corneal rotation [46], drug design [71, 82], vaccine formularies [40] and query optimization in database integration [103]. Marchuk [74] develops mathematical models of infectious diseases, antiviral immune response and antibacterial response. These models are formulated as systems of delay-differential equations and applied to the study of hepatitis B, influenza A, infectious bacterial pneumonia, and mixed infections. For future research directions and open problems in other fields of biomedicine and computational biology, the reader is referred to [39].

4 Concluding remarks

In this paper, we summarize some of the recent studies on biomedicine that utilize optimization theory. This review is not exhaustive in that, we explore some techniques in depth and give references for other studies. Applications of optimization already improve quality of biomedical treatment but there are still many open problems. As computer capabilities improve and novel implementations are developed with higher level of interdisciplinary effort, optimization will continue to provide ongoing revelations in the constantly growing field of biomedicine.

Acknowledgments

This work is partially supported by NSF and Air Force grants.

References

[1] R. Acosta, M. Ehrgott, A. Holder, D. Nevin, J. Reese, and B. Salter, *Optimization in medicine*, ch. The influence of dose grid resolution on beam selection strategies in radiotherapy treatment design, pp. 1–23, Springer, 2008.

[2] M. A. Aizerman, E. M. Braverman, and L. I. Rozonoer, *The probability problem of pattern recognition learning and the method of potential functions*, Automation and Remote Control **25** (1964), 1175–1190.

[3] S. Alexe, E. Blackstone, P. Hammer, H. Ishwaran, M. Lauer, and C. Snader, *Coronary risk prediction by logical analysis of data*, Annals of Operations Research **119** (2003), 15–42.

[4] S. Andrews, I. Tsochantaridis, and T. Hofmann, *Support vector machines for multiple-instance learning*, Advances in Neural Information Processing Systems 15, 2003, pp. 561–568.

[5] P. Auer, *On learning from multi-instance examples: Empirical evaluation of a theoretical approach*, Proceedings 14th International Conference on Machine Learning, 1997, pp. 21–29.

[6] B. Balasundaram, S. Butenko, and S. Trukhanov, *Novel approaches for analyzing biological networks*, Journal of Combinatorial Optimization **10** (2005), no. 1, 23–39.

[7] A. Ben-Dor, L. Bruhn, N. Friedman, I. Nachman, M. Schummer, and Z. Yakhini, *Tissue classification with gene expression profiles*, RECOMB '00: Proceedings of the fourth annual international conference on Computational molecular biology (New York, NY, USA), ACM Press, 2000, pp. 54–64.

[8] A. Ben-Dor, B. Chor, R. Karp, and Z. Yakhini, *Discovering local structure in gene expression data: The order-preserving submatrix problem*, Proc. Sixth Intl Conf. Computational Biology (RECOMB 02), 2002, pp. 49–57.

[9] A. Ben-Dor, N. Friedman, and Z. Yakhini, *Class discovery in gene expression data*, RECOMB '01: Proceedings of the fifth annual international conference on Computational biology (New York, NY, USA), ACM Press, 2001, pp. 31–38.

[10] K. Bennet and C. Campbell, *Support vector machines: Hype or hallelujah?*, SIGKDD Explorations **2** (2000), no. 2, 1–13.

[11] D. Bertsimas and R. Shioda, *Classification and regression via integer optimization*, Operations Research **55** (2007), no. 2, 252–271.

[12] V. Bevilacqua, G. Mastronardi, and G. Piscopo, *Evolutionary approach to inverse planning in coplanar radiotherapy*, Image and Vision Computing **25** (2007), no. 2, 196–203.

[13] S. Billups and J. Kennedy, *Minimum-support solutions for radiotherapy planning*, Annals of Operations Research **119** (2001), 229–245.

[14] Avrim Blum and Adam Kalai, *A note on learning from multiple-instance examples*, Machine Learning **30** (1998), no. 1, 23–29.

[15] M. L. Brandeau, F. Sainfort, and W. P. Pierskalla (eds.), *Handbook of operations research and health care: Methods and applications*, Kluwer Academic Publishers, 2004.

[16] S. Busygin, N. Boyko, P. M. Pardalos, M. Bewernitz, and G. Ghacibeh, *Data mining, systems analysis and optimization in biomedicine*, ch. Biclustering EEG data from epileptic patients treated with vagus nerve stimulation, pp. 220–231, American Institute of Physics, 2007.

[17] S. Busygin, G. Jacobsen, and E. Kramer, *Double conjugated clustering applied to leukemia microarray data*, Proc. Second SIAM Intl Conf. Data Mining, Workshop Clustering High Dimensional Data, 2002.

[18] S. Busygin, O. A. Prokopyev, and P. M. Pardalos, *Feature selection for consistent biclustering via fractional 0–1 programming*, Journal of Combinatorial Optimization **10** (2005), no. 1, 7–21.

[19] ———, *An optimization-based approach for data classification*, Optimization Methods & Software **22** (2007), no. 1, 3–9.

[20] ———, *Biclustering in data mining*, Computers & Operations Research **35** (2008), no. 8, 2964–2987.

[21] J. D. Carew and M. Yuan, *Data mining, systems analysis and optimization in biomedicine*, ch. Nonparametric smoothing and its applications in biomedical imaging, pp. 85–105, American Institute of Physics, 2007.

[22] Y. Censor, T. Bortfeld, B. Martin, and A. Trofimov, *A unified approach for inversion problems in intensity-modulated radiation therapy*, Physics in Medicine and Biology **51** (2006), no. 10, 2353–2365.

[23] W. Chaovalitwongse, P. M. Pardalos, L. D. Iasemidis, W. Suharitdamrong, D.-S. Shiau, L. K. Dance, O. A. Prokopyev, V. L. Boginski, P. R. Carney, and J. C. Sackellares, *Data mining in biomedicine*, Optimization and Its Applications, vol. 7, ch. Data mining in EEG: Application to epileptic brain disorders, pp. 459–481, Springer, 2007.

[24] W. Chaovalitwongse, O. Prokopyev, and P. M. Pardalos, *Electroencephalogram (EEG) time series classification: Applications in epilepsy*, Annals of Operations Research **148** (2006), no. 1, 227–250.

[25] D. L. Craft, T. F. Halabi, H. A. Shih, and T. R. Bortfeld, *Approximating convex pareto surfaces in multiobjective radiotherapy planning*, Medical Physics **33** (2006), no. 9, 3399–3407.

[26] N. Cristianini and J. Shawe-Taylor, *An introduction to support vector machines*, Cambridge University Press, Cambridge, UK, 2000.

[27] B. Dariush, *Human motion analysis for biomechanics and biomedicine*, Machine Vision and Applications **14** (2003), no. 4, 202–205.

[28] T. G. Dietterich, R. H. Lathrop, and T. Lozano-Perez., *Solving the multiple instance problem with axis-parallel rectangles*, Artificial Intelligence **89** (1997), 31–71.

[29] F. Divina and J. S. Aguilar-Ruiz, *Biclustering of expression data with evolutionary computation*, IEEE Transactions on Knowledge and Data Engineering **18** (2006), no. 5, 590–602.

[30] M. M. Donahue, W. Zhang, M. L. Harrison, J. Hu, and A. E. Rundell, *Data mining, systems analysis and optimization in biomedicine*, ch. Employing optimization and sensitivity analyses tools to generate and analyze mathematical models of T-cell signaling events, pp. 43–63, American Institute of Physics, 2007.

[31] S. Dube, J. J. Corso, T. F. Cloughesy, S. El-Saden, A. L. Yuille, and U. Sinha, *Data mining, systems analysis and optimization in biomedicine*, ch. Automated MR image processing and analysis of malignant brain tumors: enabling technology for data mining, pp. 64–84, American Institute of Physics, 2007.

[32] M. Ehrgott, H. W. Hamacher, and M. Nußbaum, *Optimization in medicine*, ch. Decomposition of matrices and static multileaf collimators: a survey, pp. 25–46, Springer, 2008.

[33] M. Ferris, J. Lim, and D. Shepard, *Radiosurgery treatment planning via nonlinear programming*, Annals of Operations Research **119** (2001), 247–260.

[34] H. K. Fung, S. Rao, C. A. Floudas, O. Prokopyev, P. M. Pardalos, and F. Rendl, *Computational comparison studies of quadratic assignment like formulations for the in silico sequence selection problem in de novo protein design*, Journal of Combinatorial Optimization **10** (2005), no. 1, 41–60.

[35] Thomas Gartner, Peter A. Flach, Adam Kowalczyk, and Alex J. Smola, *Multi instance kernels*, Proceedings of the 19th International Conference on Machine Learning, 2002, pp. 179–186.

[36] A. Genkin, C. A. Kulikowski, and Ilya Muchnik, *Set covering submodular maximization: An optimal algorithm for data mining in bioinformatics and medical informatics*, Journal of Intelligent & Fuzzy Systems **12** (2002), 5–17.

[37] A. B. Geva and D. H. Kerem, *Forecasting generalized epileptic seizures from the EEG signal by wavelet analysis and dynamic unsupervised fuzzy clustering*, Ieee Transactions on Biomedical Engineering **45** (1998), no. 10, 1205–1216.

[38] T. R. Golub, D. K. Slonim, P. Tamayo, C. Huard, M. Gaasenbeek, J. P. Mesirov, H. Coller, M. L. Loh, J. R. Downing, M. A. Caligiuri, C. D. Bloomfield, and E. S. Lander, *Molecular classification of cancer: Class discovery and class prodiction by gene expression monitoring*, Science **286** (1999), no. 5439, 531–537.

[39] H. Greenberg, W. Hart, and G. Lancia, *Opportunities for combinatorial optimization in computational biology*, INFORMS Journal on Computing **16** (2004), 211–231.

[40] S. N. Hall, S. H. Jacobson, and E. C. Sewell, *Optimization in medicine and biology*, ch. Optimizing Pediatric Vaccine Formularies, pp. 117–146, Auerbach, 2008.

[41] T. Huttunen, J. P. Kaipio, and M. Malinen, *Optimization in medicine*, ch. Optimal control in high intensity focused ultrasound surgery, pp. 169–195, Springer, 2008.

[42] L. Iasemidis, J. Principe, and C. Sackellares, *Nonlinear signal processing in medicine*, vol. 2, ch. Measurement and quantification of spatiotemporal dynamics of human epileptic seizures, pp. 294–318, IEEE press, 2000.

[43] A. N. Jain, T. G. Dietterich, R. H. Lathrop, D. Chapman, R. E. Critchlow, B. E. Bauer, T. A. Webster, and T. Lozano-Perez, *A shape-based machine learning tool for drug design*, Journal of Computer-Aided Molecular Design **8** (1994), no. 6, 635–652.

[44] David S. Jones, *Pharmaceutical statistics*, Pharmaceutical Press, 2002.

[45] I. Kalatzis, D. Pappas, N. Piliouras, and D. Cavouras, *Support vector machines based analysis of brain SPECT images for determining cerebral abnormalities in asymptomatic diabetic patients*, MED. INFORM. **28** (2003), no. 3, 221–230.

[46] I. Karpouzas and Y. Pouliquen, *Modelling and numerical optimization of corneal rotation*, Mathematical Medicine and Biology **8** (1991), no. 1, 73–82.

[47] Y. Klugar, R. Basri, J. T. Chang, and M. Gerstein, *Spectral biclustering of microarray data: Coclustering genes and conditions*, Genome Research **13** (2003), 703–716.

[48] Y. Klugar, R. Basri, J.T. Chang, and M. Gerstein, *Spectral biclustering of microarray data: Coclustering genes and conditions*, Genome Research **13** (2003), 703–716.

[49] H. M. Kluger, B. M. Kacinski, Y. Kluger, O. Mironenko, M. Gilmore-Hebert, A. S. Perkins, and E. Sapi, *Microarray analysis of invasive and metastatic phenotypes in a breast cancer model*, Lab Invest **84** (2004), no. 3, 320–331.

[50] G. Kochenberger, F. Glover, B. Alidaee, and H. Wang, *Clustering of microarray data via clique partitioning*, Journal of Combinatorial Optimization **10** (2005), no. 1, 7–21.

[51] O. E. Kundakcioglu and P. M. Pardalos, *A branch and bound algorithm for multiple instance classification*, Proceedings of the 2008 International Conference on Machine Learning; Models, Technologies and Applications (MLMTA), (H.R. Arabnia and Y. Mun, eds.), 2008, pp. 862–866.

[52] ———, *Clustering challenges in biological networks*, ch. The complexity of feature selection for consistent biclustering, World Scientific, (W. A. Chovalitwongse, S. Butenko, and P. M. Pardalos, eds.), 2009, pp. 257–266.

[53] M. Lahanas, D. Baltas, and N. Zamboglou, *A hybrid evolutionary multiobjective algorithm for anatomy based dose optimization algorithm in high-dose-rate brachytherapy*, Physics in Medicine and Biology **48** (2003), no. 3, 399–415.

[54] M. Lahanas, E. Schreibmann, and D. Baltas, *Multiobjective inverse planning for intensity modulated radiotherapy with constraint-free gradient-based optimization algorithms*, Physics in Medicine and Biology **48** (2003), no. 17, 2843–2871.

[55] L. Lazzeroni and A. Owen, *Plaid models for gene expression data*, Statistical Sinica **12** (2002), no. 1, 61–86.

[56] C.-H. Lee, M. Schmidt, A. Murtha, A. Bistritz, J. Sander, and R. Greiner, *Segmenting brain tumors with conditional random fields and support vector machines.*, CVBIA, 2005, pp. 469–478.

[57] E. K. Lee, *Optimization in medicine*, ch. Optimization-based predictive models in medicine and biology, pp. 127–151, Springer, 2008.

[58] E. K. Lee and T.-L. Wu, *Data mining, systems analysis and optimization in biomedicine*, ch. Classification and disease prediction via mathematical programming, pp. 1–42, American Institute of Physics, 2007.

[59] E.K. Lee, T. Fox, and I. Crocker, *Integer programming applied to intensity-modulated radiation treatment planning optimization*, Annals of Operations Research **119** (2001), 165–181.

[60] E.K. Lee and M. Zaider, *Mixed integer programming approaches to treatment planning for brachytherapy – application to permanent prostate implants*, Annals of Operations Research **119** (2003), 147–163.

[61] J. K. Lee, P. D. Williams, and S. Cheon, *Data mining in genomics*, Clinics in Laboratory Medicine **28** (2008), no. 1, 145–166.

[62] G.-Z. Li, T.-Y. Liu, and V. S. Cheng, *Classification of brain glioma by using SVMs bagging with feature selection.*, BioDM, 2006, pp. 124–130.

[63] Guo-Zheng Li, Jie Yang, Chen-Zhou Ye, and Dao-Ying Geng, *Degree prediction of malignancy in brain glioma using support vector machines*, Computers in Biology and Medicine **36** (2006), 313–325.

[64] S. Lloyd, *Least squares quantization in PCM*, Information Theory, IEEE Transactions on **28** (1982), no. 2, 129–137.

[65] W. A. Lodwick, S. McCourt, F. Newman, and S. Humphries, *Computational radiology and imaging: Therapy and diagnostics*, ch. Optimization Methods for Radiation Therapy Plans, pp. 229–248, Springer, 1999.

[66] P. M. Long and L. Tan, *PAC learning axis aligned rectangles with respect to product distributions from multiple instance examples*, Machine Learning **30** (1998), 7–22.

[67] A. K. Louis, *Optimization in medicine*, ch. Optimal reconstruction kernels in medical imaging, pp. 153–168, Springer, 2008.

[68] Y. Lu, S. Y. Lu, F. Fotouhi, Y. P. Deng, and S. J. Brown, *Incremental genetic k-means algorithm and its application in gene expression data analysis*, BMC bioinformatics **5** (2004), no. 172.

[69] S. C. Madeira and A. L. Oliveira, *Biclustering algorithms for biological data analysis: A survey*, IEEE Transactions on Computational Biology and Bioinformatics **1** (2004), 24–45.

[70] M. Mammadov, A. Rubinov, and J. Yearwood, *Data mining in biomedicine*, Optimization and Its Applications, vol. 7, ch. An optimization approach to identify the relationship between features and output of a multi-label classifier, pp. 141–167, Springer, 2007.

[71] M. A. Mammadov, A. M. Rubinov, and J. Yearwood, *The study of drug-reaction relationships using global optimization techniques*, Optimization Methods & Software **22** (2007), no. 1, 99–126.

[72] O. L. Mangasarian, W. N. Street, and W. H. Wolberg, *Breast cancer diagnosis and prognosis via linear programming*, Operations Research **43** (1995), 570–577.

[73] O. L. Mangasarian and E. W. Wild, *Multiple instance classification via successive linear programming*, Journal of Optimization Theory and Applications **137** (2008), no. 1, to appear.

[74] Guri I. Marchuk, *Mathematical modelling of immune response in infectious diseases*, Kluwer Academic Publishers, 1997.

[75] M. Martinez-Ramon, V. Koltchinskii, G. L. Heileman, and S. Posse, *fMRI pattern classification using neuroanatomically constrained boosting*, NeuroImage **31** (2006), 1129–1141.

[76] S. R. McAllister, R. Rajgaria, and C. A. Floudas, *Global pairwise sequence alignment through mixed-integer linear programming: a template-free approach*, Optimization Methods & Software **22** (2007), no. 1, 127–144.

[77] C. Men, H. E. Romeijn, Z. C. Taşkın, and J. F. Dempsey, *An exact approach to direct aperture optimization in IMRT treatment planning*, Physics in Medicine and Biology **52** (2007), 7333–7352.

[78] C. N. Meneses, C. A. S. Oliveira, and P. M. Pardalos, *Data mining in biomedicine*, Optimization and Its Applications, vol. 7, ch. Mathematical programming formulations for problems in genomics and proteomics, pp. 275–290, Springer, 2007.

[79] B. H. Menze, M. P. Lichy, P. Bachert, B. M. Kelm, H.-P. Schlemmer, and F. A. Hamprecht, *Optimal classification of long echo time in vivo magnetic resonance spectra in the detection of recurrent brain tumors.*, NMR Biomed **19** (2006), no. 5, 599–609.

[80] S. Mika, G. Rätsch, J. Weston, B. Schölkopf, and K. Müller, *Fisher discriminant analysis with kernels*, Proceedings of IEEE Neural Networks for Signal Processing Workshop, 1999.

[81] A. Nahapetyan, S. Busygin, and P. M. Pardalos, *An improved heuristic for consistent biclustering problems*, BIOMAT 2006, 2006.

[82] P. M. Pardalos, V. L. Boginski, O. A. Prokopyev, W. Suharitdamrong, P. R. Carney, W. Chaovalitwongse, and A. Vazacopoulos, *Essays and surveys in global optimization*, ch. Optimization techniques in medicine, pp. 211–232, Springer, 2005.

[83] P. M. Pardalos, W. Chaovalitwongse, L. D. Iasemidis, J. C. Sackellares, D.-S. Shiau, P. R. Carney, O. A. Prokopyev, and V. A. Yatsenko, *Seizure warning algorithm based on optimization and nonlinear dynamics*, Mathematical Programming **101** (2004), no. 2, 365–385.

[84] A. Pascual-Montano, P. Carmona-Saez, M. Chagoyen, F. Tirado, J. M. Carazo, and R. D. Pascual-Marqui, *bioNMF: a versatile tool for non-negative matrix factorization in biology*, BMC Bioinformatics **7** (2006), no. 366.

[85] D. Pflugfelder, J. J. Wilkens, and U. Oelfke, *Worst case optimization: a method to account for uncertainties in the optimization of intensity modulated proton therapy*, Physics in Medicine and Biology **53** (2008), 1689–1700.

[86] J. Platt, *Using sparseness and analytic QP to speed training of support vector machines*, Advances in Neural Information Processing Systems (M. S. Kearns, S. A. Solla, and D. A. Cohn, eds.), MIT Press, 1999.

[87] A. Quddus, P. Fieguth, and O. Basir, *Adaboost and support vector machines for white matter lesion segmentation in MR images*, Proceedings of the 2005 IEEE Engineering in Medicine and Biology 27th Annual Conference, 2005.

[88] M. A. Ragle, J. C. Smith, and P. M. Pardalos, *An optimal cutting-plane algorithm for solving the non-unique probe selection problem*, Annals of Biomedical Engineering **35** (2007), no. 11, 2023–2030.

[89] S. Ray and M. Craven, *Supervised versus multiple instance learnining: An emperical comparison*, Proceedings of the 22nd International Conference on Machine Learning, 2005.

[90] L. Rinaldi, P. Gallo, F. Ranzato, D. Luise, D. Colavito, M. Motta, A. Guglielmo, E. Del Giudice, C. Romualdi, E. Ragazzi, A. Darrigo, M. Dalle Carbonare, B. Leontino, and A. Leon, *Longitudinal analysis of immune cell phenotypes in early stage multiple sclerosis: distinctive patterns characterize MRI-active patients*, Brain **129** (2006), 1993–2007.

[91] S. Sabesan, L. Good, N. Chakravarthy, K. Tsakalis, P. M. Pardalos, and L. Iasemidis, *Optimization in medicine*, ch. Global optimization and spatial synchronization changes prior to epileptic seizures, pp. 103–125, Springer, 2008.

[92] B. Schölkopf, A. J. Smola, and K. R. Müller, *Kernel principal component analysis*, Advances in kernel methods: support vector learning (1999), 327–352.

[93] E. Segal, B. Taskar, A. Gasch, N. Friedman, and D. Koller, *Rich probabilistic models for gene expression*, Bioinformatics **17** (2001), 243–252.

[94] O. Seref, C. Cifarelli, O. E. Kundakcioglu, P. M. Pardalos, and M. Ding, *Detecting categorical discrimination in a visuomotor task using selective support vector machines*, Proceedings of BIOCOMP '07 (H. R. Arabnia, M. Q. Yang, and J. Y. Yang, eds.), 2007, pp. 580–590.

[95] O. Seref, O. E. Kundakcioglu, and M. Bewernitz, *Encyclopedia of healthcare information systems*, ch. Support Vector Machines in Neuroscience, p. to appear, Medical Information Science Reference (N. Wickramasinghe and E. Geisler, eds., IDEA Group Inc.), 2008, pp. 1283–1293.

[96] O. Seref, O. E. Kundakcioglu, and P. M. Pardalos, *Selective linear and nonlinear classification*, CRM Proceedings and Lecture Notes (P. M. Pardalos and P. Hansen, eds.), vol. 45, 2008, pp. 211–234.

[97] O. Seref, O. E. Kundakcioglu, O. Prokopyev, and P. M. Pardalos, *Selective support vector machines*, Journal of Combinatorial Optimization **17** (2009), no. 1, 3–20.

[98] J. Shawe-Taylor and N. Cristianini, *Kernel methods for pattern analysis*, Cambridge University Press, 2004.

[99] Q. Sheng, Y. Moreau, and B. De Moor, *Biclustering microarray data by Gibbs sampling*, Bioinformatics **19** (2003), 196–205.

[100] D. M. Shepard, M. C. Ferris, G. H. Olivera, and T. R. Mackie, *Optimizing the delivery of radiation therapy to cancer patients*, SIAM Review **41** (1999), no. 4, 721–744.

[101] W. N. Street, O. L. Mangasarian, and W. H. Wolberg, *An inductive learning approach to prognostic prediction*, in Machine Learning: Proceedings of the Twelfth International Conference, Morgan Kaufmann, 1995, pp. 522–530.

[102] S. S. Strickler, A. V. Gribenko, A. V. Gribenko, T. R. Keiffer, J. Tomlinson, T. Reihle, V. V. Loladze, and G. I. Makhatadze, *Protein stability and surface electrostatics: A charged relationship*, Biochemistry **45** (2006), no. 9, 2761–2766.

[103] W. Sujansky, *Heterogeneous database integration in biomedicine*, Journal of Biomedical Informatics **34** (2001), no. 4, 285–298.

[104] A. Tanay, R. Sharan, and R. Shamir, *Discovering statistically significant biclusters in gene expression data*, Bioinformatics **18** (2002), 136–144.

[105] ———, *Biclustering algorithms: A survey*, Tech. report, Tel Aviv University, 2004.

[106] Q. Tao, S. Scott, N. V. Vinodchandran, and T. T. Osugi, *SVM-based generalized multiple-instance learning via approximate box counting*, ICML '04: Proceedings of the Twenty-first International Conference on Machine Learning (New York, NY, USA), ACM Press, 2004, pp. 799–806.

[107] M. T. Thai, Z. Cai, and D. Z. Du, *Genetic networks: processing data, regulatory network modelling and their analysis*, Optimization Methods & Software **22** (2007), no. 1, 169–185.

[108] M. T. Thai, P. Deng, W. Wu, and T. Znati, *Data mining, systems analysis and optimization in biomedicine*, ch. Approximation algorithms of non-unique probes selection for biological target identification, pp. 174–184, American Institute of Physics, 2007.

[109] O. Ugur and G. W. Weber, *Optimization and dynamics of gene-environment networks with intervals*, Journal of Industrial and Management Optimization **3** (2007), no. 2, 357–379.

[110] B. Uthman, M. Bewernitz, C.-C. Liu, and G. Ghacibeh, *Data mining, systems analysis and optimization in biomedicine*, ch. Optimization of epilepsy treatment with vagus nerve stimulation, pp. 308–315, American Institute of Physics, 2007.

[111] V. Vapnik, *The nature of statistical learning theory*, Springer Verlag, New York, 1995.

[112] J. Wang and J.-D. Zucker, *Solving the multiple instance problem: A lazy learning approach*, Proceedings of the Seventeenth International Conference on Machine Learning (ICML-2000), 2000, pp. 1119–1125.

[113] S. Webb, *Optimization by simulated annealing of three-dimensional conformal treatment planning for radiation fields defined by a multileaf collimator*, Physics in Medicine and Biology **36** (1991), no. 9, 1201–1226.

[114] J. Weston, S. Mukherjee, O. Chapelle, M. Pontil, T. Poggio, and V. Vapnik, *Feature selection for SVMs*, NIPS, 2000, pp. 668–674.

[115] A. Wolf, J. B. Swift, H. L. Swinney, and J. A. Vastano, *Determining lyapunov exponents from a time series*, Physica D Nonlinear Phenomena **16** (1985), 285–317.

[116] E.P. Xing and R.M. Karp, *CLIFF: Clustering of high-dimensional microarray data via iterative feature filtering using normilized cuts*, Bioinformatics Discovery Note **17** (2001), 306–315.

[117] V. G. Zarnitsyn and M. R. Prausnitz, *Physical parameters influencing optimization of ultrasound-mediated DNA transfection*, Ultrasound in Med. & Biol. **30** (2004), no. 4, 527–538.

[118] Q. Zhang and S. Goldman, *EM-DD: An improved multiple-instance learning technique*, Neural Information Processing Systems, vol. 14, 2001.

Fields Institute Communications
Volume **55**, 2009

Software Development for Global Optimization

János D. Pintér
Pintér Consulting Services, Inc.
129 Glenforest Drive
Halifax, Nova Scotia, B3M 1J2 Canada
and
Bilkent University, Turkey
janos.d.pinter@gmail.com, www.pinterconsulting.com

Abstract. The theory and methods of finding the best possible solution in multi-extremal optimization models have become of growing interest in recent decades. The key theoretical results regarding important model types and basic algorithmic frameworks have been followed by software implementations that are now used to handle an increasing variety of applications. This work is focused on the practice of global optimization: key solution approaches, modeling environments, software implementations, and numerical examples are highlighted, with further pointers towards real-world applications.

1 Global optimization: problem statement and solution strategies

1.1 Introduction. Nonlinear phenomena and processes play a fundamental role in the development of great many natural and man-made systems. Consequently, nonlinear system models are often a key component of quantitative scientific studies. For related discussions that amply illustrate this point, consult Bracken and McCormick (1968), Rich (1973), Mandelbrot (1983), Murray (1983), Beck (1985), Casti (1990), Hansen and Jørgensen (1991), Schroeder (1991), Aris (1999), Gershenfeld (1999), Jacob (2001), Schittkowski (2002), Wolfram (2002), Diwekar (2003), Stojanovic (2003), Fritzson (2004), Bartholomew-Biggs (2005), Lopez (2005), as well as many other topical works.

Optimization models that incorporate a nonlinear system description frequently have multiple – local and global – optima. The objective of global optimization (GO) is to find the best possible solution under such circumstances. We shall consider the continuous global optimization (CGO) model defined by the following ingredients:

2000 *Mathematics Subject Classification.* Primary 65K05; Secondary 90C26, 90C31.
Key words and phrases. Nonlinear systems analysis and management; global optimization; exact and heuristic solution strategies; modeling environments and global solver implementations; illustrative examples; current applications and future perspectives.

- x decision vector, an element of the real Euclidean n-space $\mathbf{R}^n$
- ℓ, u explicit, finite n-vector bounds of x that define a "box" in $\mathbf{R}^n$
- $f(x)$ continuous objective function, $f : \mathbf{R}^n \to \mathbf{R}$
- $g(x)$ m-vector of continuous constraint functions, $g : \mathbf{R}^n \to \mathbf{R}^m$

Applying this notation, the CGO model can be concisely stated as follows.

$$\min f(x) \tag{1.1}$$

$$x \in D \qquad D := \{x : \ell \leq x \leq u \quad g(x) \leq 0\} \tag{1.2}$$

In (1.2) all vector inequalities are interpreted component-wise (ℓ, x, u, are n-component vectors and the zero denotes an m-component vector). The set of the additional constraints g could be empty, leading to a box-constrained GO model. Formally more general optimization models that include also $=$ and $\geq$ constraint relations and/or explicit lower bounds on the constraint function values can be simply reduced to the CGO model form (1.1)–(1.2).

The CGO model is very general: in fact, it evidently subsumes linear and convex nonlinear programming models, under corresponding additional specifications. It is also easy to see that CGO formally subsumes the entire class of pure integer (combinatorial) and mixed integer optimization problems. (This is evidently true, since bounded integer variables can be represented by a corresponding set of binary variables, and then every binary variable $y \in \{0,1\}$ can be equivalently represented by its continuous extension $0 \leq y \leq 1$, and the added non-convex constraint $y(1-y) \leq 0$.) Of course, this direct reformulation approach may not be best – or even suitable – for all (pure or mixed) integer optimization models: however, it certainly shows the generality of the model as stated above.

Ideally, we want to find all global solutions to CGO model. The more practical (numerical) objective of GO is to find suitable approximations of the set of global optima and the corresponding optimum value. This objective has to be attained by carrying out a finite number of search steps (model objective and constraint function evaluations) at algorithmically selected points.

Let us observe that if D is non-empty, then the above key analytical assumptions already guarantee (by the classical theorem of Weierstrass) that the optimal solution set X^* in the CGO model is non-empty. For reasons of numerical tractability, the following additional requirements are also often postulated: D is a full-dimensional subset ("body") in $\mathbf{R}^n$; the set X^* is at most countable; and f and g (component-wise) are Lipschitz-continuous functions on $[l, u]$. Without going into details, notice that the first of these assumptions (the set D is the closure of its non-empty interior) makes algorithmic search easier – or at all possible – within D. The second assumption supports theoretical convergence results (in most well-posed practical GO problems the set of global optimizers consists either of a single point x^* or at most of several points). The third assumption is a sufficient condition for estimating $f(x^*)$ on the basis of a finite set of generated feasible search points.

Based on these remarks, all models defined by continuously differentiable functions f and g belong to the CGO or even to the Lipschitz model class. As a consequence of this generality, the CGO model family includes also instances that are tremendously difficult to handle numerically. For instance, D could be disconnected, and even some of its component sets could be non-convex; furthermore, f could be highly multiextremal. Therefore CGO model-instances can have an unknown (possibly huge) number of global and local solutions. To illustrate this point, consider Trefethen's (2002) Problem 4 – a merely two-variable, box-constrained GO model – based on the formulation (1.3) shown below.

$$\min \frac{1}{4}(x_1^2 + x_2^2) + e^{\sin(50x_1)} + \sin(60e^{x_2}) + \sin(70\sin(x_1)) + \sin(\sin(80x_2))$$
$$- \sin(10x_1 + 10x_2) \qquad -3 \leq x_1 \leq 3, \ -3 \leq x_2 \leq 3. \tag{1.3}$$

Let us remark that Trefethen's original model statement did not include the variable bounds $[-3, 3]$: however, it is easy to deduct analytically that these bounds are valid for the objective function in (1.3), without missing the global optimum. Notice the quadratic term in (1.3) that will dominate all other terms as the Euclidean norm of the vector (x_1,x_2) grows: the function looks approximately quadratic "from a distance". However, a closer look reveals a massively multimodal structure: see Figure 1. This figure also hints at the fact that model complexity could increase exponentially as the model size, expressed by the number of variables and constraints, grows. One can also think of higher-dimensional extensions of (1.3), perhaps with added nonlinear constraints, to visualize the tremendous possible complexity of GO models.

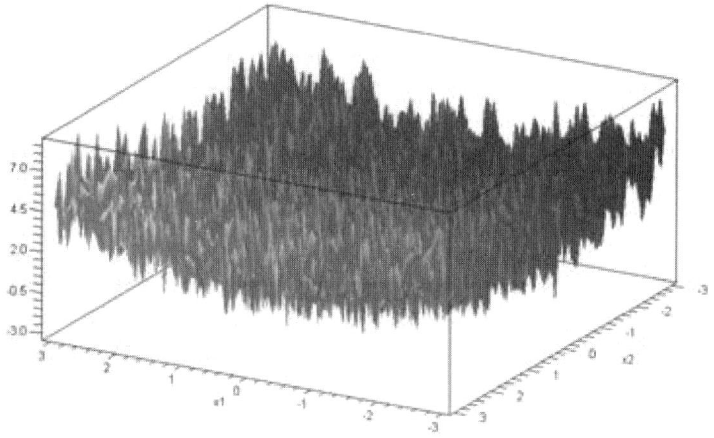

Figure 1 The objective function in Trefethen's Problem 4

This "simple" example implies that certain GO models could become very difficult to handle numerically. Let us also point out that a direct analytical solution approach is viable only in very special cases, since in general there is no straightforward algebraic characterization of global optimality. By contrast, in traditional local (nonlinear) optimization, most algorithms attempt to solve the Karush-Kuhn-Tucker system of (necessary) optimality conditions. The corresponding system of equations and inequalities in the present framework becomes another GO problem, often at least as difficult as the original model. To enumerate and investigate all KKT points of the model shown in Figure 1 is obviously a daunting task.

Arguably, not all GO models are as difficult as (1.3): however, in many cases, we don't have the possibility to directly inspect or estimate the numerical difficulty of a general nonlinear optimization model. A practically important case is when one has to optimize the parameters of a model that has been developed by someone else. The model may be confidential, or just complicated enough to prevent its easy analysis by inspection; it could even be presented to the optimization method (software) as a compiled (object, dynamic link library, shared library, or similar) module. In such situations, direct model inspection and structural verification are frequently not possible. In other practically relevant cases, the evaluation of the optimization model functions may require the numerical solution of a system of embedded differential and/or algebraic equations, the evaluation of special functions, integrals, the execution of certain deterministic computational procedures or stochastic simulation modules, and so on. Again, model structure analysis by inspection may not be an option in such cases. Hence, to verify model convexity could remain elusive, and one should attempt to solve such models as a GO problem – if aiming at the best possible solution is important.

Most traditional nonlinear optimization methods discussed in prominent textbooks such as Bazaraa, Sherali, and Shetty (1993), Bertsekas (1999), Chong and Zak (2001), Boyd and Vandenberghe (2004), Hillier and Lieberman (2005), Nocedal and Wright (2006) search only for local optima. This approach is based on the key assumption that a "sufficiently good" initial solution (that is located in the region of attraction of the "true" global solution) is easily available. An essentially convex model structure guarantees that local search indeed suffices; otherwise, to supply that good initial solution guess is essential. Needless to say, one should always take advantage of such a situation since it will enable the use of well-developed and efficient local optimization strategies and software. However, Figure 1 and the practical situations mentioned above imply that to postulate or to verify a convex or similarly "easy" model structure may not always be a realistic assumption. Nonlinear models with perhaps less dramatic difficulty, but in (much) higher dimensions and with added constraints also require a global optimization based approach. For instance, in various areas of engineering design optimization models with tens, hundreds, thousands, or more variables and constraints are analyzed and need to be solved. In such cases, even an approximately executed, but genuine global scope search strategy could – and typically will – yield better results than the most sophisticated local optimization approach "started from the wrong valley"... (Again, recall Figure 1 to illustrate this point.)

The mentioned facts have motivated the design of GO algorithms, with a view towards their computer-based implementations. In this article, we only highlight several broad families of exact and heuristic strategies that are currently considered most promising: further details regarding these are found in related references.

1.2 Exact GO methods. Exact methods possess proven deterministic or stochastic global convergence properties. That is, if such a search method could be carried out completely as an infinite iterative process, then the generated limit point(s) would belong to the set of global solutions X^*. In the case of stochastic GO methods, the above convergence statement is valid "only" with probability one. In practice – after a finite number of search steps – one can only expect a numerically validated (deterministic) or estimated (stochastic) lower bound for the global optimum value $f(x^*)$, as well as a best feasible or near-feasible global solution estimate to approximate one (or several) element(s) of X^*. The following list summarizes the most prominent exact strategies.

- Branch-and-bound methods, including interval arithmetic based strategies
- Homotopy, path following, deformation, continuation, trajectory, and similar methods

- Implicit enumeration techniques (e.g., vertex enumeration in concave minimization)
- Passive set covering approaches such as grid search, quasi-random sequences, pure random search, and their variants
- Stochastically convergent methods: adaptive random searches, Bayesian optimization strategies, and their combinations

For detailed expositions related to exact deterministic GO strategies, consult, e.g., Horst and Tuy (1990), Horst and Pardalos (1995), Kearfott (1996), Pintér (1996a), Pardalos and Romeijn (2002), Tawarmalani and Sahinidis (2002), Neumaier (2004), and Nowak (2005). Regarding stochastic GO strategies, see, e.g., Zhigljavsky (1991), Boender and Romeijn (1995), Pintér (1996a), and Zabinsky (2003).

1.3 Heuristic GO methods. Heuristic methods typically do not have provable convergence guarantees. However, they have a promising – even if sometimes over-estimated – track record, and may offer practical tools to handle models that are (at least currently) out of the reach of theoretically correct, rigorous methodology. These features make such approaches popular among practitioners, especially in the non-mathematician (engineering and scientist) community. Let us point out at the same time that many of the heuristic methods are "only" meta-heuristics: that is, they may need considerable problem-specific tailoring to become competitively efficient. In addition, there are GO model classes that truly need a formal optimization based approach (or component). This fact is in contrast with some of the overly optimistic claims regarding the "universal" applicability of certain heuristics approaches. The following list summarizes the most prominent heuristic strategies.

- Ant colony optimization
- Basin-hopping strategies
- Convex underestimation heuristics
- Evolutionary search methods
- Genetic algorithms
- Greedy adaptive search strategies
- Memetic algorithms
- Neural networks
- Response surface and surrogate model based methods
- Scatter search
- Simulated annealing
- Tabu search
- Tunneling, filled function methods, and similar approaches

For discussions of these methods, consult, e.g., Goldberg (1989), Michalewicz (1996), Osman and Kelly (1996), Glover and Laguna (1997), Rudolph (1997), Voss, Martello, Osman and Roucairol (1999), Edgar, Himmelblau and Lasdon (2001), Ferreira (2002), Rothlauf (2002), Jones and Pevzner (2004). The topical chapters in Pardalos and Resende (2002) also offer expositions related to both exact and heuristic GO approaches. For further discussions regarding GO software development, consult, e.g., Pintér (2002) and Liberti (2006), in addition to the cited topical references.

This section is concluded by proposing some general guidelines that can also drive practically useful GO software development. Exact deterministic algorithms offer a rigorous guarantee for finding at least one or all global solutions: however, the associated computational burden can become excessive (theoretically increasing at an exponential rate, as the model size increases). Therefore even the seemingly "unlimited" increase of computational power will not resolve the genuine intractability of arbitrary size GO models, in a strict deterministic sense. For this reason, in higher dimensions and/or in the presence of complicated constraints, and without an exploitable special model structure, there is perhaps more practical hope in exact stochastic algorithms, and in intelligent heuristic methods which have a well-designed stochastic global search component. This is not to say that exact deterministic methods can not be combined with the other methods: in fact, we are convinced that combining various (perhaps radically different) strategies can help to find high-quality solutions to difficult GO models. Given the expectable computational demand of all exact global scope algorithms (if these are used stand-alone), such methods should be carefully combined with efficient local optimization procedures whenever this is possible. Global convergence, however, can be guaranteed only by the global scope component algorithm. This controversy between theoretical rigor and numerical efficiency poses an inherent challenge in the development of theoretically sound, robust, and efficient GO software.

2 Software development: User needs and current implementations

In solving (numerically) an optimization model, a key aspect to consider is the choice of software. Ideally, we want to use "the most suitable" software to solve the problem at hand. However, the word "suitability" aggregates different criteria for different people.

For example, a theoretically motivated GO study invites rigorous global optimization methodology that guarantees the accuracy of the solution obtained, while perhaps paying only secondary (or no) attention to time and resource demands and efficiency issues. A good – and fully meritorious – example to illustrate this approach is the recent work of Markót and Csendes (2005): they provide tight, rigorous enclosures for the optimal configuration of 28, 29, and 30 identical circles in the unit square. The associated computational work and resource needs can be very significant, however. As Csendes (2007) states, "Standard interval optimization algorithms could not solve the problems. A careful problem decomposition, and custom made built in acceleration devices based on the understanding of the problem structure enabled the successful solution. The running times were below 3 days, and ca. one million subintervals were stored during the solution process. The uncertainty in the position of the circles has been decreased by more than 700 orders of magnitude in each case. Due to the controlled outward rounding mode applied, the obtained results are reliable even in the sense of a rigorous mathematical theorem."

In many other cases – typically motivated by practical considerations and requirements – software users (engineers and research scientists) place the key emphasis on the "acceptable" execution speed of software. In other words, they want "only" a quickly obtained high-quality global search based solution, even though the solution found is not rigorously guaranteed to be globally optimal. In professional (commercial) environments, acceptance by the organization, compatibility with the hardware and operating system platform as well as with other software, ease of use, quality of software documentation and user support may dominate all other (perhaps more objective and technical) aspects. Hence, the choice of "best" tools greatly depends on specific user criteria: the modeling environments and software revised below evidently attempt to serve a broad range of users and their optimization needs.

To start with the "lowest level" (core) optimization software implementations, there exists a range of compiler platform-based systems, with more or less built-in model development functionality. Typically, the core solver engine is developed in C or Fortran, but the user interface and other features are implemented using one of C++, C#, Delphi, Java, Visual Basic, etc. Such solver implementations are suitable to build directly into various decision support systems and applications, due to their customizability and small hardware + software "footprint". At the same time, these implementations may require the most expertise to use – unless set up as a fully or nearly automated "push button" solver option.

In the context of operations research, management science, industrial engineering and related disciplines, the core solvers referred to above are frequently linked to algebraic (optimization) modeling languages. Examples of prominent modeling systems include AIMMS (Paragon Decision Technology, 2006), AMPL (Fourer, Gay, and Kernighan, 1993), the Excel Premium Solver Platform (Frontline Systems, 2006), GAMS (Brooke, Kendrick, and Meeraus, 1988), ILOG OPL Studio (ILOG, 2004), the LINDO Solver Suite (LINDO Systems, 2006), and MPL (Maximal Software, 2006). For current information, visit the corresponding websites listed in the references; see also Kallrath (2004).

There has been also remarkable development in relation to integrated scientific and technical computing (ISTC) systems such as Maple (Maplesoft, 2006), *Mathematica* (Wolfram Research, 2006), Mathcad (Mathsoft, 2006) and MATLAB (The MathWorks, 2006). These systems offer also optimization features, either as built-in functions or as add-on products.

The modeling environments listed above meet the requirements and needs of different types of users. Key categories include educational users (instructors and students); research scientists, engineers, consultants, and other practitioners (possibly, but not necessarily equipped with an in-depth optimization related background); optimization experts, software application developers, and other "power users". (These user categories are not necessarily disjoint.) The pros and cons of the individual software products – in terms of hardware and software platform demands, ease of use, model prototyping options, detailed code development and maintenance features, optimization model checking and processing tools, availability of solver options and other auxiliary tools, program execution speed, overall level of system integration, quality of related documentation and technical support, customization options, and communication with end users – make the corresponding modeling and solver approaches more or less attractive for the various user groups.

Next we present a concise summary of the currently available modeling platforms and solvers for doing global optimization. The early software review (Pintér, 1996b; available also at the website of Mittelmann, 2007) listed a few dozens of individual software products, with further website links. Neumaier's (2007) webpage lists more than 100 software development projects; see also Fourer (2007) and Mittelmann (2007). These reviews include general purpose solvers, as well as application-specific products. (A word of warning: quite a few of the software links are now obsolete, or have been changed; and there are also quite a few other software development projects that are not included in these reviews.)

The user's preference obviously depends on another key factor: whether one prefers to use "free" (non-commercial, research, or even open source) code, or looks for a "ready to use" professionally supported commercial product. There is a significant body of freely available GO solvers, although their quality arguably varies. (Of course, this remark could well apply also to commercial products.) Instead of trying to impose personal judgment, the reader is encouraged to do some web browsing and experimentation, as his/her time and resources allow. Both Mittelmann and Neumaier provide far more extensive information on non-commercial systems. Therefore we will focus here on software products that are part of commercial systems, typically as an add-on option, but in some cases as a built-in option. We list in alphabetic order the currently available products that are explicitly targeted towards global optimization, as stated by the corresponding vendors. For this reason, stand-alone local solvers are not listed here; furthermore, we will not mention modeling environments that currently have no global solver options. Most of the listed solvers implement exact deterministic or stochastic methods, but there are also a few heuristic solvers. (The software package list is provided to our best knowledge: further information is always welcome.)

AIMMS, by Paragon Decision Technology (www.aimms.com): the BARON and LGO global solvers are offered with this modeling system as add-on options. AIMMS also has an Outer Approximation (AOA) module to assist the solution of mixed integer nonlinear programming (MINLP) problems.

The **Excel Premium Solver Platform (PSP)** is developed by Frontline Systems (www.solver.com). The PSP offers a stochastic global pre-solver option that can be used with several local optimization engines: these currently include KNITRO, LSGRG, and LSSQP. Frontline Systems also offers an Interval Global Solver, an Evolutionary Solver, and OptQuest as genuine global solvers.

GAMS, by the GAMS Development Corporation (www.gams.com): AlphaECP, BARON, DICOPT, LGO, LINDOGlobal, MSNLP, OQNLP, SBB are offered as solver options that are aimed at the solution of continuous global optimization and/or MINLP models.

LINDO, by LINDO Systems (www.lindo.com): both the LINGO modeling environment and What'sBest! (the company's spreadsheet solver) have built-in global solver functionality.

Maple, by Maplesoft (www.maplesoft.com) offers the Global Optimization Toolbox as an add-on product.

Mathematica, by Wolfram Research (www.wolfram.com) has a built-in function (called NMinimize) for numerical global optimization. In addition, there are several third-party GO packages that can be directly linked to Mathematica: these are Global Optimization, MathOptimizer, and MathOptimizer Professional.

MPL, by Maximal Software (www.maximal-usa.com): the LGO solver engine is offered as an add-on.

TOMLAB, by TOMLAB Optimization AB (www.tomopt.com) is an optimization platform for solving MATLAB models. The TOMLAB global solvers currently include CGO, LGO, MINLP, and OQNLP. Let us note here that MATLAB's own Genetic Algorithm and Direct Search Toolboxes also have heuristic global solver capabilities.

For further up-to-date information on these GO solver products, visit the websites shown above. In addition, e.g., recent works by Neumaier (2004), Nowak

(2005), Liberti and Maculan (2006), Li and Sun (2006), Csendes (2007) also discuss several notable software development projects: to our best knowledge, these are not publicly available commercial products, however.

3 LGO solver suite for global and local nonlinear optimization

In order to illustrate the functionality and actual use of global optimization software, we will briefly review (and then use) the Lipschitz Global Optimizer (LGO) solver package. LGO has been continuously developed since the late 1980s, and offered as a professional software package since the mid-nineties. It is now available for a range of compiler platforms, modeling languages, and ISTCs.

3.1 Key features. The overall design of LGO is based on the combination of theoretically convergent global and efficient local search strategies. Currently, LGO offers the following standard solver options:

- Adaptive partition and search (branch-and-bound) based global search (BB)
- Adaptive global random search (single-start) (GARS)
- Adaptive global random search (multi-start) (MS)
- Constrained local search by the generalized reduced gradient (GRG) method (LS)

In a typical global optimization run, the user selects one of the global (BB, GARS, MS) solver options; the corresponding global search phase is automatically followed by the local search option. It is also possible to apply only the LS option, making use of an automatically set (default) or a user-supplied initial solution point.

The global search methodology implemented in LGO is based on the detailed exposition in (Pintér, 1996a), with many added numerical features. The well-known GRG method is discussed in numerous articles and textbooks, consult for instance Edgar, Himmelblau, and Lasdon (2001). Therefore only a brief overview of these solver components is provided here.

BB, GARS, and MS are all based on globally convergent search methods. Specifically, in the case of Lipschitz-continuous models (assuming the availability of suitable Lipschitz-constant overestimates for all model functions) BB theoretically generates a sequence of search points that will converge to the global solution point. If there is a countable set of such optimal points, then a convergent search point sequence will be generated in association with each of these.

For a GO model with a continuous structure (but without postulating access to valid Lipschitz information), both GARS and MS are theoretically globally convergent, with probability one (w.p.1). In other words, the sequence of points that is associated with the generated sequence of global optimum estimates will converge to a point which belongs to X^*, w.p.1. (If several such convergent point sequences are generated by a suitable stochastic search procedure, then each of these sequences has a corresponding limit point in X^*, w.p.1.)

The LS method (GRG) is aimed at finding a locally optimal solution that satisfies the Karush-Kuhn-Tucker system of necessary local optimality conditions, assuming standard model smoothness and regularity conditions.

In all three LGO global search modes the model functions are aggregated by an exact penalty (merit) function. In the local search phase all model functions are considered and handled individually. The global search phases incorporate both deterministic and stochastic sampling procedures: the latter support the use of statistical bound estimation methods, under basic continuity assumptions. In numerical practice – time-limited runs, with user-defined or default option settings – the LGO global solver options generate a global solution estimate that is refined by the seamlessly following local search mode. If the LS mode is used without a preceding global search phase, then LGO serves as a general purpose local solver.

All LGO component algorithms require only model function return values $f(x)$ and $g(x)$, at each algorithmically chosen input vector x. The global search options BB, GARS, and MS are genuinely based on direct sampling information; in the LS phase central differences are used to approximate function gradients (assuming a locally smooth model structure). This approach reflects our design objective to handle (also) models that are defined by merely computable, continuous functions, including completely "black box" systems. This design is markedly different from GO software packages that require explicit (analytical) model information that supports model parsing and model function component specific operations (such as function specific bounding procedures). Let us emphasize that we do not claim that one design is superior to the other. The model function decomposition and parser based approach supports the efficient and precise solution of a range of GO models which are defined by a predefined set of component functions. However, the design of LGO allows the handling of problems that will remain outside of the scope of the analytical model decomposition based approach.

The expected practical result of using LGO – barring numerical problems caused by a "tricky" multimodal objective and/or feasible set, poor model scaling, ill-conditioning, etc. (which could impede any nonlinear optimization method) – is a global search based high-quality feasible solution that meets at least the local optimality conditions. Extensive numerical tests and a range of practical applications demonstrate that LGO can find a close numerical approximation of the global solution not only when using the standard (simple and arguably over-used) academic test problems, but also in more complicated, sizeable GO models. This point will be illustrated in the next section. At the same time, one should keep in mind that no global solver will "always" work satisfactorily, with default settings and under resource limitations (time, model function evaluation, or other preset limits).

3.2 LGO implementations. The current platform-specific implementations include the following:

- LGO with a calling (driver) program option, or as a stand-alone executable
- LGO with a text input/output interface, for C and Fortran compiler platforms
- LGO integrated development environment with a Microsoft Windows style menu interface, for C and Fortran compiler platforms
- AIMMS /LGO solver engine
- AMPL/LGO solver engine
- GAMS /LGO solver engine
- Global Optimization Toolbox for Maple (LGO solver engine with a link to Maple)
- MathOptimizer Professional (LGO solver engine with a link to Mathematica)
- MPL /LGO solver engine
- TOMLAB /LGO, for MATLAB users

Technical descriptions of these software implementations – including detailed numerical tests and a range of applications – have appeared elsewhere: consult Pintér (1996a, 1997, 2001a,b, 2003, 2005, 2006a,b, 2007a,b), as well as Stortelder, de Swart, and Pintér (2001), Isenor, Pintér and Cada (2003), Tervo et al. (2003), Pintér and Kampas (2003, 2005), Pintér, Holmström, Göran and Edvall (2004), Kampas and Pintér (2006), Pintér, Linder, and Chin (2006), Castillo, Kampas, and Pintér (2008).

The compiler platform based LGO solver suite can be used in stand-alone mode, and also as a solver option in decision support systems and other customized applications. In its text input/output based) implementation version, LGO reads an input text file that contains application-specific (model descriptor) information, as well as several solver options (such as the global solver type, precision settings, resource and time limits). During the program run, LGO iteratively calls the application-specific model function file that returns function values for the algorithmically chosen sequence of arguments. Upon completing the LGO run, automatically (optionally) generated summary and detailed report files are available. As it can be expected, this LGO implementation version has the lowest hardware demands, and it runs fastest. This version can be directly embedded into various decision support systems, including proprietary user applications. The same core LGO system is also available in directly callable form, without reading and writing text files: this version is frequently used as a built-in solver module in other customized modeling and optimization systems. LGO can also be equipped – as a readily available option – with a Windows graphical user interface: this version is referred to as the LGO Integrated Development Environment (IDE). The LGO IDE supports model development, compilation, linking, execution, and the inspection of results, together with built-in basic help facilities.

In the two LGO implementations mentioned above, models can be connected to LGO using any one of several programming languages that are available on personal computers and workstations. Currently supported platforms include, in principle, "all" professional Fortran 77/90/95 and C/C++ compilers. Examples of supported compilers include Compag, Intel, Lahey, and Salford Fortran, as well as g77 and g95, and Borland and Microsoft C/C++ for Windows, and gcc/g77, Intel Fortran for Linux. Other LGO versions (to use with other compilers or software applications) are made available upon request.

In the optimization modeling language (AIMMS, AMPL, GAMS, and MPL) or ISTC (Maple, Mathematica, and TOMLAB) environments the core LGO solver engine is linked to the corresponding modeling platform, as a dynamically callable or shared library, or as an executable program. The key advantage of using LGO within a modeling or ISTC environment is the combination of modeling system specific features – model prototyping and detailed development, model consistency checking, integrated documentation, visualization, and other platform-specific features – with a numerical performance comparable to that of the stand-alone LGO solver suite.

Let us mention that a size limited demo version of LGO serves to illustrate global optimization software – in connection with a demo version of the MPL modeling system – in the prominent O.R. textbook by Hillier and Lieberman (2005). Recently, we also have made available two stand-alone demo versions of LGO that can be used in conjunction with the C# or the Tiny C (freely available) compilers:

this work was done in cooperation with Frank Kampas who developed the interface links to LGO.

4 Illustrative applications

4.1 Trefethen's Problem 4. This model (recall the problem statement (1.3) and Figure 1) is one of the standard test problems set up to test current LGO implementations. Here we cite the result obtained by LGO with a text I/O interface. The MS+LS search option is used, and the number of global search phase function evaluations is limited by 10^5. These are direct function evaluations, without gradient or higher order calculations: hence, they are rather quick unless the model function evaluations themselves become expensive.

The global optimum value reported by LGO equals -3.3068686475: this is identical to all 10 decimal digits (standard LGO output precision) to the value reported by Trefethen at web.comlab.ox.ac.uk/oucl/work/nick.trefethen/hundred.html and elsewhere. The corresponding LGO runtime is 0.16 seconds (for a little less than 10^5 function evaluations, due to other stopping criteria) on an AMD Athlon 64 3,200+ 2GHz processor based desktop machine, running under the Windows XP (Version 2002, Service Pack 2) operating system. In this numerical example, LGO is used in conjunction with the Lahey Fortran 95 (LF95) compiler.

4.2 Electrical circuit design problem. Following the classical study of Ebers and Moll (1954), a bipolar transistor can be modeled by an electrical circuit. This model leads to a square system of highly nonlinear equations in 9 variables that has been studied by numerous researchers, in attempts to solve it, and then to prove the correctness of the suggested solution. The system of equations is defined by

$$a_k(x) = 0 \quad k = 1, \ldots, 4; \quad b_k(x) = 0 \quad k = 1, \ldots, 4; \quad c(x) = 0. \tag{4.1}$$

Here (using for better readability the $\exp(x)$ notation for e^x)

$$\begin{aligned} a_k(x) &= (1 - x_1x_2)x_3\{\exp[x_5(g_{1k} - g_{3k}x_7 - g_{5k}x_8)] - 1\} \\ &\qquad -g_{5k} + g_{4k}x_2 \quad k = 1, \ldots, 4; \\ b_k(x) &= (1 - x_2x_2)x_4\{\exp[x_6(g_{1k} - g_{2k} - g_{3k}x_7 + g_{4k}x_9)] - 1\} \\ &\qquad -g_{5k}x_1 + g_{4k} \quad k = 1, \ldots, 4; \\ c(x) &= x_1x_3 - x_2x_4. \end{aligned}$$

This is a well-known (and arguably rather difficult) global optimization test problem, frequently cited in the literature. Consult, for example, Dimmer and Cutteridge (1980); for a more recent detailed discussion and historical notes, see Ratschek and Rokne (1993).

Using the standard model formulation, by assumption the vector variable belongs to the box region $[0, 10]^9$. The numerical values of the constants g_{ik} $i = 1, \ldots, 5$, $k = 1, \ldots, 4$ are listed in the paper of Ratschek and Rokne, and will not be repeated here. (In order to make the model functions more readable, several constants are simply aggregated in the above formulae, when compared to that paper.) To solve this model rigorously, Ratschek and Rokne applied a combination of interval arithmetic, subdivision and branch-and-bound strategies. Their (expert) solution process required billions of function evaluations and several months of coordinated computational effort, using a sizeable farm of Sparc 1 workstations (state-of-the-art hardware at the time). Their study resulted in a guaranteed interval estimate that is component-wise within at least 10^{-4} precision of the postulated (unique, approximate) global solution:

$$x* = (0.9, 0.45, 1.0, 2.0, 8.0, 8.0, 5.0, 1.0, 2.0).$$

By taking, e.g., the Euclidean norm of the overall error in the equations, the problem of finding the global solution can be formulated as a standard GO model. Alternatively, one can use a dummy (constant valued) objective function and then just solve the system of nonlinear equation constraints. Setting up and solving this model in the latter fashion (again, using the core LGO implementation with the LF95 compiler), the numerical solution found by LGO – directly imported from the answer report – is shown below.

$$\begin{aligned} x^* \; = \; & (0.8999997968, 0.4500000151, 0.9999998620, 1.9999992055, 7.9999999686, \\ & 8.0000025641, 4.9999999577, 1.0000000010, 1.9999992108). \end{aligned}$$

All constraint equations (4.1) are satisfied by this solution, at least with 10^{-10} precision. The LGO solution time is 1.31 seconds. As this example shows, one can produce a high-precision numerical solution in a matter of seconds using LGO, using a few hundred thousand function evaluations. Here LGO is used again in its default MS+LS search mode, and the solution is already found around 122000 function evaluations, but the search continues as per the given option settings.

Please note (again) that our result is not to be interpreted as a dismissive comparison to rigorous global optimization approaches. Instead, we wish to illustrate that a careful combination of global and local search strategies (as implemented in LGO) can also lead to high-quality solutions to difficult global optimization problems, within a practically affordable timeframe.

4.3 Circle and sphere packing examples. In close cooperation with Frank Kampas, we have been using various circle, sphere and other packing problems to test our MathOptimizer Professional (LGO linked to Mathematica) software package (Pintér and Kampas, 2003) as well as the core LGO software implementation. In the examples reviewed here, we seek high-quality (in theory, globally optimized) solutions to the problem of finding a minimal size circle (or sphere) that contains a given collection of non-uniform size circles (spheres) in a tight, but non-overlapping arrangement.

Let us remark that such general packing problems have been considered much less frequently than the packing of identical circles, and that we are not aware of other global optimization studies regarding the packing a general collection of spheres. Al Zimmermann's recent non-uniform size circle packing contest (see http://recmath.org/contest/CirclePacking) has received considerable attention: 27490 putative solutions were submitted. The (GO expert) winning team of Addis, Locatelli and Schoen (2006) applied a mixture of global and local optimization strategies with random search and local moves.

In our recent circle and sphere packing study (Kampas and Pintér, 2007) first we perform an initial global optimization procedure, and then we use various local search techniques to improve the initial solution. These local search techniques involve swapping the position of the circles or spheres in order to decrease the radius of the enclosing circle or sphere. The initial global optimization and all local searches are performed by MathOptimizer Professional (MOP), taking advantage of its ability to reuse machine compiled code for model function evaluations. The reuse of machine compiled code enables the circle or sphere swapping to be performed several thousand times faster than the equivalent calculation performed in Mathematica, using its built-in FindMinimum function. MathOptimizer Professional has been described in other works that review its implementation details, and the solution of a large variety of test problems. The latter also include the solution of various circle packing problems and their potential industrial applications. For additonal details not discussed here, consult Kampas and Pintér (2004, 2006), Pintér and Kampas (2003, 2005), Castillo, Kampas and Pintér (2007).

In the first example, we consider 10 circles with given radii $r_i = i^{-1/2}$ $i = 1, \ldots, 10$. If the circumscribed circle has its center at the origin, then the objective function (circumscribing radius r) value equals the maximum outer distance of the list of circles with radii r_i. The variables to optimize are the positions c_i of the circle centers within the circumscribed circle. Considering also the pair-wise "no overlap" condition, the general 10-circle packing global optimization problem is stated as

$$\begin{aligned} &\text{minimize } r \\ &r_i + \|c_i\| \le r \qquad i = 1, \ldots, 10 \\ &r_i + r_j \le \|c_i - c_j\| \qquad 1 \le i < j \le 10. \end{aligned} \tag{4.2}$$

Notice that the number of model variables increases linearly with the number of circles considered, but the number of non-convex ("no overlap") constraints grows quadratically. This way, even moderate size model instances lead to non-trivial GO test problems. In the 10-circle instance, the number of decision variables is 21, and the number of constraints is 55 (of which 45 are non-convex). The global solution estimate (minimized radius of the circumscribed circle) found in our study is $r = 1.8259562513496725$; the corresponding circle configuration is shown by Figure 2.

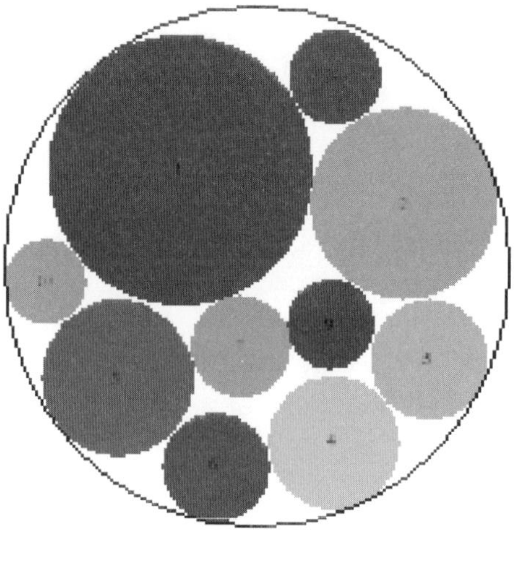

Figure 2 The best 10-circle configuration found

The runtime for the initial (generic) global optimization is less than a second. The subsequently applied local search strategies take from less than a second (single swaps) to a few minutes (multiple swaps) for each search strategy applied. The direct total runtime of the entire study is less than an hour on a desktop PC.

Let us consider now the three-dimensional variant of the same model. Given 10 spheres with radii $r_i = i^{-1/3}$ for $i = 1, \ldots, 10$, we want to find the smallest circumscribed sphere that contains their non-overlapping arrangement. Under the conditions defined above, our model is identical to (4.2), except that now the Euclidean norm in $\mathbf{R}^3$ is used. This model has 31 decision variables and 55 constraints. The best result (with minimal radius of the embedding sphere) found is $r = 1.9148313247862758$; the corresponding sphere configuration is shown by Figure 3.

Figure 3 The best 10-sphere configuration found

The runtime to execute the initial global search is again less than a second, and the individually tested swapping strategies take from a second to a few minutes. The total net runtime of using all strategies (that we tried) is still less than an hour. In our numerical study, Mathematica version 6, and the corresponding current MOP version has been used, in conjunction with the Salford Fortran (FTN95) compiler. Further details, including a discussion of the strategies used and an extensive set of results, will appear elsewhere.

4.4 Global optimization with embedded computable functions: A parametric integral example. In the last numerical example presented here, we will use the Global Optimization Toolbox (GOT) for Maple: this is the LGO implementation linked to Maple (Maplesoft, 2004). Let us consider the parametric integral expression $int(a)$ defined by

$$int(a) = 2 \int_{-1.5a}^{2a} [0.01x + \sin(5x)\cos(3x) - \sin(3x)\cos(2x) + \sin(x)\cos(7x)]dx. \quad (4.3)$$

In other words, for every given parameter value a, we can calculate the corresponding value of the integral shown in (4.3): for example, $int(-2) \sim 1.050870659$, and *int*(9) $\sim$ 1.129267105. For our present discussion, it is secondary how the integration itself is done (whether analytically or numerically). We only wish to highlight a combined feature of a computing system (Maple) that serves to evaluate a "black box" function and of a global solver (LGO) that is able to operate directly based on such information. Our illustrative problem is the following: find the parameter value a that minimizes the integral in the parameter range $-3 \leq a \leq 10$. Using the GOT with Maple, this model can be stated very concisely: first we define *int*(a) by (4.3), and then we can essentially follow the GO model statement

$$\min int(a) \qquad -3 \leq a \leq 10. \tag{4.4}$$

Applying the GOT with its default settings in this implementation, we obtain the numerical solution $a^* = 4.92574563473295957$, $int(a^*) = -1.86463327469610008$. The runtime is 15 seconds, using 1351 model function (integral) evaluations: the corresponding program run has been done on the desktop machine mentioned earlier. Figure 4 shows the parametric integral values in the range $[-3, 10]$, and visually confirms the numerical global optimality of our result.

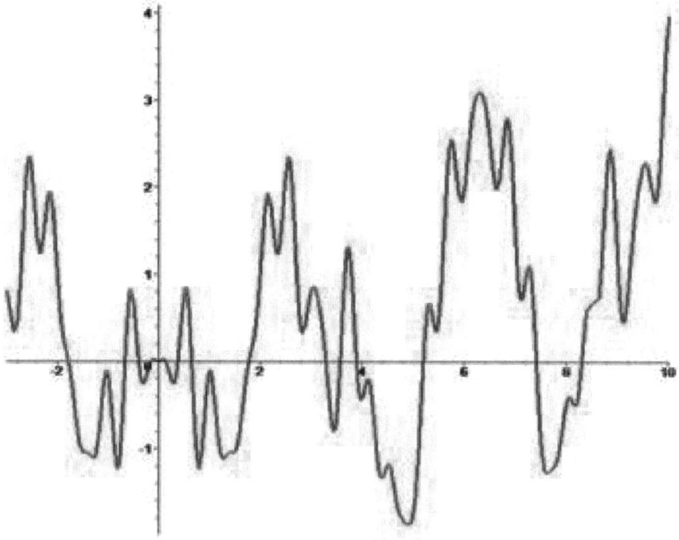

Figure 4 The optimization model defined by (4.3) and (4.4)

5 Application areas

Global optimization is used across many fields in the sciences and engineering. Without going into details, here we only list a number of prominent and broad application areas. Many of the application areas listed below are based on information received from users of LGO (in its various implementations); some others are cited from the rapidly growing GO literature.

- Acoustics
- Aerospace engineering
- Agro-ecosystem management
- Animal husbandry operations management

- Assembly line design
- Automotive (system component) design
- Cancer therapy planning
- Chemical and process engineering
- Computational chemistry and molecular biology: molecular structure and protein folding models
- Computational finance, financial engineering
- Computational fluid mechanics
- Data classification, pattern recognition
- Dynamic resource management problems
- Electrical engineering (nonlinear circuits and other design problems)
- Energy production and distribution systems
- Experimental design
- Facility layout, location and related problems
- Finite element methods
- Identification of dynamical systems
- Image reconstruction
- Laser equipment design
- Manufacturing, product design
- Medical and pharmaceutical research
- Mechanical engineering design
- Mobile network design
- Model fitting to data (calibration, nonlinear regression, inverse problems, parameter estimation)
- Object configuration design, including stock cutting and packing problems
- Optimal control problems
- Risk analysis and management
- Robotics
- Sensor networks and telecommunication
- Simulation and optimization
- Stochastic (non-convex) optimization problems
- Supply chain reliability management
- Systems of nonlinear equations and inequalities
- VLSI design

For detailed discussions of a large variety of GO applications and perspectives, consult, e.g., Mockus, Eddy, Mockus, Mockus and Reklaitis (1996), Grossmann (1996), Pardalos, Shalloway and Xue (1996), Pintér (1996a), Bomze, Csendes, Horst and Pardalos (1997), Migdalas, Pardalos and Värbrand (1997), Mistakidis and Stavroulakis (1997), Hendrix (1998), Corliss and Kearfott (1999), Floudas et al. (1999), Floudas and Pardalos (2000), Papalambros and Wilde (2000), Atkinson, Bogacka and Zhigljavsky (2001), Edgar, Himmelblau and Lasdon (2001), Gao, Ogden and Stavroulakis (2001), Pintér (2001), Pardalos and Resende (2002), Tawarmalani and Sahinidis (2002), Diwekar (2003), Zabinsky (2003), Neumaier (2004), Bartholomew-Biggs (2005), Nowak (2005), Liberti and Maculan (2006), Li and Sun (2006), and Pintér (2006a, 2006b). In addition to these (and other) books typically with a strong GO content, science, engineering, economics and finance journals and books often publish work that directly calls for GO methodology.

As of today, global optimization software is used around the world at industrial and consulting organizations, research and educational institutions. The number of such institutions is arguably in the thousands, and the range and depth of GO applications is steadily increasing. Without doubt, the future of global optimization will get even brighter.

Acknowledgements

I wish to thank Panos Pardalos for his kind invitation to contribute to this volume. The review presented in this work has benefited from information provided by colleagues, collected from books, articles, and Web sites, as reflected also by the list of references. The growing worldwide user base of the LGO software implementations – and the feedback received from these users – has motivated my views on the practice of global optimization. I also wish to thank Frank Kampas and a large number of other colleagues (as shown also by the co-authored references) for their cooperation and contributions to joint software development projects.

References

[1] Addis, B., Locatelli, M., and Schoen, F. (2006) Efficiently packing unequal disks in a circle: A computational approach which exploits the continuous and combinatorial structure of the problem. Preprint available at www.optimization-online.org/DB_FILE/2006/03/1343.pdf. Submitted for publication,

[2] Aris, R. (1999) *Mathematical Modeling: A Chemical Engineer's Perspective.* Academic Press, San Diego, CA.

[3] Atkinson, A., Bogacka, B. and Zhigljavsky, A.A., Eds. (2001) *Optimum Design 2000.* Kluwer Academic Publishers, Dordrecht.

[4] Bartholomew-Biggs, M. (2005) *Nonlinear Optimization with Financial Applications.* Kluwer Academic Publishers, Dordrecht.

[5] Bazaraa, M.S., Sherali, H.D. and Shetty, C.M. (1993) *Nonlinear Programming: Theory and Algorithms.* Wiley, New York.

[6] Beck, M.B. (1985) *Water Quality Management: A Review of the Development and Application of Mathematical Models.* Springer, Berlin / Heidelberg / New York.

[7] Bertsekas, D.P. (1999) *Nonlinear Programming.* (2nd Edition) Athena Scientific, Cambridge, MA.

[8] Boender, C.G.E. and Romeijn, H.E. (1995) Stochastic methods. In: Horst and Pardalos, Eds. *Handbook of Global Optimization. Volume 1*, pp. 829-869. Kluwer Academic Publishers, Dordrecht.

[9] Bomze, I.M., Csendes, T., Horst, R., and Pardalos, P.M., Eds. (1997) *Developments in Global Optimization.* Kluwer Academic Publishers, Dordrecht

[10] Boyd, S. and Vandenberghe, L. (2004) *Convex Optimization.* Cambridge University Press, Cambridge, UK.

[11] Bracken, J. and McCormick, G.P. (1968) *Selected Applications of Nonlinear Programming.* Wiley, New York.

[12] Brooke, A., Kendrick, D. and Meeraus, A. (1988) *GAMS: A User's Guide.* The Scientific Press, Redwood City, CA. (Revised versions are available from the GAMS Corporation.) See also www.gams.com.

[13] Casti, J.L. (1990) *Searching for Certainty.* Morrow & Co., New York.

[14] Castillo, I., Kampas, F.J., and Pintér, J.D. (2008) Solving circle packing problems by global optimization: numerical results and industrial applications. *European Journal of Operations Research* 16, 165-178.

[15] Chong, E.K.P. and Zak, S.H. (2001) *An Introduction to Optimization.* (2nd Edition) Wiley, New York.

[16] Corliss, G.F. and Kearfott, R.B. (1999) Rigorous global search: industrial applications. In: Csendes, T., Ed. *Developments in Reliable Computing,* pp. 1-16. Kluwer Academic Publishers, Dordrecht.

[17] Csendes, T. (2007) *Reliable Optimization: Methods and Applications.* D.Sc. Dissertation, Hungarian Academy of Sciences.
[18] Dimmer, P.R. and Cutteridge, O.P.D. (1980) Second derivative Gauss-Newton-based method for solving nonlinear simultaneous equations. *IEE Proceedings* 127, 278-283.
[19] Diwekar, U. (2003) *Introduction to Applied Optimization.* Kluwer Academic Publishers, Dordrecht.
[20] Ebers, J. J. and. Moll,. J. L. (1954) Large signal behaviour of junction transistors, *IEE Proceedings* 42, 1761-1772.
[21] Edgar, T.F., Himmelblau, D.M., and Lasdon, L.S. (2001) *Optimization of Chemical Processes.* (2nd Edition) McGraw-Hill, New York.
[22] Ferreira, C. (2002) *Gene Expression Programming.* Angra do Heroísmo, Portugal.
[23] Floudas, C.A., Pardalos, P.M., Adjiman, C.,Esposito, W.R., Gumus, Z.H., Harding, S.T., Klepeis, J.L., Meyer, C.A., and Schweiger, C.A. (1999) *Handbook of Test Problems in Local and Global Optimization.* Kluwer Academic Publishers, Dordrecht.
[24] Fourer, R. (2007) *Nonlinear Programming Frequently Asked Questions.* Optimization Technology Center of Northwestern University and Argonne National Laboratory. See www-unix.mcs.anl.gov/otc/Guide/faq/nonlinear-programming-faq.html.
[25] Fourer, R., Gay, D.M. and Kernighan, B.W. (1993) *AMPL - A Modeling Language for Mathematical Programming.* The Scientific Press, Redwood City, CA. (Reprinted by Boyd and Fraser, Danvers, MA, 1996.) See also www.ampl.com.
[26] Frontline Systems (2006) *Premium Solver Platform - Solver Engines. User Guide.* Frontline Systems, Inc. Incline Village, NV. See www.solver.com.
[27] Fritzson, P. (2004) *Principles of Object-Oriented Modeling and Simulation with Modelica 2.1.* IEEE Press, Wiley-Interscience, Piscataway, NJ.
[28] Gao, D.Y., Ogden, R.W. and Stavroulakis, G.E., Eds. (2001) *Nonsmooth/Nonconvex Mechanics: Modeling, Analysis and Numerical Methods.* Kluwer Academic Publishers, Dordrecht.
[29] Gershenfeld, N. (1999) *The Nature of Mathematical Modeling.* Cambridge University Press, Cambridge.
[30] Glover, F. and Laguna, M. (1997) *Tabu Search.* Kluwer Academic Publishers, Dordrecht.
[31] Grossmann, I.E., Ed. (1996) *Global Optimization in Engineering Design.* Kluwer Academic Publishers, Dordrecht.
[32] Hendrix, E.M.T. (1998) *Global Optimization at Work.* Ph.D. Thesis, University of Wageningen, Netherlands.
[33] Hansen, P.E. and Jørgensen, S.E., Eds. (1991) *Introduction to Environmental Management.* Elsevier, Amsterdam.
[34] Hillier, F.J. and Lieberman, G.J. (2005) *Introduction to Operations Research.* (8th Edition) McGraw-Hill, New York.
[35] Horst, R. and Pardalos, P.M., Eds. (1995) *Handbook of Global Optimization. Volume 1.* Kluwer Academic Publishers, Dordrecht.
[36] Horst, R. and Tuy, H. (1990) *Global Optimization – Deterministic Approaches.* Springer, Berlin.
[37] ILOG (2004) ILOG OPL Studio and Solver Suite. www.ilog.com.
[38] Isenor, G., Pintér, J.D. and Cada, M (2003) A global optimization approach to laser design. *Optimization and Engineering* 4, 177-196.
[39] Jacob, C. (2001) *Illustrating Evolutionary Computation with Mathematica.* Morgan Kaufmann Publishers, San Francisco.
[40] Jones, N.C. and Pevzner, P.A. (2004) *An Introduction to Bioinformatics Algorithms.* The MIT Press, Cambridge, MA.
[41] Kallrath, J., Ed. (2004) *Modeling Languages in Mathematical Optimization.* Kluwer Academic Publishers, Dordrecht.
[42] Kampas, F.J. and Pintér, J.D. (2004) Generalized circle packings: model formulations and numerical results. *Proceedings of the International Mathematica Symposium* (Banff, AB, Canada, August 2004).
[43] Kampas, F.J. and Pintér, J.D. (2006) Configuration analysis and design by using optimization tools in *Mathematica. The Mathematica Journal* 10 (1), 128-154.

[44] Kampas, F.J. and Pintér, J.D. (2007) Testing the "No free lunch theorem": Circle and sphere packings with MathOptimizer Professional. Presented at the Wolfram Technology Conference, Champaign, IL, October 2007.

[45] Kearfott, R.B. (1996) *Rigorous Global Search: Continuous Problems.* Kluwer Academic Publishers, Dordrecht.

[46] Li, D. and Sun, X. (2006) *Nonlinear Integer Programming.* Springer Science + Business Media, New York.

[47] Liberti, L. (2006) Writing global optimization software. In: Liberti, L. and Maculan, N., Eds., *Global Optimization: From Theory to Implementation*, pp. 211-262. Springer Science + Business Media, New York.

[48] Liberti, L. and Maculan, N., Eds. (2006) *Global Optimization: From Theory to Implementation.* Springer Science + Business Media, New York.

[49] LINDO Systems (1996) *Solver Suite.* LINDO Systems, Inc., Chicago, IL. See also www.lindo.com.

[50] Lopez, R.J. (2005) *Advanced Engineering Mathematics with Maple.* (Electronic book edition.) Maplesoft, Inc., Waterloo, ON. See www.maplesoft.com/products/ebooks/AEM/.

[51] Mandelbrot, B.B. (1983) *The Fractal Geometry of Nature.* Freeman & Co., New York.

[52] Maplesoft (2004) *Global Optimization Toolbox for Maple.* Maplesoft, Inc. Waterloo, ON. See www.maplesoft.com/products/toolboxes/globaloptimization/.

[53] Maplesoft (2006) *Maple.* Maplesoft, Inc., Waterloo, ON. www.maplesoft.com.

[54] Markót, M. Cs. and Csendes, T. (2005) A new verified optimization technique for the "Packing Circles in a Unit Square" problems. *SIAM J. Optimization* 16, 193-219.

[55] Mathsoft (2006) *Mathcad.* Mathsoft Engineering & Education, Inc., Cambridge, MA.

[56] Maximal Software (2006) *MPL Modeling System.* Maximal Software, Inc. Arlington, VA. www.maximal-usa.com.

[57] Michalewicz, Z. (1996) *Genetic Algorithms + Data Structures = Evolution Programs.* (3rd Edition) Springer, New York.

[58] Mistakidis, E.S. and Stavroulakis, G.E. (1997) *Nonconvex Optimization. Algorithms, Heuristics and Engineering Applications of the F.E.M.* Kluwer Academic Publishers, Dordrecht.

[59] Mittelmann, H.D. (2007) *Decision Tree for Optimization Software.* See plato.la.asu.edu/guide.html. (This website was started and maintained jointly for several years with Peter Spellucci.)

[60] Murray, J.D. (1983) *Mathematical Biology.* Springer, Berlin.

[61] Nocedal, J. and Wright, S.J. (2006) *Numerical Optimization.* (2nd Edition) Springer Science + Business Media, New York.

[62] Neumaier, A. (2004) Complete search in continuous global optimization and constraint satisfaction. In: Iserles, A., Ed. *Acta Numerica 2004*, pp. 271-369. Cambridge University Press, Cambridge.

[63] Neumaier, A. (2007) *Global Optimization.* www.mat.univie.ac.at/~neum/glopt.html.

[64] Nowak, I. (2005) *Relaxation and Decomposition Methods for Mixed Integer Nonlinear Programming.* Birkhäuser, Basel.

[65] Osman, I.H. and Kelly, J.P., Eds. (1996) *Meta-Heuristics: Theory and Applications.* Kluwer Academic Publishers, Dordrecht.

[66] Papalambros, P.Y. and Wilde, D.J. (2000) *Principles of Optimal Design.* Cambridge University Press, Cambridge.

[67] Paragon Decision Technology (2006) AIMMS. Paragon Decision Technology BV, Haarlem, The Netherlands. See www.aimms.com.

[68] Pardalos, P.M., Shalloway, D. and Xue, G., Eds. (1996) *Global Minimization of Nonconvex Energy Functions: Molecular Conformation and Protein Folding.* DIMACS Series, Vol. 23, American Mathematical Society, Providence, RI.

[69] Pardalos, P.M. and Resende, M.G.C., Eds. (2002) *Handbook of Applied Optimization.* Oxford University Press, Oxford.

[70] Pardalos, P.M. and Romeijn, H.E., Eds. (2002) *Handbook of Global Optimization. Volume 2.* Kluwer Academic Publishers, Dordrecht.

[71] Pintér, J.D. (1996a) *Global Optimization in Action.* Kluwer Academic Publishers, Dordrecht.

[72] Pintér, J.D. (1996b) Continuous global optimization software: A brief review. *Optima* 52, 1-8. (Web version is available at plato.la.asu.edu/gom.html.)
[73] Pintér, J.D. (1997) LGO – A program system for continuous and Lipschitz optimization. In: Bomze, I.M., Csendes, T., Horst, R. and Pardalos, P.M., Eds. *Developments in Global Optimization*, pp. 183-197. Kluwer Academic Publishers, Dordrecht.
[74] Pintér, J.D. (2001a) *Computational Global Optimization in Nonlinear Systems.* Lionheart Publishing Inc., Atlanta, GA.
[75] Pintér, J.D. (2001b) Globally optimized spherical point arrangements: model variants and illustrative results. *Annals of Operations Research* 104, 213-230.
[76] Pintér, J.D. (2002) Global optimization: software, test problems, and applications. In: Pardalos and Romeijn, Eds. *Handbook of Global Optimization. Volume 2*, pp. 515-569. Kluwer Academic Publishers, Dordrecht.
[77] Pintér, J.D. (2003) Globally optimized calibration of nonlinear models: techniques, software, and applications. *Optimization Methods and Software* 18, 335-355.
[78] Pintér, J.D. (2005) Nonlinear optimization in modeling environments: Software implementations for compilers, spreadsheets, modeling languages, and integrated computing systems. In: Jeyakumar, V. and Rubinov, A.M., Eds., *Continuous Optimization: Current Trends and Modern Applications*, Chapter 5 (pp. 147-173). Springer Science + Business Media, New York.
[79] Pintér, J.D., Ed. (2006a) *Global Optimization - Scientific and Engineering Case Studies.* Springer Science + Business Media, New York.
[80] Pintér, J.D. (2006b) *Global Optimization with Maple: An Introduction with Illustrative Examples.* An electronic book published and distributed by Pintér Consulting Services, Inc. Halifax, NS, Canada and Maplesoft, Inc. Waterloo, ON, Canada.
[81] Pintér, J.D. (2007a) Nonlinear optimization with GAMS /LGO. *Journal of Global Optimization* 38, 79-101.
[82] Pintér, J.D. (2007b) *LGO - A Model Development System for Continuous Global Optimization. User's Guide.* (Current revision) Pintér Consulting Services, Inc., Halifax, NS. For summary information, see www.pinterconsulting.com.
[83] Pintér, J.D., Holmström, K., Göran, A.O., and Edvall, M.M. (2004) User's Guide for TOMLAB /LGO. TOMLAB Optimization AB, Västerås, Sweden. See www.tomopt.com/docs/TOMLAB_LGO.pdf.
[84] Pintér, J.D. and Kampas, F.J. (2003) *MathOptimizer Professional – An Advanced Modeling and Optimization System for Mathematica Users with an External Solver Link. User Guide.* Pintér Consulting Services, Inc., Halifax, NS, Canada. For a summary, see also www.wolfram.com/products/ applications/mathoptpro/.
[85] Pintér, J.D. and Kampas, F.J. (2005) Nonlinear optimization in *Mathematica* with *MathOptimizer Professional. Mathematica in Education and Research* 10, 1-18.
[86] Pintér, J.D., Linder, D., and Chin, P. (2006) Global Optimization Toolbox for Maple: an introduction with illustrative applications. *Optimization Methods and Software* 21 (4) 565-582.
[87] Ratschek, H., and Rokne, J. (1993) Experiments using interval analysis for solving a circuit design problem. *Journal of Global Optimization* 3, 501-518.
[88] Rich, L.G. (1973) *Environmental Systems Engineering.* McGraw-Hill, Tokyo.
[89] Rothlauf, F. (2002) *Representations for Genetic and Evolutionary Algorithms.* Physica-Verlag, Heidelberg.
[90] Rudolph, G. (1997) *Convergence Properties of Evolutionary Algorithms.* Verlag Dr. Kovac, Hamburg.
[91] Schittkowski, K. (2002) *Numerical Data Fitting in Dynamical Systems.* Kluwer Academic Publishers, Dordrecht.
[92] Schroeder, M. (1991) *Fractals, Chaos, Power Laws.* Freeman & Co., New York.
[93] Stojanovic, S. (2003) *Computational Financial Mathematics Using Mathematica.* Birkhäuser, Boston.
[94] Stortelder, W.J.H., de Swart, J.J.B., and Pintér, J.D. (2001) Finding elliptic Fekete point sets: two numerical solution approaches. *Journal of Computational and Applied Mathematics* 130, 205-216.

[95] Tawarmalani, M. and Sahinidis, N.V. (2002) *Convexification and Global Optimization in Continuous and Mixed-integer Nonlinear Programming.* Kluwer Academic Publishers, Dordrecht.

[96] Tervo, J., Kolmonen, P., Lyyra-Laitinen, T., Pintér, J.D., and Lahtinen, T. (2003) An optimization-based approach to the multiple static delivery technique in radiation therapy. *Annals of Operations Research* 119, 205-227.

[97] TOMLAB Optimization (2006) TOMLAB. TOMLAB Optimization AB, Västerås, Sweden. See www.tomopt.com.

[98] The MathWorks (2006) *MATLAB.* The MathWorks, Inc., Natick, MA. See www.mathworks.com.

[99] Trefethen, N.L. (2002) A hundred-dollar, hundred-digit challenge. *SIAM News* 35 (1), p. 3.

[100] Voss, S., Martello, S., Osman, I.H. and Roucairol, C., Eds. (1999) *Meta-Heuristics: Advances and Trends in Local Search Paradigms for Optimization.* Kluwer Academic Publishers, Dordrecht.

[101] Wolfram, S. (2002) *A New Kind of Science.* Wolfram Media, Champaign, IL, and Cambridge University Press, Cambridge.

[102] Wolfram, S. (2003) *The Mathematica Book.* (4th Edition) Wolfram Media, Champaign, IL, and Cambridge University Press, Cambridge.

[103] Wolfram Research (2006) *Mathematica.* Wolfram Research, Inc., Champaign, IL. www.wolfram.com.

[104] Zabinsky, Z.B. (2003) *Stochastic Adaptive Search for Global Optimization.* Kluwer Academic Publishers, Dordrecht.

[105] Zhigljavsky, A.A. (1991) *Theory of Global Random Search.* Kluwer Academic Publishers, Dordrecht.

Fields Institute Communications
Volume **55**, 2009

Connected Dominating Set in Hypergraph

Linxia Li
Department of Mathematics
Xi'an Jiaotong University
Xi'an, China
`li.lin.xia.652@stu.xjtu.edu.cn`

Xu Zhu
Department of Mathematics
Xi'an Jiaotong University
Xi'an, China
`zhuxu@mail.xjtu.edu.cn`

Ding-Zhu Du
Department of Computer Science
University of Texas at Dallas
Richardson, TX 75083 USA
`dzdu@utdallas.edu`

Panos M. Pardalos
Department of Industrial and Systems Engineering
University of Florida
Gainesville, FL 32611 USA
`pardalos@ufl.edu`

Weili Wu
Department of Computer Science
University of Texas at Dallas
Richardson, TX 75083 USA
`weiliwu@utdallas.edu`

Abstract. In this paper, we study both connected dominating node-sets and connected dominating edge-sets in hypergraphs. By extending a recent general result on greedy approximation with submodular potential functions to nonsubmodular functions, we establish the performance ratio of some greedy approximation for problems of minimizing connected dominating node-set and connected dominating edge-sets in hypergraphs.

2000 *Mathematics Subject Classification.* Primary 68Q25; Secondary 90C27.

The work of D-Z. Du was partially supported by a grant from the Research Grants Council of the Hong Kong Special Administrative Region, China [Project No. CityU 1165/04E] and also partially supported by National Science Foundation of USA under grant CCF-0728851.

Research of P.M. Pardalos was partially supported by NSF and Airforce grants.

Support of W. Wu in part by the National Science Foundation of USA under grants CCF-0514796 and CCF-0750992.

1 Introduction

Given a hypergraph $G = (V, E)$, a subset D of nodes is a *dominating node-set* if every node in V is either in D or adjacent to a node in D. Futhermore, a dominating node-set D is called a *connected* dominating node-set if D induces a connected sub-hypergraph. [1] Finding a connected dominating node-set with minimum cardinality is called the problem of CONNECTED DOMINATING NODE-SET IN HYPERGRAPH.

CONNECTED DOMINATING NODE-SET IN HYPERGRAPH is a generalization of the problem of the minimum connected dominating set in graph. The latter has been studied extensively in the literature and widely used as virtual backbone in wireless networks [7].

A subset of edges in a hypergraph $G = (V, E)$ is called a *dominating edge-set* if they cover all nodes. Furthermore, if those edges induce a connected sub-hypergraph, then we call it a *connected dominating edge-set.* The minimum connected dominating edge-set in graph is easy to be computed since every spanning tree is an optimal solution. However, computing the minimum connected dominating edge-set in hypergraph, called CONNECTED DOMINATING EDGE-SET IN HYPERGRAPHS, is NP-hard and has no polynomial-time $\rho \ln n$-approximation for any $0 < \rho < 1$ unless $NP \subseteq DTIME(n^{O(\ln n)})$ where n is the number of nodes in G.

In fact, we can reduce the following problem to CONNECTED DOMINATING EDGE-SET IN HYPERGRAPHS.

WEAKLY CONNECTED DOMINATING SET IN GRAPH: Given a connected graph G, find a weakly connected dominating set of G, with the minimum cardinality. Here, a dominating set D in a graph is a dominating node-set and furthermore is said to be *weakly connected* if all edges incident to D induce a connected subgraph.

Given a connected graph $G = (V, E)$, construct a hypergraph $G^h = (V, E^h)$ by setting $E^h = \{e_v \mid v \in V\}$ where $e_v = \{u \mid (u, v) \in E\} \cup \{v\}$. It is not hard to show that $D \subseteq V$ is a dominating node-set of G if and only if $D^h = \{e_v \mid v \in D\}$ is a dominating edge-set of G^h and D is weakly connected in G if and only if D^h is connected in G^h. Since the minimum weakly connected dominating node-set in graph is NP-hard and has no polynomial-time $\rho \ln n$-approximation for any $0 < \rho < 1$ unless $NP \subseteq DTIME(n^{O(\ln n)})$ [4], the analogous results hold for CONNECTED DOMINATING EDGE-SET IN HYPERGRAPH.

[1] showed that the minimum connected dominating edge-set can be approximated by a greedy algorithm with performance ratio $(1 + \ln n)$. In fact, it can be reduced to the well-known submodular cover problem. To explain this, let us first introduce the submodularity of functions.

Consider a set X and a function f defined on the power set 2^X, i.e., the family of all subsets of X. f is said to be *submodular* if for any two subsets A and B in 2^X,

$$f(A) + f(B) \geq f(A \cap B) + f(A \cup B).$$

f is said to be *monotone nondecreasing* if for any two subsets A and B in 2^X

$$A \subset B \Rightarrow f(A) \leq f(B).$$

---

[1]The hypergraph $G' = (D, E')$ is said to be sub-hypergragh induced by D for G where $E' = \{e \cap D | e \cap D \neq \emptyset, e \in E\}$.

f is said to be a *polymatroid function* if f is submodular, monotone nondecreasing and

$$f(\emptyset) = 0.$$

For any submodular and monotone decreasing function f on 2^X, define

$$\Omega(f) = \{A \in 2^X \mid f(A) = f(X)\}.$$

The submodular cover problem is defined as follows.

SUBMODULAR COVER: Given a polymatroid function f and a nonnegative real function c on X, find a $A \in \Omega(f)$ to minimize $c(A)$ where $c(A) = \sum_{x\in A} f(x)$.

The following is a greedy approximation for SUBMODULAR COVER.

Greedy Algorithm G

input submodular function f and cost function c;
$A \leftarrow \emptyset$;
while there exists $x \in X$ such that $\Delta_x f(A) > 0$
  do select a vertex x that maximizes $\Delta_x f(A)/c(x)$ and set
    $A \leftarrow A \cup \{x\}$;
return A.

The performance ratio of above greedy approximation is described in the following theorem.

Theorem 1.1 *If f is an integer polymatroid function, then the Greedy Algorithm G produces an approximation solution within a factor of $H(\gamma)$ from optimal for* SUBMODULAR COVER, *where $\gamma = \max_{x\in X} f(\{x\})$ and $H(\gamma)$ is the harmonic function, i.e., $H(\gamma) = \sum_{i=1}^{\gamma} \frac{1}{i}$.*

Consider a hypergraph $G = (V, E)$. For any $F \subseteq E$, define $f(A) = |V| - p(A)$ where $p(A)$ is the number of connected components of sub-hypergraph (V, A), induced by the hyperedge set A. Then f is a polymatroid function and $A \in \Omega(f)$ if and only if A is a connected dominating edge-set.

In this paper, we present some general results similar to Theorem 1.1, which extend the analysis of greedy approximation with submodular potential functions to nonsubmodular potential functions of certain type. As an application, we establish the performance ratio of Greedy Algorithm G for CONNECTED DOMINATING NODE-SET IN HYPERGRAPH.

2 Main results

We first present some general results about greedy approximation.

Consider a monotone nondecreasing function $f : 2^X \to R$ with $f(\emptyset) = 0$. Define $\Omega(f) = \{A \mid f(A) = f(X)\}$. We are interested in the following condition on function f:

(*) For any $A \in \Omega(f)$, elements of A can be arranged into an ordering $A = \{x_1, x_2, ..., x_k\}$ such that for any $B \subseteq X$,

$$\Delta_{x_i} f(B) \geq \Delta_{x_i} f(B \cup \{x_1, ..., x_{i-1}\}).$$

A function satisfying condition (*) will be called a (*)-function. Every monotone nondecreasing and submodular function f is a (*)-function. However, there exist monotone nondecreasing (*)-functions which are nonsubmodular.

With monotone nondecreasing (*)-function f, we can also study the SUBMODULAR COVER problem and its greedy approximation, the Greedy Algorithm G. Indeed, the following lemma establishes the correctness of Greedy Algorithm G.

Lemma 2.1 *Suppose F is a monotone nondecreasing (*)-function. If $f(B) < f(X)$, then there exists $x \in X - B$ such that $\Delta_x f(B) > 0$.*

Proof Choose $A = \{x_1, x_2, ..., x_k\} \in \Omega(f)$ such that $\Delta_{x_i} f(B) \geq \Delta_{x_i} f(B \cup A_{i-1})$ for every $i = 1, ..., k$, where $A_{i-1} = \{x_1, x_2, ..., x_{i-1}\}$. Note that

$$\sum_{i=1}^{k} \Delta_{x_i} f(B \cup A_{i-1}) = f(B \cup A) - f(B) = f(X) - f(B) > 0.$$

Therefore, there exists i, $1 \leq i \leq k$ such that $\Delta_{x_i} f(B \cup A_{i-1}) > 0$. □

Next, we present a result similar to Theorem 1.1, but, it can apply to some nonsubmodular potential functions.

Theorem 2.2 *Let f be a monotone nondecreasing (*)-function. Suppose in Greedy Algorithm G, selected x always satisfies*

$$\frac{\Delta_x f(A)}{c(x)} \geq 1.$$

Then, Greedy Algorithm G produces an approximation solution within a factor of

$$1 + \ln \frac{f^*}{opt}$$

from optimal for SUBMODULAR COVER, *where opt is the objective function value of optimal solution and $F^* = f(X)$.*

Proof Let $A_g = \{x_1, x_2, ..., x_g\}$ be obtained by Greedy Algorithm G in the ordering of their appearance. Denote $A_i = \{x_1, x_2, ..., x_i\}$ and $A_0 = \emptyset$. Let $A^* = \{y_1, y_2, ..., y_k\}$ be an optimal solution and its elements in the ordering for meeting condition (*).

Denote $a_i = \Delta_{A^*} f(A_i)$ for $i = 0, 1, ..., g$. Then

$$\begin{aligned} a_{i-1} - a_i &= \Delta_{A^*} f(A_{i-1}) - \Delta_{A^*} f(A_i) \\ &= f(A_{i-1} \cup A^*) - f(A_{i-1}) - f(A_i \cup A^*) - f(A_i) \\ &= \Delta_{x_i} f(A_{i-1}). \end{aligned}$$

Denote $A_j^* = \{y_1, y_2, ..., y_j\}$ and $A_0^* = \emptyset$. Then

$$a_{i-1} = \Delta_{A^*} f(A_{i-1}) = \sum_{j=1}^{k} \Delta_{y_j} f(A_{i-1} \cup A_{j-1}^*) \leq \sum_{j=1}^{k} \Delta_{y_j} f(A_{i-1}).$$

Hence

$$\begin{aligned} \frac{a_{i-1}}{opt} &\leq \frac{\sum_{j=1}^k \Delta_{y_j} f(A_{i-1})}{\sum_{j=1}^k c(y_j)} \\ &\leq \max_{1 \leq j \leq k} \frac{\Delta_{y_j} f(A_{i-1})}{c(y_j)} \\ &\leq \frac{\Delta_{x_i} f(A_{i-1})}{c(x_i)}. \end{aligned}$$

Thus,

$$\frac{a_{i-1}}{opt} \leq \frac{a_{i-1} - a_i}{c(x_i)}.$$

So, we have

$$\begin{aligned} a_i &\leq a_{i-1}\left(1 - \frac{c(x_i)}{opt}\right) \\ &\leq a_{i-1}e^{-c(x_i)/opt} \\ &\leq \cdots \\ &\leq a_0 e^{-(c(x_i)+\cdots+c(x_1))/opt}. \end{aligned}$$

Note that

$$a_g = \Delta_{A^*} f(A_g) = f(A^* \cup A_g) - f(A_g) = 0$$

and

$$\begin{aligned} a_0 &= f(A^*) - f(\emptyset) \\ &= f(A_g) - f(\emptyset) \\ &= \sum_{i=1}^{g} \Delta_{x_i} f(A_{i-1}) \\ &\geq \sum_{i=1}^{g} c(x_i) \\ &\geq opt. \end{aligned}$$

There exists i, $0 \leq i \leq g_1$ such that

$$a_{i+1} < opt \leq a_i.$$

Denote $a' = a_i - opt$ and $a'' = opt - a_{i+1}$ so that

$$a_i - a_{i+1} = \Delta_{x_{i+1}} f(A_i) = a' + a''.$$

Let $c(x_{i+1}) = c' + c''$ such that

$$\frac{a'}{c'} = \frac{a''}{c''} = \frac{\Delta_{x_{i+1}} f(A_i)}{c(x_{i+1})} \geq 1.$$

Then

$$\frac{opt - a_{i+1}}{c'} = \frac{a'}{c'} = \frac{\Delta_{x_{i+1}} f(A_i)}{c(x_{i+1})} \geq \frac{a_i}{opt}.$$

Thus

$$opt \leq a_i\left(1 - \frac{c'}{opt}\right) \leq a_0 e^{-(c'+c(x_i)+\cdots+c(x_1))/opt}.$$

Hence

$$c(x_1) + \cdots + c(x_i) + c' \leq opt \ln \frac{a_0}{opt}.$$

Moreover,

$$\begin{aligned} &c'' + c(x_{i+2}) + \cdots + c(x_g) \\ &\leq a'' + \Delta_{x_{i+2}} f(A_{i+1}) + \cdots + \Delta_{x_g} f(A_{g-1}) \\ &= (opt - a_{i+1}) + (a_{i+1} - a_{i+2}) + \cdots + (a_{g-1} - a_g) \\ &= opt. \end{aligned}$$

Therefore,

$$c(A_g) \leq opt \cdot \left(1 + \ln \frac{f^*}{opt}\right).$$

□

A monotone nondecreasing function $f : 2^X \to R$ is called a d-shifting (*)-function if it satisfies the following two conditions:

(a) For any $A \in \Omega(f)$, its elements can be arranged into an ordering $A = \{x_1, x_2, ..., x_k\}$ such that for any $B \subseteq X$,

$$\Delta_{x_i} f(B) \geq \Delta_{x_i} f(B \cup \{x_1, x_2, ..., x_{i-1}\}) - d.$$

(b) For any B with $f(B) < f(X)$, there exists $x \in X - B$ such that $f(B) < f(B \cup \{x\})$.

About d-shifting (*)-function, we have the following result:

Theorem 2.3 *Let f be an integer monotone nondecreasing d-shifting (*)-function. Then, Greedy Algorithm G produces an approximation solution within a factor of*

$$1 + d + \ln(d + \frac{d + f^*}{opt})$$

from optimal for Submodular Cover *with cost function $c(x) = 1$ for all $x \in X$, where opt is the objective function value of optimal solution.*

Proof Let $A_g = \{x_1, x_2, ..., x_g\}$ be obtained by Greedy Algorithm G in the ordering of their appearance. Denote $A_i = \{x_1, x_2, ..., x_i\}$ and $A_0 = \emptyset$. Let $A^* = \{y_1, y_2, ..., y_{opt}\}$ be an optimal solution and its elements in the ordering for meeting condition (a).

By the greedy rule,

$$\Delta_{x_{i+1}} f(A_i) \geq \Delta_{y_j} \text{ for all } 1 \leq j \leq opt.$$

Thus,

$$\begin{aligned} \Delta_{x_{i+1}} f(A_i) &\geq \frac{\sum_{j=1}^{opt} \Delta_{y_j} f(A_i)}{opt} \\ &\geq \frac{-(opt-1)d + \sum_{j=1}^{opt} \Delta_{y_j} f(A_i \cup A^*_{j-1})}{opt} \\ &= \frac{-(opt-1)d + f(A_i \cup A^*) - f(A_i)}{opt} \\ &= \frac{-(opt-1)d + f^* - f(A_i)}{opt}. \end{aligned}$$

Therefore,

$$f(A_{i+1}) \geq f(A_i) + \frac{-(opt-1)d + f^* - f(A_i)}{opt}.$$

Denote

$$a_i = -(opt-1)d + f^* - f(A_i).$$

Then

$$a_{i+1} \leq a_i \left(1 - \frac{1}{opt}\right) \leq a_0 \left(1 - \frac{1}{opt}\right)^{i+1} \leq a_o e^{-(i+1)/opt}.$$

Choose i to satisfy

$$a_{i+1} < opt \leq a_i.$$

(If such i does not exist, i.e., $a_0 < opt$, then $f^* \leq d \cdot opt$ and hence $g \leq f^* \leq d \cdot opt$.) Then

$$g - i \leq d(opt - 1) + opt$$

and

$$opt \leq a_0 e^{-i/opt}.$$

Thus,

$$g \leq d(opt-1) + opt + i \leq opt\left(d + 1 + \ln \frac{a_o}{opt}\right),$$

where

$$\frac{a_0}{opt} = \frac{-(opt-1)d + f^*}{opt} \leq -d + \frac{d+f^*}{opt}.$$

□

Now, we come back to connected dominating node-set in hypergraphs.

For input hypergraph $G = (V, E)$, we always assume that G is connected; otherwise no connected dominating node-set exists. For any subset A of V, define $p(A)$ to be the number of connected components of the sub-hypergraph induced by A and $q(A)$ the number of connected components of the sub-hypergraph with node set V and edges set $\{e \mid e \cap A \neq \emptyset, e \in E\}$. Define $g(A) = |V| - p(A) - q(A)$.

Lemma 2.4 *For $n \geq 3$, A is a connected dominating node-set if and only if $g(A) = n - 2$ where $n = |V|$.*

Proof If A is a connected dominating node-set, then $p(A) = q(A) = 1$ and hence $g(A) = n - 2$. Conversely, if $g(A) = n - 2$, then $A \neq \emptyset$ since $f(\emptyset) = 0$. Note that for nonempty A, $p(A) \geq 1$ and $q(A) \geq 1$. Moreover, it follows from $g(A) = n - 2$ that $p(A) + q(A) = 2$. Therefore, $p(A) = q(A) = 1$. $q(A) = 1$ implies that A is a dominating node-set and $p(A) = 1$ implies that A induces a connected sub-hypergraph. □

Lemma 2.5 *A is a connected dominating node-set if and only if $g(A \cup \{u\}) = g(A)$ for every $u \in V$.*

Proof It is easy to prove when there is one or two nodes in G. Then we consider there are three nodes in G. If A is a connected dominating set, then $g(A) = n - 2$, which reaches the maximum value. Therefore, $g(A \cup \{u\}) = g(A)$.

Conversely, suppose $g(A \cup \{u\}) = g(A)$ for every $u \in V$. First, A cannot be the empty set. In fact, for contradiction, suppose $A = \emptyset$. Since G is a connected hypergraph, we consider two cases:

Case 1. There is just one hyperedge that is incident to all nodes in G. If we can choose any one node u, then $g(\{u\}) = |V| - p(\{u\}) - q(\{u\}) = n - 2$. But $g(\emptyset) = 0 < f(\{u\})$, a contradiction.

Case 2. There are more than one hyperedges in G. There must exist a node u with degree at least two and for such a node u, $p(\{u\}) = p(\emptyset) + 1$, $q(\{u\}) \leq q(\emptyset) - 2$, $g(\{u\}) \geq g(\emptyset) + 1 > g(\emptyset)$, also a contradiction.

Now, we may assume $A \neq \emptyset$. Consider a connected component of the sub-hypergraph induced by A. Let B denote its node set which is a subset of A. For every grey node v adjacent to B, if v is adjacent to a black node not in B, then $p(A \cup \{v\}) < p(A)$ and $q(A \cup \{v\}) \leq q(A)$; otherwise, if v is adjacent to a white node or a grey node not adjacent to B, then we must have $p(A \cup \{v\}) \leq p(A)$ and $q(A \cup \{v\}) < q(A)$; hence, in all cases $g(A \cup \{v\}) > g(A)$, a contradiction. Therefore, every grey node adjacent to B cannot be adjacent to any node neither in B nor adjacent to B. Since G is connected, it follows that every node of G must belong to B or adjacent to B. That is, $B = A$ is a connected dominating node-set. □

Lemma 2.6 *g is a monotone nondecreasing 1-shifting (*)-function.*

Proof First, we show that g is monotone nondecreasing. Consider a node-subset A and a node $v \notin A$.

If v is adjacent to a node in A, then $q(A) = q(A \cup \{v\})$. Moreover, it is clear that $p(A) \geq p(A \cup \{v\}$. Therefore, $g(A) \leq g(A \cup \{v\})$.

If v is not adjacent to any node in A, then $q(A \cup \{v\}) = q(A)+1$. However, in this case, v itself is a connected component in the sub-hypergraph with nodee set V and edge set $\{e \in E \mid e \cap A \neq \emptyset\}$. Thus, there is at least one hyper-edge at v connecting it to some other connected component. This means that $p(A \cup \{v\}) \leq p(A) - 1$. It follows that $g(A) \leq g(A \cup \{v\})$.

Next, we show that g is a 1-shift (*)-function.

Consider a connected dominating node-set A, we can always arrange elements of A into an ordering $v_1, v_2, \ldots, v_{|A|}$ such that for $i = 1$, v_1 is chosen arbitrarily; for $i \geq 2$, v_i is chosen to be adjacent to a node in $A_{i-1} = \{v_1, v_2, ..., v_{i-1}\}$ Then, for any node-subset B, v_i can dominate at most one additional connected component in the sub-hypergraph induced by B than in the sub-hypergraph induced by $B \cup A_{i-1}$, since $v_1, \ldots, v_{j-1}$ are connected. Thus,

$$\Delta_{v_i} q(B) \leq \Delta_{v_i} q(B \cup A_{i-1}) + 1.$$

Note that $n - p(A)$ is submodular with respect to A so that

$$\Delta_{v_i} p(B) \leq \Delta_{v_i} p(B \cup A_{i-1}).$$

Therefore,

$$\Delta_{v_i} g(B) \geq \Delta_{v_i} g(B \cup A_{i-1}) - 1.$$

This means that g satisfies condition (a). Moreover, by Lemma 2.5, g satisfies condition (b). Therefore, g is a 1-shifting (*)-function. □

Theorem 2.7 *Greedy Algorithm G is a $\big(2 + \ln(\delta(\mu - 1))\big)$-approximation for* CONNECTED DOMINATING NODE-SET IN HYPERGRAPH, *where μ is the maximum number of nodes in a hyperedge, δ is the maximum node-degree in input hypergraph.*

Proof It follows immediately from Lemma 2.6 and Theorem 2.3. □

Corollary 2.8 [7] *When G is a graph, Greedy Algorithm G is a $(2 + \ln \delta)$-approximation for* CONNECTED DOMINATING NODE-SET IN GRAPH, *where δ is the maximum node-degree in input graph.*

3 Discussion

Consider a finite set V and a collection $\mathcal{I}$ of subsets of X. $(X, \mathcal{I})$ is called a matroid if

(M1) $A \subset B \in \mathcal{I} \Rightarrow A \in \mathcal{I}$, and

(M2) $|A| > |B|$ for $A, B \in \mathcal{I} \Rightarrow$ there exist $x \in A$ such that $B \cup \{x\} \in \mathcal{I}$.

All subsets in $\mathcal{I}$ are called *independent sets.* For every $A \subseteq X$, define the rank of A to be $r(A) = \max\{|I| \mid I \in \mathcal{I}, I \subseteq A\}$. It is well-known that $r(A)$ is a polymatroid function satisfying $r(x) = 1$ for every $x \in X$. Therefore, Theorem 1.1 indicates that using the rank function r as potential function f, Greedy Algorithm G would produces an optimal solution for SUBMODULAR COVER problem.

Greedoid is a generalization of matroid [5]. $(X, \mathcal{I})$ is called a *greedoid* if
(G1) any $A \in \mathcal{I}$ contains an element x such that $A - \{x\} \in \mathcal{I}$, and
(G2) $|A| > |B|$ for $A, B \in \mathcal{I} \Rightarrow$ there exist $x \in A$ such that $B \cup \{x\} \in \mathcal{I}$.
Similarly, we can define the rank function of greedoid as follows

$$r(A) = \max\{|I| \mid I \in \mathcal{I}, I \subseteq A\}.$$

It is also known that r is the rank function of some greedoid if and only if
(R1) for every $A \subseteq X$, $r(A) \leq |A|$,
(R2) r is monotone nondecreasing,
(R3) for any $A \subseteq X$ and $x, y \in X$, $r(A) = r(A \cup \{x\})$ and $r(A \cup \{y\})$ imply $r(A) = r(A \cup \{x, y\})$.

Let us call a function satisfying (R2) and (R3) a polygreedoid function. Then a polygreedoid function may not be submodular since $\Delta_x r(A)$ may be smaller than $\Delta_x r(B)$ for $A \subset B$ when set $\mathcal{I} = \{A_i \mid i = 1, ..., 5\}$ where $A_i = \{1, 2, ..., i\}$, and consider $A = \{1, 2\}$, $B = \{1, 2, 3, 4\}$ and $x = 5$.

It is well-known that greedoid generalizes the relationship between matroid and greedy algorithms, that is, there are some results on greedoid for Grredy Algorithm G to produce an optimal solution for SUBMODULAR COVER problem. However, no generalization of Theorem 1.1 has been found on greedoids. Is there any relationship between polygreedoid and (*)-functions? Those research topics will bring more understanding on the analysis of greedy approximation with non-submodular potential functions [2].

References

[1] G. Baudis, C. Gröpl, S. Hougardy, T. Nierhoff, H.J. Prömel, Approximating Minimum Spanning Sets in Hypergraphs and Polymatroids, technical report, Humboldt-Universitt zu Berlin, 2000.

[2] Ding-Zhu Du, R.L. Graham, P.M. Pardalos, P.-J. Wan, W. Wu and W. Zhao, Analysis of Greedy Approximations with Nonsubmodular Potential Functions, *Proceedings of the 19th Annual ACM-SIAM Symposium on Dicrete Algorithms (SODA)*, San Francisco, USA; January 20-22, 2008, pp. 167-175.

[3] T. Fujito. Approximation algorithms for submodular set cover with application, *IEICE Trans. Inf. Syst*, E83-D(3):480–487, 2000.

[4] S. Guha and S. Khuller, Approximation algorithms for connected dominating sets, *Algorithmca*, 20 (1998) 374-387.

[5] B. Korte, L. Lovász, and R. Schrader, *Greedoids*, Springer-Verlag, Berlin Heidelberg 1991.

[6] L. A. Wolsey, An analysis of the greedy algorithm for submodular set covering problem, *Combinatorica*, vol 2 no 4 (1982) 385-393.

[7] Lu Ruan, Hongwei Du, Xiaohua Jia, Weili Wu, Yingshu Li, Ker-I Ko, A greedy approximation for minimum connected dominating sets, *Theoretical Computer Science* 329 (2004) 325-330.

Fields Institute Communications
Volume **55**, 2009

Global Optimisation of Pessimistic Bi-Level Problems

Angelos Tsoukalas
Department of Computing
Imperial College of Science, Technology and Medicine
London SW7 2RH, UK
at102@doc.ic.ac.uk

Wolfram Wiesemann
Department of Computing
Imperial College of Science, Technology and Medicine
London SW7 2RH, UK
wwiesema@doc.ic.ac.uk

Berç Rustem
Department of Computing
Imperial College of Science, Technology and Medicine
London SW7 2RH, UK
br@doc.ic.ac.uk

Abstract. We study the solution of non-convex, pessimistic bi-level problems. After providing several motivating examples, we relate the problem to existing research in optimisation. We analyse key properties of the optimisation problem, such as closedness of the feasible region and computational complexity. We then present and investigate a semi-infinite solution approach that solves ϵ-approximations of the problem. To the best of our knowledge, this represents the first solution technique proposed for this problem class. We close with numerical results and a discussion on fruitful directions for future research.

1 Introduction

This chapter is concerned with problems of the following type:

$$\min_{x \in X} \; f(x) \tag{1.1a}$$

subject to

$$g(x, y) \leq 0 \qquad \forall\, y \in \arg\max_{y' \in Y} \; h(x, y'), \tag{1.1b}$$

where $X \subseteq \mathbb{R}^n$, $Y \subseteq \mathbb{R}^m$, $f : X \mapsto \mathbb{R}$ and $g, h : X \times Y \mapsto \mathbb{R}$. We assume that X and Y are non-empty and compact, while f, g and h are continuous in their arguments. In case multiple constraints $g_1, \ldots, g_p : X \times Y \mapsto \mathbb{R}$ should be considered, they

2000 *Mathematics Subject Classification.* Primary 90C26, 90C34, 91-08.

can be condensed to $g(x,y) = \max\{g_i(x,y) : i = 1,\dots,p\}$. Similarly, the case of multiple constraints g_i with individual second-stage objectives h_i can be handled by a straightforward extension of what follows. We assume that the feasible region of the subordinate optimisation problem does not depend on x, the variables of the master problem.

Model (1.1) is understood best in the context of non-cooperative game theory: two players A and B participate in a hierarchic game where A (the leader) acts first and B (the follower) responds. In the game, information is assumed to be complete (i.e., the players' payoff functions f_A and f_B, as well as the constraint sets X_A and X_B, are common knowledge) and perfect (i.e., the follower knows the leader's decision). Having observed the leader's action x_A, the follower chooses a feasible decision $x_B \in X_B$ that maximises his payoff function $f_B(\cdot)$ (i.e., his 'best response'). Knowing that the follower does so, the leader chooses an admissible decision $x_A \in X_A$ which maximises his payoff function $f_A(\cdot)$, taking into account that the follower will choose his best response afterwards. The setting just described is known as 'Stackelberg leader-follower game' [1] and can be formalised as follows.

$$\max_{x_A \in X_A} \min_{x_B \in X_B} \left\{ f_A(x_A, x_B) : x_B \in \operatorname*{arg\,max}_{x'_B \in X_B} f_B(x_A, x'_B) \right\}. \tag{1.2}$$

Problem (1.2) is readily identified as a special case of model (1.1): the feasible region of the first-stage problem becomes $x = (x_A, \tau) \in X = X_A \times \mathbb{R}$, while the feasible region of the second-stage problem is described by $y = x_B \in Y = X_B$. The objective function is represented by $f((x_A, \tau)) = -\tau$. The constraint becomes $g((x_A, \tau), x_B) = \tau - f_A(x_A, x_B)$, which has to be less than or equal to zero for all maxima of $h((x_A, \tau), x'_B) = f_B(x_A, x'_B)$.

If the follower's decision problem is convex but its objective function h lacks the property of strict convexity, the follower can possess several different globally optimal responses. If the follower's decision problem is non-convex, the set of optimal follower responses might even be disconnected. In either situation, the follower is indifferent between several equally beneficial responses. Multiple optimal responses pose a problem to the leader since he cannot anticipate which of these optima will be chosen by the follower, and the corresponding optimal leader decisions can differ significantly. In the optimistic version of the Stackelberg leader-follower game, the leader anticipates that the follower chooses a best response that is most suitable for the leader. In essence, this means that the leader can decide which of the best responses is implemented by the follower. In (1.1) and (1.2), on the other hand, the leader does not have this possibility. Instead, he prepares for the worst case, that is, he chooses a decision $x_A \in X_A$ which performs best in view of the 'worst' (globally optimal) follower response. Because of this, (1.1) and (1.2) are called pessimistic versions of the bi-level problem and Stackelberg game, respectively.

Stackelberg games and, more generally, bi-level problems naturally arise in economics, engineering, optimisation under uncertainty and several other application areas [8, 9, 33]. In the following, we consider two examples that illustrate some of the difficulties which have to be addressed by solution procedures.

Production Planning. Two companies A and B produce the same product and need to decide on their production quantities. Company A is the market leader and has to decide first on a quantity $x_A \in [0, 200]$. Company B, on the other hand, is a follower and decides on a quantity $x_B \in [0, 100]$ after observing A's decision. The market price for the product depends on the aggregated supply

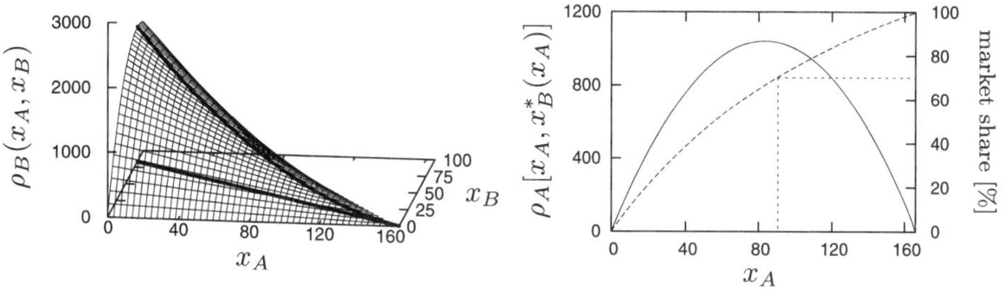

Figure 1 Production planning without investment option. The left graph visualises the follower's optimal response to the leader's production quantity, while the right graph shows the leader's profit (straight line) and market share (dashed line) as functions of the leader's production output. $x^*_B(x_A)$ denotes the set of optimal responses of B to a production quantity x_A of A.

$x_\Sigma = x_A + x_B$ and is determined by $p(x_\Sigma) = 100 - 0.3x_\Sigma$, while the (constant) unit production costs are 50 for both companies. A and B want to maximise their profits. Additionally, company A wishes to preserve its market dominance by providing at least 70% of the aggregated supply.

The profit of company A and B amounts to $\rho_i(x_A, x_B) = (p(x_A + x_B) - 50)\, x_i$, $i \in \{A, B\}$. Company A's decision problem can be written as follows.

$$\max_{x_A \in [0,200]} \ \min_{x_B \in x^*_B(x_A)} \ \{(50 - 0.3(x_A + x_B))\, x_A\}$$

subject to

$$\frac{x_A}{x_A + x_B} \geq 0.7 \qquad \forall x_B \in x^*_B(x_A),$$

where

$$x^*_B(x_A) = \underset{x_B \in [0,100]}{\arg\max} \ \{(50 - 0.3(x_A + x_B))\, x_B\}$$

denotes B's set of optimal responses to a given production quantity of A. The objective is to maximise company A's worst-case profit. The constraint ensures that A preserves a market share of at least 70%, no matter which optimal response is chosen by B. It is easy to rewrite this model into the form of (1.1).

Figure 1 (left) illustrates company B's optimal decision x^*_B as function of company A's production quantity. As expected, B's optimal output is monotonically decreasing in A's production quantity. Figure 1 (right) shows A's profit and market share as a function of A's production quantity, anticipating B's optimal response from Figure 1 (left). One can see that it would be optimal for A to produce ≈ 83.33 units (with a profit of ≈ 1041.67), which would cause B to produce ≈ 41.66 units (with a profit of ≈ 520.87). Since A's market share would be $\approx 66.66\%$ in this case, however, A instead has to produce the higher quantity of ≈ 89.75 units (with a lower profit of ≈ 1035.49), which causes B to produce ≈ 38.45 units (resulting in a profit of ≈ 443.71).

Let us now assume that B has the option to invest in new production facilities (at a cost of 500) that reduce B's per-unit production costs to 38. In this case, ρ_B changes to

$$\rho_B(x_A, x_B) = \max\{(p(x_A + x_B) - 50)\, x_B, (p(x_A + x_B) - 38)\, x_B - 500\},$$

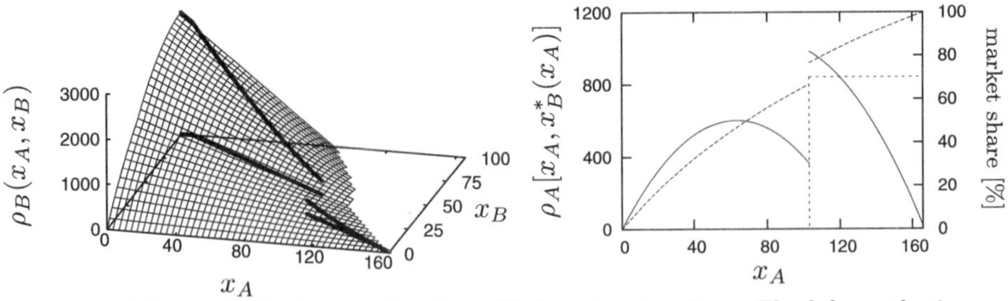

Figure 2 Production planning with investment option. The left graph visualises the follower's optimal response to the leader's production quantity, while the right graph shows the leader's profit (straight line) and market share (dashed line) as functions of the leader's production output. The follower invests if the leader produces less than ≈ 103.33 units. As in Figure 1, $x_B^*(x_A)$ denotes the set of optimal responses of B to a production quantity x_A of A.

while the other problem parameters remain the same. Figure 2 (left) illustrates B's new response function. B's optimisation problem is clearly non-convex. Furthermore, B's response function has a discontinuity at $x_A \approx 103.33$: if A produces more than this quantity, the investment is not beneficial for B. If A produces less than this quantity, on the other hand, B should invest in new facilities. In this case, however, the investment costs have to be amortised by a considerably higher production quantity. Note that at $x_A \approx 103.33$, B might or might not invest since he is indifferent between both choices. Figure 2 (right) shows how A's profit and market share depend on A's production quantity in the new situation. Note that due to the discontinuity in B's response function, both A's profit and his market share possess similar discontinuities at $x_A \approx 103.33$. With and without the market penetration constraint, A's optimal production quantity is 'just above' this discontinuity. Since A optimises in view of the worst of B's globally optimal responses, the point of discontinuity itself is *not* optimal: at this point, A has to consider the possibility that B might indeed invest and thus decrease the market price (by producing a large quantity in order to amortise his investment costs). As a result, A's decision problem contains a discontinuous objective function and a non-closed feasible region (due to the market penetration constraint). A's optimal production quantity is ≈ 103.34 units (resulting in a profit of ≈ 981.69 and a market share of $\approx 76.54\%$), while B's optimal response is ≈ 31.67 units (leading to a profit of 360.84).

It is interesting to note that B's additional investment option results in a smaller profit for B, even though B does not exercise this option in the realised solution. This seems to contradict the principles of decision theory, where the presence of additional alternatives should not decrease the achievable profit. The reason for this surprising (but well-known) effect is that A is aware of B's investment opportunity. Hence, A knows that the previously optimal production quantity of ≈ 89.75 would cause B to exercise his investment option and produce a higher amount. This, in turn, would result in a lower market price, lower profits and a violation of the market penetration constraint. Thus, A is forced to produce more, which results in lower profits for both companies but nevertheless constitutes the best achievable result for A in this new setting.

Principal-agent problem. The principal-agent problem illustrates the difficulty to design contracts under incomplete and asymmetric information [16]. Assume that agent A is supposed to perform a task for principal P. The generated profit ρ of that task is a function of the effort $\phi \in [0,1]$ that A puts into the task. Since P cannot observe ϕ, he designs an incentive-based contract, that is, apart from a fixed salary, he pays A a fraction β of the observable task profit $\rho(\phi)$. P wants to choose β in such a way that his part of the profit is maximised. The setting can be cast as a Stackelberg game: P, the leader, moves first by deciding on the contract parameter β. After he has observed β, A (the follower) chooses the effort $\phi^*(\beta)$ that maximises his utility $u(\beta,\phi)$. Anticipating A's decision, P wants to choose a contract parameter β that maximises his part of the profit, that is, $(1-\beta)\rho[\phi^*(\beta)]$.

In the following, assume that A's utility is determined by $u(\beta,\phi) = s(\beta,\phi) - r(\phi)$, where $s(\beta,\phi) = 10 + \beta\rho(\phi)$ denotes his salary and $r(\phi) = \phi + (5[\phi - 1/2]^+)^2$ represents A's reluctance to work. The profit generated by the task outcome is determined by $\rho(\phi) = 5\phi^2$. With this notation, P's optimisation problem can be described as follows.

$$\max_{\beta\in[0,1]} \min_{\phi\in\phi^*(\beta)} \left\{5(1-\beta)\phi^2\right\},$$

where

$$\phi^*(\beta) = \arg\max_{\phi\in[0,1]} \left\{10 + 5\beta\phi^2 - \phi - (5[\phi - 1/2]^+)^2\right\}$$

denotes the set of optimal efforts for a given contract design β. The model maximises P's worst-case profit, anticipating that A will choose any of his optimal responses. Again, it is straightforward to rewrite the model into the form of (1.1).

Figure 3 (left) illustrates A's optimal effort levels for different contract designs β. One can see that A's optimal strategy is to put no effort into the task if the profit share β is less than $\approx 39.2\%$. At $\beta \approx 39.2\%$, A is indifferent between $\phi = 0$ and $\phi \approx 0.520$. Above this margin, A's optimal response increases with β. Figure 3 (right) shows P's profit as a function of β, anticipating the optimal response of A from Figure 3 (left). As in the production planning example with investment opportunity, the leader's objective function is discontinuous. The optimal contract design is 'just above' this discontinuity at $\beta \approx 39.2\%$, causing A to put an effort of $\phi \approx 0.521$ into the task. Again, the point of discontinuity itself is not optimal since P has to consider that A puts no effort into the task.

In the examples, we have encountered two cases where the follower's optimal response function is discontinuous. Such discontinuities lead to set-valued response functions since at the points of discontinuity, the follower is indifferent between two or more globally optimal responses. Furthermore, we have seen that quite naturally, the leader's optimal strategy can consist in 'pushing' the follower as close as possible to a particular side of such a discontinuity. In Stackelberg games of type (1.2), this results in (leader) optimisation problems whose projections on the x_A-space possess discontinuous objective functions. More generally, in pessimistic bi-level problems, this results in non-closed feasible regions. Hence, one can expect (1.1) to be generically very difficult to solve.

The remainder of this text is organised as follows. In the next section, we relate formulation (1.1) to existing research in optimisation. In particular, we will highlight its connection with variants of the minimax problem, bi-level optimisation and mathematical programming with equilibrium constraints. In Section 3, we

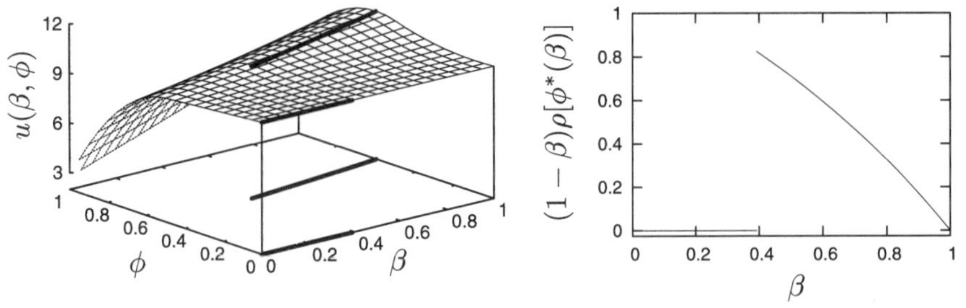

Figure 3 Principal-agent problem. The left and right graph visualise the agent's optimal response and the principal's profit, both as functions of the contract design chosen by the principal. The agent does not put any effort into the task if the profit share is below $\approx 39.2\%$.

analyse properties of problem (1.1), such as closedness of the feasible region and the computational complexity under various assumptions. In Section 4, we present a solution procedure based on semi-infinite programming. Numerical results are provided in Section 5. We conclude in Section 6. Throughout this paper, the discussion will take place on an 'intuitive' level. Formal justification for the claims made here is provided in the accompanying paper [31].

2 Relation to existing research

In this section, we relate model (1.1) to existing research. In particular, we consider various minimax formulations, bi-level models and equilibrium constraint problems. It will become apparent that our problem formulation fills a gap in the literature on optimisation under uncertainty and hierarchic games.

2.1 Minimax problems. Minimax problems [11, 25] are very similar to the Stackelberg game discussed in Section 1. Like Stackelberg games, minimax problems can be regarded as hierarchic games with two players A (the leader) and B (the follower). Contrary to Stackelberg games, however, minimax problems constitute constant-sum games, i.e., both players possess the same objective function, which is to be minimised by player A and maximised by player B. Apart from game-theoretic applications similar to those outlined in Section 1, minimax problems arise naturally in optimisation under uncertainty. In such settings, player B plays the role of 'nature' which reacts to player A's decision in the most destructive way. Hence, player A (the decision maker) is interested in a strategy which performs best in view of the worst possible future contingency.

Based on the involved constraints, one distinguishes between uncoupled and coupled minimax problems. The uncoupled minimax problem is defined as:

$$\min_{x \in X} \max_{y \in Y} f(x, y). \tag{2.1}$$

The characteristic feature of uncoupled minimax problems is that the feasible regions X and Y are independent. Note, however, that the optimal strategy of player B nevertheless depends on the decision of player A since they are coupled by the objective function. Similar to Stackelberg games, a purely local optimum with respect to both x and y is typically not meaningful for practical applications. Although suboptimal solutions with respect to x are acceptable, a global maximum with respect to y is needed in order to correctly anticipate player B's response. As a result, solution procedures for (2.1) typically determine global optima with respect to y.

Depending on the optimality guarantees provided with respect to x, one distinguishes between local and global optimisation approaches. Local optimisation algorithms for (2.1) have been proposed in [19, 22, 25, 26, 28]. A global optimisation procedure is described in [23]. Problem (2.1) can furthermore be regarded as a special case of semi-infinite programming, which implies that the respective solution procedures are applicable as well [5, 6, 15]. Due to their favourable convergence properties, however, specialised local optimisation procedures for (2.1) are preferable to semi-infinite programming techniques.

Contrary to (2.1), the coupled minimax problem contains coupled constraints which involve the decisions of both players. Coupled constraints can arise in game-theoretic settings (both players have to obey certain regulations, the leader wants to retain a certain market share regardless of the follower's response, etc.) and in optimisation under uncertainty (a constraint needs to be satisfied under any possible future contingency). In the following, we distinguish between two different types of coupled constraints: type A-constraints restrain player A, whereas type B-constraints restrain (primarily) player B.

A type A-constraint $g_A : X \times Y \mapsto \mathbb{R}$ can be included in (2.1) as follows.

$$\min_{x \in X} \max_{y \in Y} f(x, y) \tag{2.2a}$$

subject to

$$g_A(x, y') \leq 0 \qquad \forall y' \in Y. \tag{2.2b}$$

Hence, player A has to ensure that the coupled constraint g_A is satisfied for all of player B's possible responses. Note that g_A does not couple player A's decision to player B's eventual response. Instead, any possible response of the follower is considered. Hence, g_A can be considered as part of the description of X, player A's feasible region, and the term 'coupled constraint' is misleading. Type A-constraints can be regarded as a natural extension of the uncoupled minimax-problem (2.1) and have various applications in game theory (similar to our discussion in Section 1) and optimisation under uncertainty (e.g., a constraint has to be satisfied under any possible future contingency). Problem (2.2) is typically solved as a semi-infinite program [34].

Often, player B does not represent nature (whose most destructive responses should be anticipated), but rather a rational player whose objective is to maximise f. In this case, type A-constraints are unnecessarily restrictive. Indeed, in such settings player A merely has to ensure that all *optimal* responses of player B satisfy the coupled constraint. The resulting formulation is a special case of (1.2) and hence represents an instance of the pessimistic bi-level problem (1.1).

Contrary to type A-constraints, a type B-constraint $g_B : X \times Y \mapsto \mathbb{R}$ restricts the admissible decisions of player B:

$$\min_{x \in X} \max_{y \in Y} f(x, y) \tag{2.3a}$$

subject to

$$g_B(x, y) \leq 0. \tag{2.3b}$$

More precisely, (2.3) is to be understood as

$$\min_{x \in X} \max_{y \in Y} \{f(x,y) \,:\, g_B(x,y) \leq 0\}.$$

Hence, player A can choose any strategy $x \in X$, while player B's response $y \in Y$ additionally has to satisfy the coupled constraint $g_B(x,y) \leq 0$. Contrary to type A-constraints, type B-constraints indeed couple the decisions of the players.

Note that (2.3) implicitly assumes the existence of a $y \in Y$ that satisfies $g_B(x^*,y) \leq 0$ for the x^* chosen by player A since otherwise, the objective value would be undefined. If non-emptyness of $\{y \in Y \,:\, g_B(x,y) \leq 0\}$ is not vacuously satisfied for every $x \in X$, player A needs to ensure that player B has a feasible response that satisfies the coupled constraint:

$$\min_{x \in X} \Big\{ \max_{y \in Y} \{f(x,y) \,:\, g_B(x,y) \leq 0\} \,:\, \exists\, y \in Y \,.\, g_B(x,y) \leq 0 \Big\}.$$

This problem type seems less common in practice, however, since it enforces cooperation (in view of the satisfaction of the coupled constraint) among players that otherwise behave as perfect competitors (due to the objective function). Furthermore, this 'cooperation' is assumed to take place sequentially since player A decides before player B responds. A local optimisation procedure for minimax problems with type B-constraints can be found in [24].

2.2 Bi-level problems. Minimax problems are special instances of bi-level problems. Structurally, a bi-level problem is an optimisation problem that contains a constraint which requires a subset of the variables to optimise a subordinate optimisation problem. The optimistic (cooperative) bi-level problem can be defined as follows [9].

$$\min_{x,y} \; f(x,y) \tag{2.4a}$$

subject to

$$y \in \arg\min_{y' \in Y} \big\{h(x,y') \,:\, q(x,y') \leq 0\big\} \tag{2.4b}$$

$$g(x,y) \leq 0 \tag{2.4c}$$

$$x \in X. \tag{2.4d}$$

Again, the problem can be interpreted in a game-theoretic setting: for a fixed leader decision $x^* \in X$, player B (the follower) implements a decision

$$y^*(x^*) \in \arg\min_{y' \in Y} \big\{h(x^*,y') \,:\, q(x^*,y') \leq 0\big\}.$$

Player A (the leader), on the other hand, solves the problem

$$\min_{x \in X} \big\{f(x,y^*(x)) \,:\, g(x,y^*(x)) \leq 0\big\}$$

for *some* optimal follower response $y^*(x)$. Problem (2.4) is optimistic in the sense that the leader can choose which of the optimal responses $y^*(x)$ is implemented by the follower. While this might reflect reality in some settings, it is difficult to justify in a game-theoretic context. One could argue that the leader is allowed to cooperate with the follower in order to decide which of his optimal responses is implemented. Note, however, that absence of cooperation is a fundamental assumption of the bi-level problem. Indeed, if cooperation was allowed, the two-level structure would be inappropriate and the players would rather agree on a Pareto-optimal solution obtained by means of a bargaining process.

Alternatively, one could keep up the assumption of perfect competition but postulate that binding contracts are feasible. In this case, the leader can pay an arbitrarily small amount ϵ to the follower if the latter implements an optimal response that is favourable for the leader. If transfer payments are allowed, however, the leader would have an incentive to exploit their full potential and rather solve the following optimisation problem:

$$\min_{x,\theta(\cdot),y} \; \{f(x,y) + \theta(y)\}$$

subject to

$$\begin{aligned} y \in \arg\min_{y' \in Y} \; &\{h(x,y') - \theta(y') \,:\, q(x,y') \leq 0\} \\ g(x,y) &\leq 0 \\ \theta(y) &\geq 0 \qquad \forall\, y \in Y \\ x &\in X. \end{aligned}$$

Here, $\theta(y)$ represents the amount of money that the leader pays to the follower in case the latter implements decision $y \in Y$.

If there is a unique globally optimal follower response for every leader decision $x \in X$, the optimistic bi-level problem (2.4) is equivalent to (1.1).[1] This is the case, for example, if Y is convex and $h(x,\cdot)$ and $q(x,\cdot)$ are strictly convex and quasi-convex in their second arguments, respectively, for every $x \in X$. If the follower's optimal response is not guaranteed to be unique, the interpretation of the obtained solution needs to be carefully considered. In any case, the optimistic bi-level problem provides a lower (optimistic) bound for (1.1).

Another difference between (2.4) and (1.1) deserves attention: in the former problem, the feasible region of the second-stage problem depends on the solution chosen in the first-stage problem through the constraint '$q(x,y') \leq 0$'. This is not the case in (1.1) since the feasible region of the second-stage problem is defined through '$y \in Y$' only, i.e., it is independent of the first-stage decision x. In the following, we refer to the former problem type as 'dependent' bi-level problems, while the latter problem class is termed 'independent' bi-level problems. In this paper, we analyse and solve independent pessimistic bi-level problems, see (1.1).

If the second-stage problem is convex, optimistic bi-level problems can be expressed as mathematical programs with equilibrium constraints. Hence, the solution approaches described in the next section can be used for the local optimisation of (2.4). For problems with linear and quadratic constraints, global optimisation procedures have been suggested in [12, 13]. Global optimisation procedures for general instances of (2.4) have been proposed in [21, 30].

An alternative to (2.4) is the pessimistic bi-level problem presented in [21]:

$$\min_{x \in X} \; \max_{y \in \mathcal{Y}(x)} \; f(x,y), \tag{2.5a}$$

where for $x \in X$,

$$\mathcal{Y}(x) = \Big\{ y \in \arg\min_{y' \in Y} \{h(x,y') \,:\, q(x,y') \leq 0\} \,:\, g(x,y) \leq 0 \Big\}. \tag{2.5b}$$

---

[1] More precisely the *independent* version of (2.4) is equivalent to (1.1), see below.

Contrary to (2.4), the leader cannot decide anymore which optimal response is implemented by the follower. A subtle but important problem arises when interpreting '$g(x, y) \leq 0$', which implicitly represents a 'cooperative' constraint in (2.5). Clearly, it restricts the set of admissible follower responses. Since the objective value is undefined for leader decisions $x \in X$ for which there is no admissible follower response, however, the constraint implicitly restricts the leader as well. This entails two problems. Firstly, we already saw that 'cooperative' constraints are difficult to justify in a two-stage game-theoretic setting. Secondly, (2.5) could be regarded as inconsistent since its objective function anticipates the (non-cooperative) worst case, whereas its constraint assumes (some form of) cooperation between the players.

It seems more reasonable to assume that cautious leaders want to implement a decision x for which *every* optimal follower response y vacuously satisfies the constraint '$g(x, y) \leq 0$'. This would result in the following formulation:

$$\min_{x \in X} \left\{ \max_{y \in \mathcal{Y}'(x)} \{f(x, y)\} \; : \; g(x, y) \leq 0 \;\; \forall \, y \in \mathcal{Y}'(x) \right\},$$

where for $x \in X$,

$$\mathcal{Y}'(x) = \arg\min_{y' \in Y} \left\{ h(x, y') \; : \; q(x, y') \leq 0 \right\}.$$

The independent version of this problem is readily identified as a special case of (1.1). Indeed, assume that there is a function $\delta : Y \mapsto \mathbb{R}$ such that $q(x, \cdot) = \delta(\cdot)$ for all $x \in X$. Then the problem can be reformulated as follows.

$$\min_{(x, \tau) \in X \times \mathbb{R}} \quad \tau$$

subject to

$$\max \left\{ f(x, y) - \tau, g(x, y) \right\} \leq 0 \qquad \forall \, y \in \arg\min_{y' \in Y} \left\{ h(x, y') \; : \; \delta(y') \leq 0 \right\}.$$

2.3 Mathematical programs with equilibrium constraints. Mathematical programming with equilibrium constraints (MPEC) is concerned with problems of the following type [20].

$$\min_{x, y} \; f(x, y) \tag{2.6a}$$

subject to

$$y \in \left\{ z \in C(x) \; : \; F(x, z)^{\mathrm{T}}(z' - z) \geq 0 \;\; \forall \, z' \in C(x) \right\} \tag{2.6b}$$

$$(x, y) \in Z. \tag{2.6c}$$

Here, C is a set-valued function that maps vectors x to closed convex subsets of admissible vectors y. Constraint (2.6b) is a parametric variational inequality. Its value becomes apparent when recalling the first-order optimality conditions for optimisation problems: for a convex problem of type

$$\min_{y \in C(x)} \; \theta(x, y),$$

$y^* \in C(x)$ is an optimal solution if and only if

$$\left[\nabla_y \theta(x, y^*) \right]^{\mathrm{T}} (y - y^*) \geq 0 \qquad \forall \, y \in C(x). \tag{2.7}$$

For general non-convex problems, this variational inequality represents a necessary (but not sufficient) optimality condition.

Employing variational inequality (2.7), MPECs of type (2.6) can be used to solve the optimistic bi-level problem (2.4) if the follower's optimisation problem is convex. In case the follower's objective function is furthermore strictly convex, the optimal solution of (2.6) coincides with the solutions obtained from the pessimistic bi-level formulations (1.1) and (2.5). If the follower's objective function is convex (but not strictly convex), MPECs are valuable if the leader can choose which optimal response is implemented by the follower. As we discussed in the previous section, however, this seems difficult to justify in game-theoretic settings. If the follower's optimisation problem is non-convex, (2.6) allows the leader to choose any local optimum of the follower's optimisation problem. While this can provide useful lower (optimistic) bounds for the optimal value of (1.1) and (2.5), the obtained solution itself seems to be of little interest for both game-theoretic applications and optimisation under uncertainty. Indeed, any rational follower has an incentive to deviate from strategies prescribed by non-global optima of the follower's decision problem. Summing up, for the application areas envisioned in this paper, MPECs are primarily of interest if the follower's optimisation problem is convex and possesses a strictly convex objective function. These assumptions are rather restrictive; indeed, they are not even fulfilled by linear programs in general.

MPECs are generically non-convex and non-smooth optimisation problems. Without further assumptions, the feasible region of (2.6) is not even guaranteed to be closed. Hence, MPECs are very difficult to solve in general. In recent years. a variety of local optimisation procedures have been proposed that efficiently find feasible (but in general suboptimal) solutions for (2.6), see [20].

3 Problem analysis

In this section, we analyse the theoretical properties of the independent pessimistic bi-level problem (1.1):

$$\min_{x \in X} \; f(x)$$

subject to

$$g(x, y) \leq 0 \qquad \forall\, y \in \arg\max_{y' \in Y} \; h(x, y'),$$

where $X \subseteq \mathbb{R}^n$, $Y \subseteq \mathbb{R}^m$, $f : X \mapsto \mathbb{R}$ and $g, h : X \times Y \mapsto \mathbb{R}$. As stated in Section 1, we assume that X and Y are non-empty and compact, while f, g and h are continuous in their arguments. We also relate our findings to other variants of the bi-level problem (see Section 2.2).

In the first part of this section, we analyse the closedness of the feasible region of (1.1). Afterwards, we study the computational complexity of the problem. We close with a discussion of an ϵ-approximation of the problem. This approximation will provide the basis for our solution procedure in the next section.

3.1 Closedness of the feasible region. We denote the point-to-set mapping that relates a first-stage decision $x \in X$ to the set of optimal second-stage decisions $y \in Y$ by

$$\mathcal{Y}(x) = \arg\max_{y \in Y} \; h(x, y).$$

Note that the maximum $h^*(x) = \max_{y \in Y} h(x, y)$ exists for $x \in X$ since Y is compact and h continuous. For the same reasons, $\mathcal{Y}(x) = \{y' \in Y : h(x, y') \geq h^*(x)\}$ is

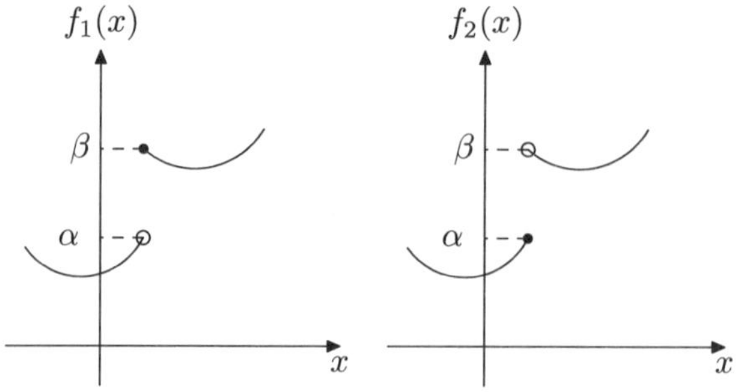

Figure 4 Semicontinuity of constraints determines closedness of the feasible region. Function f_1 in the left graph is upper semicontinuous, while function f_2 in the right graph is lower semicontinuous. $\{x : f_1(x) \leq \delta\}$ is only closed for $\delta < \alpha$ and $\delta \geq \beta$, whereas $\{x : f_1(x) \geq \delta\}$ is closed for all δ. Similarly, $\{x : f_2(x) \geq \delta\}$ is only closed for $\delta \leq \alpha$ and $\delta > \beta$, whereas $\{x : f_2(x) \leq \delta\}$ is closed for all δ.

compact for every $x \in X$. Thus, we can denote the maximum value of $g(x, \cdot)$ over all optimal second-stage solutions $y \in \mathcal{Y}(x)$ by $g^*(x)$:

$$g^*(x) = \max_{y \in \mathcal{Y}(x)} g(x, y).$$

With this notation, the independent pessimistic bi-level problem (1.1) equals

$$\min_{x \in X} \left\{ f(x) : g^*(x) \leq 0 \right\}.$$

This problem is closed if g^* is lower semicontinuous, see Figure 4. Closedness of the feasible region is a crucial property of optimisation problems. Indeed, the minimum or maximum of a continuous function over a non-closed domain might not be attained. The following example illustrates that the feasible region of (1.1) is generically non-closed:

$$\max_{x \in [-1,1]} \quad x$$

subject to

$$x \leq y \qquad \forall y \in \operatorname*{arg\,min}_{y' \in [-1,1]} \left\{ xy' \right\}.$$

The model is readily identified as an instance of (1.1). Note that points $x > 0$ are infeasible since the constraint becomes '$x \leq -1$' in that case. Similarly, $x = 0$ is infeasible since the constraint becomes '$0 \leq y \; \forall y \in [-1, 1]$' in that case. For points $x < 0$ the constraint becomes '$x \leq 1$' and is as such vacuously satisfied. As a result, the admissible solutions are $x \in [0, 1)$, that is, the feasible region is not closed. Furthermore, since the objective is to maximise x, the optimum is undefined. Note that optimal solutions do exist for other objective functions, such as 'max $-x$'.

The non-closedness of the feasible region of (1.1) can be further analysed by considering Berge's Maximum theorem. For this we need the notions of hemicontinuity and closedness of point-to-set mappings. A point-to-set mapping $\gamma : X \mapsto \mathcal{P}(Y)$ maps points $x \in X$ to subsets of Y. γ is upper hemicontinuous at $x \in X$ if $\gamma(x) \subseteq S$ for some open set $S \subseteq Y$ implies that $\gamma(x') \subseteq S$ for all x' in a neighbourhood of x. Similarly, γ is lower hemicontinuous at $x \in X$ if $\gamma(x) \cap S \neq \emptyset$ for some open set $S \subseteq Y$ implies that $\gamma(x') \cap S \neq \emptyset$ for all x' in a neighbourhood of x. γ is continuous at $x \in X$ if it is lower and upper hemicontinuous at x. The concepts of upper and

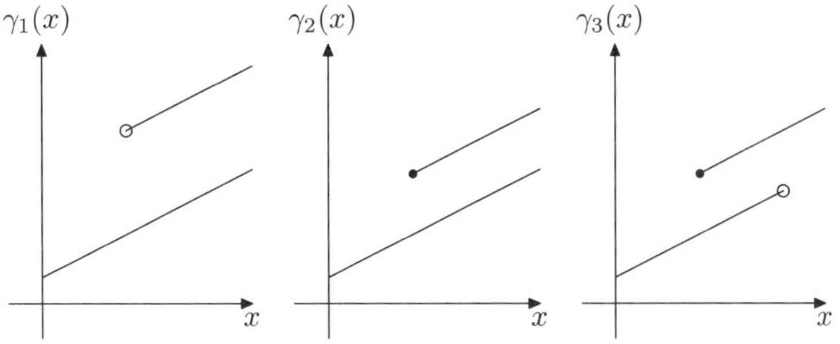

Figure 5 Upper and lower hemicontinuity. Point-to-set mapping γ_1 is lower, but not upper hemicontinuous. Point-to-set mapping γ_2, on the other hand, is upper, but not lower hemicontinuous. Point-to-set mapping γ_3, finally, is neither upper nor lower hemicontinuous.

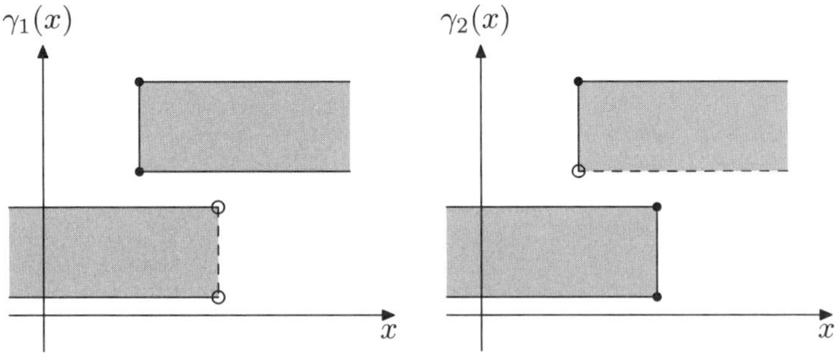

Figure 6 Closed and closed-valued point-to-set mappings. Point-to-set mapping γ_1 is closed-valued, but not closed. Point-to-set mapping γ_2 is neither closed-valued nor closed.

lower hemicontinuity are illustrated in Figure 5. γ is closed at $x \in X$ if $x_n \to x$ and $y_n \to y$ for $y_n \in \gamma(x_n)$ imply that $y \in \gamma(x)$. Note that closedness of the point-to-set mapping γ at $x \in X$ (as just described) implies closedness of the set $\gamma(x)$, but not vice versa. In order to avoid confusion, we will refer to the latter concept as closed-*valued* (or, in connection with boundedness, as compact-*valued*) in the following. Both concepts are illustrated in Figure 6. We say that γ is lower/upper hemicontinuous or closed if it is lower/upper hemicontinuous or closed at all $x \in X$.

With this notation, Berge's Maximum theorem can be stated as follows.

Theorem 3.1 (Berge's Maximum Theorem [3, 7]) *For $X \subseteq \mathbb{R}^n$ and $Y \subseteq \mathbb{R}^m$, let $\gamma : X \mapsto \mathcal{P}(Y)$ be a compact-valued point-to-set mapping. Furthermore, for a continuous function $\eta : X \times Y \mapsto \mathbb{R}$, let*

$$\Gamma(x) = \underset{y \in \gamma(x)}{\arg\max}\ \eta(x, y).$$

Then Γ is compact-valued for all $x \in X$. Furthermore, if γ is continuous at $x \in X$, then Γ is closed and upper hemicontinuous at x.

In our setting, X and Y coincide with the respective sets of (1.1), $\gamma(x) = Y$ for all $x \in X$, $\eta(x, y) = h(x, y)$ and $\Gamma(x) = \mathcal{Y}(x)$. Since Y is compact, γ is compact-valued in our case. Continuity of γ is vacuously satisfied. Hence, Berge's Maximum

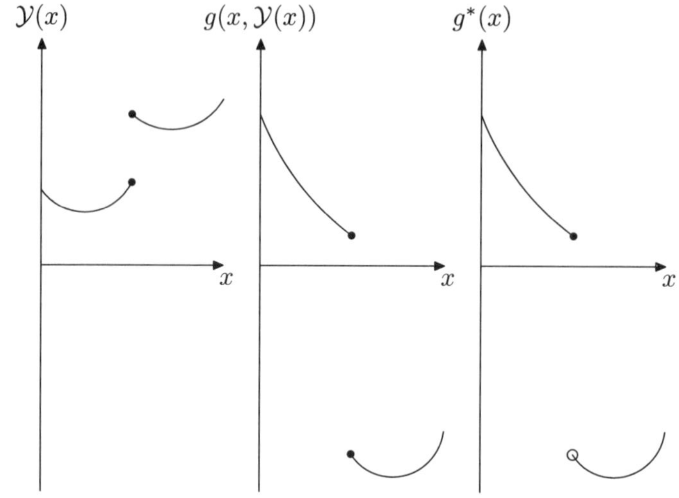

Figure 7 Closedness and upper hemicontinuity of $\mathcal{Y}$ do not imply lower semicontinuity of g^*. For the point-to-set mapping $\mathcal{Y}$ in the left graph, the middle and right graphs visualise $x \mapsto \{g(x,y) : y \in \mathcal{Y}(x)\}$ and $x \mapsto g^*(x)$, respectively. Although $\mathcal{Y}$ is closed and upper hemicontinuous, g^* fails to be lower semicontinuous.

theorem is applicable and $\mathcal{Y}$ is closed and upper hemicontinuous. These properties do *not* imply lower semicontinuity of g^*, however. This is illustrated in Figure 7.

Let us contrast this with the independent *optimistic* version of the bi-level problem. Here, the optimisation problem is

$$\min_{x \in X} \left\{ f(x) : g'(x) \leq 0 \right\} \quad \text{with} \quad g'(x) = \min_{y \in \mathcal{Y}(x)} g(x,y).$$

Again, we can employ Berge's Maximum theorem to show that $\mathcal{Y}$ is closed and upper hemicontinuous. With the new choice of functions, however, this implies that g' is lower semicontinuous. Hence, the independent optimistic bi-level problem has a closed feasible region. Note that the same reasoning can be employed to show the *upper* semicontinuity of g^* in the case of the independent pessimistic bi-level problem.

Let us compare this result with the well-known fact that *dependent* optimistic bi-level problems have non-closed feasible regions in general. This can be illustrated by the following example, which is borrowed from [29]:

$$\min_{\substack{x \in \mathbb{R}, \\ y \in [0,2]}} \left\{ y - x \right\}$$

subject to

$$x \leq 1$$
$$y \in \underset{y' \in [0,2]}{\arg\max} \left\{ x + y' : xy' \geq 0 \right\}.$$

One can readily verify that the feasible region of this problem is described by the non-closed set $(x,y) \in (-\infty,0) \times \{0\} \cup [0,1] \times \{2\}$. In the terminology of Theorem 3.1, the point-to-set mapping γ is not continuous at zero and Berge's Maximum theorem is as such not applicable.[2] Hence, the added generality of the dependent optimistic bi-level problem over its independent counterpart comes at the price of generically non-closed feasible regions. Of course the dependent pessimistic bi-level problem has a generically non-closed feasible region as well since it encompasses its independent counterpart as a special case.

Summing up, we have seen that the pessimistic bi-level problem is more involved than its optimistic counterpart. The feasible region of both problems can be expressed as $\{x \in X : g(x) \leq 0\}$ for some function $g : X \mapsto \mathbb{R}$. For independent optimistic bi-level problems, we can employ Berge's Maximum theorem to show that this g is lower semicontinuous, which in turn implies that the feasible region is closed. Although Berge's Maximum theorem is applicable to independent pessimistic bi-level problems as well, this would only allow us to prove upper semicontinuity of g. Indeed, a simple example demonstrated that the feasible region of independent pessimistic bi-level problems is generically non-closed. Dependent bi-level problems, finally, have generically non-closed feasible regions, no matter whether the optimistic or pessimistic variant is considered.

It turns out that even in benign cases where the feasible region is closed, bi-level problems are difficult to solve in general. In the next section, we examine the computational complexity of several variants of the bi-level problem.

3.2 Computational complexity. In this section, we consider the computational complexity of optimistic and pessimistic bi-level problems. Clearly, these problems are difficult to solve in general, i.e., without further assumptions about the feasible regions and involved functions. In the following, we will look at special cases and show that even in seemingly benign settings, bi-level problems remain difficult to solve. We assume some familiarity with the basic notions of complexity theory, in particular with $\mathcal{NP}$-hardness. For a textbook introduction to complexity theory, the reader is referred to [14]. Complexity results for dependent optimistic bi-level problems are compiled in [10, 33]; we will mention some of them further below.

We will base our discussion on the Knapsack problem: imagine n different goods with weights w_i and utilities u_i, $i = 1, \ldots, n$. The goal is to find a utility-maximising subset of goods whose collective weight does not exceed W:

$$\max_{x \in \{0,1\}^n} \left\{ \sum_{i=1}^{n} u_i x_i \; : \; \sum_{i=1}^{n} w_i x_i \leq W \right\}. \tag{3.1}$$

The Knapsack problem is known to be $\mathcal{NP}$-hard [14]. Intuitively speaking, this is caused by the binarity requirement, which forces (current) solution procedures to consider many of the elements of $\{0,1\}^n$ in the worst case. Since this set is of size 2^n, this results in exponential solution procedures. For the remainder of this section, we will be interested in the following reformulation:

$$\max_{x \in [0,1]^n} \left\{ \sum_{i=1}^{n} u_i x_i - M \sum_{i=1}^{n} \min\{x_i, 1 - x_i\} \; : \; \sum_{i=1}^{n} w_i x_i \leq W \right\}. \tag{3.1'}$$

Here, the binarity requirement has been dropped in favour of the penalty function $M \sum_{i=1}^{n} \min\{x_i, 1 - x_i\}$. Note that this function vanishes at binary solutions, while other x are penalised with a multiplicity of the constant M. It can be shown that

[2]More precisely, γ is not lower hemicontinuous in this case. To see this, consider the open set $S = (1, 2)$: $\gamma(x) \cap S \neq \emptyset$ for $x \geq 0$, but $\gamma(x) \cap S = \emptyset$ for $x < 0$.

the penalty function is exact for finite M, i.e., for a suitably large (but polynomial) M, the optimal solution of the latter problem is guaranteed to be binary [2, 4]. For such M, the optimal solutions of (3.1) and (3.1') coincide. Due to the penalty term, (3.1') remains a non-convex optimisation problem.

Let us now consider linear bi-level problems:

$$\min_{x \in \mathbb{R}^n_+} \; c^{\mathrm{T}} x \tag{3.2a}$$

subject to

$$Ax + By \leq b \qquad \forall\, y \in \arg\max_{z \in \mathbb{R}^m_+} \left\{ d^{\mathrm{T}} x + f^{\mathrm{T}} z \,:\, Cx + Dz \leq g \right\}. \tag{3.2b}$$

Model (3.2) represents a dependent pessimistic bi-level problem. If the '$\forall$' in the first constraint is replaced by an '$\exists$', we obtain the corresponding dependent optimistic bi-level problem. Let us first assume that $C = 0$. In this case, we consider independent bi-level problems, i.e., the set of admissible second-stage decisions does not depend on the first-stage variables. Moreover, since the contribution of the first-stage variables to the second-stage objective is constant, the set of optimal second-stage decisions does not depend on the first-stage variables either. In itself, however, this does not imply that (3.2) can be solved efficiently. Indeed, it could well be that the (vertex) set of optimal second-stage solutions has exponential cardinality, and the mere enumeration of its elements would be a difficult problem. It turns out, however, that one can employ duality arguments to show that both the optimistic and the pessimistic version of the linear independent bi-level problem can be solved in polynomial time [31].

Let us now assume that $C \neq 0$. In [10] it has been shown that the linear dependent optimistic bi-level problem can be solved in polynomial time if the number of second-stage variables is regarded as a constant. The authors show that the result can be extended to the linear dependent pessimistic bi-level problem. If we drop the assumption of a constant number of second-stage variables, however, the linear dependent optimistic bi-level problem becomes strongly $\mathcal{NP}$-hard [18]. In the following, we show that the linear dependent pessimistic bi-level problem is $\mathcal{NP}$-hard as well, which is an immediate consequence of [2]. Consider the following formulation:

$$\max_{(x,\tau) \in \mathbb{R}^{n+1}_+} \left\{ \sum_{i=1}^n u_i x_i - M\tau \right\}$$

subject to

$$\sum_{i=1}^n w_i x_i \leq W$$

$$\tau \geq \sum_{i=1}^n y_i \qquad \forall\, y \in \arg\max_{z \in [0,1]^n} \left\{ \sum_{i=1}^n z_i \,:\, z \leq x,\; z \leq e - x \right\},$$

where e denotes the vector of all ones. Note that for a given first-stage solution x, the optimal second-stage solution is y^* with $y_i^* = \min\{x_i, 1 - x_i\}$. Thus, this problem is equivalent to (3.1'). As a result, the linear dependent pessimistic bi-level problem is indeed $\mathcal{NP}$-hard.

Let us now consider nonlinear versions of the bi-level problem. Clearly, non-linear dependent bi-level problems are $\mathcal{NP}$-hard since they encompass the linear dependent bi-level problems as special cases. In the following, we show that contrary its linear counterpart, the nonlinear independent pessimistic bi-level problem is $\mathcal{NP}$-hard as well. More precisely, we consider

$$\min_{x \in X} f(x) \tag{3.3a}$$

subject to

$$g(x,y) \geq 0 \qquad \forall\, y \in \arg\max_{z \in Y} h(x,z) \tag{3.3b}$$

with f, g linear, X and Y polyhedral and h concave and smooth. Let us take a look at the following problem:

$$\max_{(x,\tau) \in \mathbb{R}_+^{n+1}} \left\{ \sum_{i=1}^{n} u_i x_i - M\tau \right\}$$

subject to

$$\sum_{i=1}^{n} w_i x_i \leq W$$

$$\tau \geq \sum_{i=1}^{n} y_i^2 \qquad \forall\, (y^1, y^2) \in \arg\max_{(z^1,z^2) \in \mathbb{R}_+^{2n}} \left\{ -\left\|z^1 - x\right\|_2 \,:\, z^2 \leq z^1,\, z^2 \leq e - z^1 \right\},$$

where $\|\cdot\|_2$ denotes the Euclidean norm. Clearly, the problem is an instance of (3.3). For a given first-stage solution x, the optimal second-stage solution is (y^1, y^2) with $y_i^1 = x_i$ and $y_i^2 \in \left[0, \min\left\{x_i, 1 - x_i\right\}\right)$. Since τ has to be greater than or equal to the sum of y^2's components for *every* optimal y^2, this problem is equivalent to (3.1'), too. We note that the problem remains $\mathcal{NP}$-hard if we allow the second-stage problem to be solved only with (polynomially bounded) accuracy ϵ. The corresponding proof can be found in [31].

Summing up, linear independent bi-level problems are efficiently solvable. The nonlinear independent pessimistic bi-level problem, however, is $\mathcal{NP}$-hard even in very elementary settings. Dependent bi-level problems are in general more difficult than their independent counterparts. Indeed, the only efficiently solvable case we are aware of are linear dependent bi-level problems under the additional assumption of a constant number of second-stage variables.

3.3 ϵ-approximations. We have seen in Section 3.1 that the feasible region of the independent pessimistic bi-level problem (1.1),

$$\min_{x \in X} f(x)$$

subject to

$$g(x,y) \leq 0 \qquad \forall\, y \in \arg\max_{y' \in Y} h(x,y'),$$

is generically non-closed. In this section, we construct an ϵ-approximation of (1.1) that is closed and, under certain technical assumptions, 'converges' (with respect to the set of optimal solutions) to problem (1.1) as $\epsilon \to 0$. This approximation serves as the basis for the solution procedure which we will discuss in Section 4.

Using the notation from Section 3.1, we can reformulate (1.1) as follows.

$$\min_{x \in X} \left\{ f(x) \,:\, g(x,y) \leq 0 \;\; \forall\, y \in \mathcal{Y}(x) \right\},$$

where

$$\mathcal{Y}(x) = \left\{ y \in Y \,:\, h(x,y) \geq h(x,y') \;\; \forall\, y' \in Y \right\}.$$

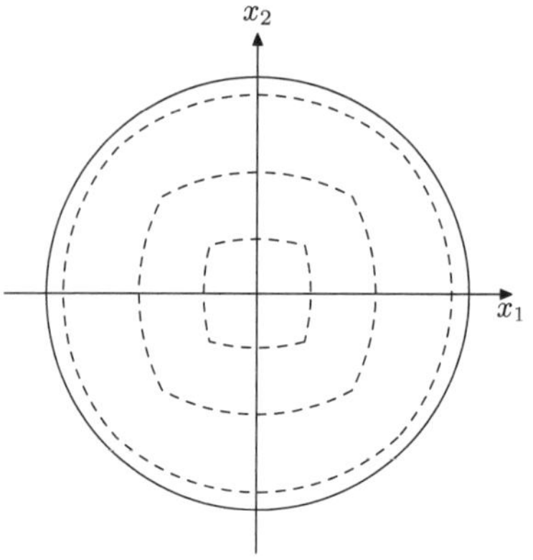

Figure 8 Feasible region of P and $P(\epsilon)$ for different values of ϵ. Shown is the feasible region (with respect to the X-space) induced by $g(x,y) \leq 0$ for $y \in \mathcal{Y}(x)$ (straight line) and $y \in \mathcal{Y}_\epsilon(x)$ for different ϵ (dashed lines). In this example, $x, y \in \mathbb{R}^2$, $g(x,y) = \|x\|_2^2 + \|y\|_2^2 - 50$ and $h(x,y) = \|x - y\|_2$.

In the remainder of this section, we refer to this problem as P. Let us denote by θ^* the tightest bound on the objective value of P that we can obtain:

$$\theta^* = \inf_{x \in X} \left\{ f(x) \,:\, g(x,y) \leq 0 \;\; \forall y \in \mathcal{Y}(x) \right\}.$$

Since P is not guaranteed to be closed, θ^* might or might not be attainable.

We consider the following approximation $P(\epsilon)$ of P, $\epsilon > 0$:

$$\min_{x \in X} \left\{ f(x) \,:\, g(x,y) \leq 0 \;\; \forall y \in \mathcal{Y}_\epsilon(x) \right\}, \tag{3.4a}$$

where

$$\mathcal{Y}_\epsilon(x) = \left\{ y \in Y \,:\, h(x,y) > h(x,y') - \epsilon \;\; \forall y' \in Y \right\}. \tag{3.4b}$$

We denote the optimal objective value of this problem by $\theta(\epsilon)$ if it exists and set $\theta(\epsilon) = \infty$ otherwise. Note that $\mathcal{Y}_\epsilon(x)$ is a superset of $\mathcal{Y}(x)$, $x \in X$. Hence, $P(\epsilon)$ enforces more '$g(x,y) \leq 0$'-constraints and as such constitutes a conservative approximation of P, see Figure 8. This, together with the fact that the objective functions of both problems coincide, implies that $\theta(\epsilon) \geq \theta^*$ for all $\epsilon > 0$. Decreasing the value of ϵ leads to better approximations of P in the following sense:

$$\forall \kappa > 0 \; \exists \lambda > 0 \,:\, \forall x \in X, \, \epsilon \in (0, \lambda], \, y \in \mathcal{Y}_\epsilon(x) \;\; \exists y' \in \mathcal{Y}(x) \,:\, \|y - y'\| \leq \kappa. \tag{3.5}$$

Hence, the distance between every $y \in \mathcal{Y}_\epsilon(x)$ and its 'closest' $y' \in \mathcal{Y}(x)$ can be made arbitrarily small by decreasing ϵ. The nature of $\mathcal{Y}_\epsilon$ and the approximation thereby obtained is further illustrated in Figure 9.

Contrary to P, $P(\epsilon)$ has a closed feasible region for every $\epsilon > 0$. For this property, the strict inequality in the definition of $\mathcal{Y}_\epsilon$ turns out to be crucial. To see this, note that (3.4) can be equivalently expressed as

$$\min_{x \in X} \left\{ f(x) \,:\, g^*_\epsilon(x) \leq 0 \right\} \quad \text{with} \quad g^*_\epsilon(x) = \sup_{y \in \mathcal{Y}_\epsilon(x)} g(x,y). \tag{3.4'}$$

Figure 10 (left) shows that $\mathcal{Y}_\epsilon$ is 'open from the sides' in the following sense: if $y \in \mathcal{Y}_\epsilon(x)$, then there is a neighbourhood $N(x)$ of x such that $y \in \mathcal{Y}_\epsilon(x')$ for all $x' \in N(x)$. This property is a direct consequence of the strict inequality in the definition of $\mathcal{Y}_\epsilon$. As Figure 10 (middle) illustrates, this property is passed on to

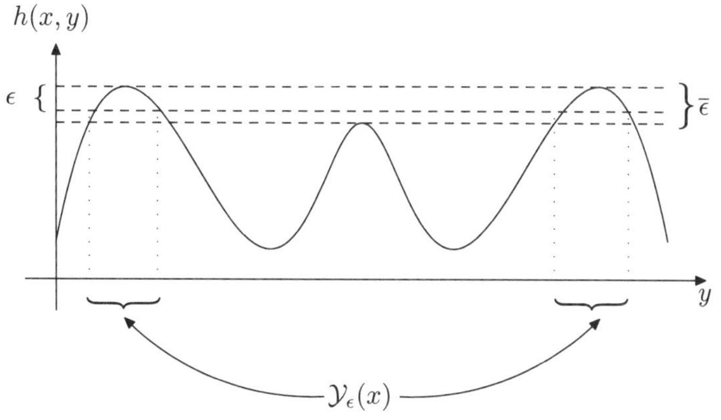

Figure 9 Convergence of $\mathcal{Y}_\epsilon(x)$ to $\mathcal{Y}(x)$ for given x. For ϵ small enough (in this case, for $\epsilon < \bar{\epsilon}$), $\mathcal{Y}_\epsilon(x)$ contains only second-stage solutions 'close' to global maxima of h.

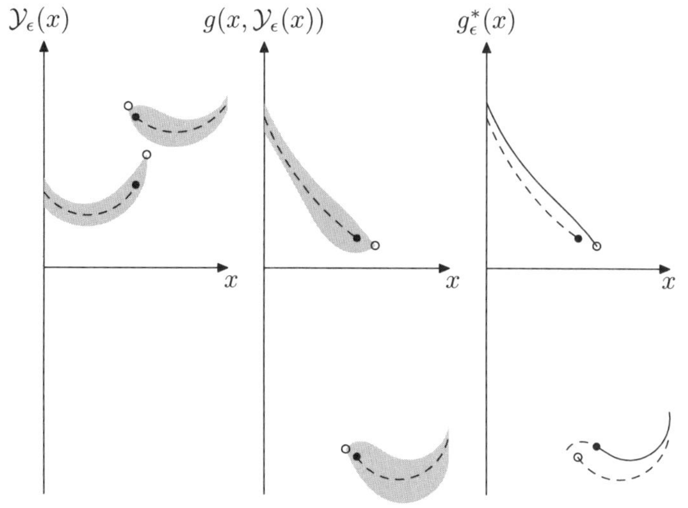

Figure 10 Employing $\mathcal{Y}_\epsilon$ instead of $\mathcal{Y}$ leads to closed, conservative approximations of P. The graphs illustrate the consequences of using $\mathcal{Y}_\epsilon$ (shaded area) instead of $\mathcal{Y}$ (dashed lines) for the example from Figure 7. Function g^*_ϵ is lower semicontinuous and bounds g^* from above.

the mapping $x \mapsto \{g(x, y) : y \in \mathcal{Y}_\epsilon(x)\}$. Figure 10 (right), finally, shows that this property ensures lower semicontinuity of g^*_ϵ. Lower semicontinuity of g^*_ϵ, however, guarantees that (3.4′) (and thus, (3.4)) has a closed feasible region.

The difference between $\theta(\epsilon)$ and θ^* depends on the value of ϵ. From the relation between $\mathcal{Y}_\epsilon$ and $\mathcal{Y}$, one expects that smaller values of ϵ lead to smaller gaps between $\theta(\epsilon)$ and θ^*. It can happen, however, that $P(\epsilon)$ is infeasible for every $\epsilon > 0$, whereas P contains feasible solutions.[3] Similarly, optimal solutions of $P(\epsilon)$ might 'substantially deviate' from those of P for every $\epsilon > 0$. In the following, we discuss a technical condition under which such problems cannot arise. More precisely, we

[3]This can happen, for example, if the feasible region of P contains no interior points.

give a condition under which

$$\lim_{\epsilon \to 0} \theta(\epsilon) = \theta^*.$$

By construction, $\theta(\epsilon) \geq \theta^*$ for all $\epsilon > 0$. Moreover, feasible solutions of P have objective values of θ^* or larger. Since P is not necessarily closed, however, θ^* might not be achievable. Hence, we cannot hope for more than convergence of $\theta(\epsilon)$ to θ^* for $\epsilon \to 0$. In order to show such a convergence, we introduce another approximation $P'(\delta)$ of P, $\delta > 0$:

$$\min_{x \in X} \left\{ f(x) \,:\, g(x,y) \leq -\delta \;\; \forall\, y \in \mathcal{Y}(x) \right\}.$$

Let us denote the optimal objective value of $P'(\delta)$ by $\psi(\delta)$; in case $P'(\delta)$ is infeasible, we set $\psi(\delta) = \infty$. For every $\delta > 0$, we can find an $\epsilon(\delta) > 0$ such that $\theta(\epsilon(\delta)) \leq \psi(\delta)$. Indeed, let $x \in X$ be feasible for $P'(\delta)$, that is, $g(x,y) \leq -\delta$ for all $y \in \mathcal{Y}(x)$. We want to show that x is feasible for some $P(\epsilon)$ as well, that is, $g(x,y) \leq 0$ for all $y \in \mathcal{Y}_\epsilon(x)$. Due to the continuity of g, we can find a $\phi > 0$ such that $g(x,y') \leq 0$ for all $y' \in Y$ with $\|y - y'\| \leq \phi$. Hence, we need to choose ϵ such that all $y' \in \mathcal{Y}_\epsilon(x)$ have a distance of at most ϕ to some $y \in \mathcal{Y}(x)$. This is exactly what property (3.5) allows us to do. Hence, $P'(\delta)$ is indeed a conservative approximation of some $P(\epsilon(\delta))$, and we have shown that

$$x \text{ is feasible for } P'(\delta) \;\rightsquigarrow\; x \text{ is feasible for } P(\epsilon(\delta)) \;\rightsquigarrow\; x \text{ is feasible for } P.$$

Furthermore, the objective functions of all three problems coincide.

It remains to be shown that under a certain condition,

$$\lim_{\delta \to 0} \psi(\delta) = \theta^*.$$

Tuy [17, 32] calls such problems P 'stable'. A sufficient condition for the stability of P is that the function g^* from Section 3.1,

$$g^*(x) = \max_{y \in \mathcal{Y}(x)} g(x,y),$$

has no local minimum of value zero, see [17, 32]. A tighter description can be found in [31], where it is shown that P is stable as long as not all of its global optima (if such global optima exist) are attained at local minima of g^* with value zero. The stability condition is illustrated in Figure 11. Since $\theta(\epsilon(\delta))$ is 'sandwiched' between $\psi(\delta)$ and θ^*, stability of P implies convergence of $\theta(\epsilon)$ to θ^* for $\epsilon \to 0$.

Summing up, we have discussed a conservative approximation $P(\epsilon)$ of the independent pessimistic bi-level problem P. Contrary to the latter problem, this approximation is closed and as such amenable to optimisation techniques. In the next section, we will present a semi-infinite solution procedure that allows us to solve $P(\epsilon)$ for fixed ϵ. The resulting solution represents a feasible but in general suboptimal solution to P. Better solutions can be found for smaller values of ϵ, and if a technical assumption is satisfied, these solutions are guaranteed to converge to the global optima of P (if existent) for $\epsilon \to 0$.

Throughout this section, our key motivation for the ϵ-approximation (3.4) was computational tractability. Indeed, contrary to the original problem (1.1), the approximation (3.4) has a closed feasible region. Apart from computational aspects, there are at least two further reasons why the ϵ-approximation is a valuable tool in the setting of Stackelberg games. Firstly, it is very likely that the leader does not know the follower's optimisation problem exactly, that is, he might not be sure about the precise shape of all involved functions. In this setting, which is

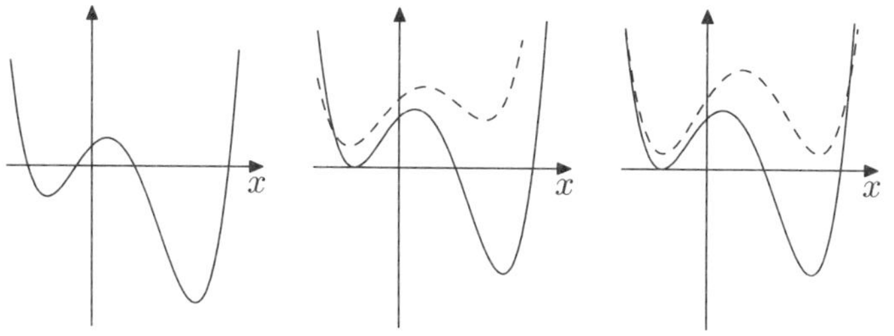

Figure 11 Technical condition for the convergence of $P(\epsilon)$ to P, $\epsilon \to 0$. The straight and dashed lines represent g^* and f, respectively. In the first graph, g^* does not possess local minima of value zero and hence, convergence can be guaranteed for any f. In the second graph, the unique global optimum of f is located at a solution where g^* attains a local minimum of value zero. As a result, convergence cannot be ensured in this case. In the third graph, finally, there are global optima of f where g^* does not attain a local minimum of value zero and hence, convergence is guaranteed.

termed 'imperfect information' in game theory, the leader might want to include a 'safety margin' by solving the ϵ-approximation (3.4) instead of the original problem (1.1). Secondly and closely related, real-world decision makers suffer from 'bounded rationality'. In our setting, this means that the follower might not possess the resources to solve his optimisation problem to global optimality. Instead, he might settle for slightly suboptimal, e.g., ϵ-optimal decisions. By considering not only optimal, but also ϵ-optimal follower decisions, the leader can hedge against followers with potentially bounded rationality.

4 Semi-infinite solution procedure

In this section, we present a solution procedure for the ϵ-approximation (3.4) of problem (1.1) with $\epsilon > 0$ fixed:

$$\min_{x \in X} \left\{ f(x) \, : \, g(x,y) \leq 0 \;\; \forall \, y \in \mathcal{Y}_\epsilon(x) \right\},$$

where

$$\mathcal{Y}_\epsilon(x) = \left\{ y \in Y \, : \, h(x,y) > h(x,y') - \epsilon \;\; \forall \, y' \in Y \right\}.$$

We will employ semi-infinite programming principles [15] to solve this problem. Note that the constraint '$g(x,y) \leq 0 \;\; \forall \, y \in \mathcal{Y}_\epsilon(x)$' is equivalent to

$$\begin{aligned} & \left[y \in \mathcal{Y}_\epsilon(x) \;\Rightarrow\; g(x,y) \leq 0 \right] && \forall \, y \in Y \\ \Leftrightarrow \quad & \left[y \notin \mathcal{Y}_\epsilon(x) \right] \;\vee\; \left[g(x,y) \leq 0 \right] && \forall \, y \in Y \\ \Leftrightarrow \quad & \left[\exists \, y' \in Y \, : \, h(x,y) \leq h(x,y') - \epsilon \right] \;\vee\; \left[g(x,y) \leq 0 \right] && \forall \, y \in Y, \end{aligned}$$

which bears structural similarity to the discussion in [21].[4] We are thus lead to the following reformulation of (3.4):

$$\min_{(x,y') \in X \times Y} \;\; f(x)$$

subject to

$$\left[g(x,y) \leq 0 \right] \;\vee\; \left[h(x,y) \leq h(x,y') - \epsilon \right] \qquad \forall \, y \in Y.$$

[4]In the context of [21], however, disjunctive constraints are needed to enforce dependent second-stage constraints, whereas we employ disjunctive constraints in order to enforce $g(x,y) \leq 0$ for all second-stage optimisers, i.e., to reflect the 'pessimism' of the leader.

Note that this problem contains an infinite number of constraints in general, namely one for every $y \in Y$. In the spirit of semi-infinite programming, we solve this problem iteratedly for finite subsets Y_K of Y. More precisely, at the Kth iteration, we consider

$$\min_{(x,y')\in X\times Y} f(x) \tag{4.1a}$$

subject to

$$\big[g(x,y_k) \le 0\big] \vee \big[h(x,y_k) \le h(x,y') - \epsilon\big] \qquad \forall\, y_k \in Y_K \tag{4.1b}$$

for a finite subset $Y_K = \{y_1,\dots,y_K\} \subseteq Y$. Clearly, if some $x \in X$ is feasible for (3.4), then it is also feasible for (4.1) and the objective values coincide. If $x \in X$ is feasible for (4.1), however, we cannot conclude that x is also feasible for (3.4). Hence, we refer to (4.1) as Y_K-relaxation of (3.4). The optimal solution of (4.1), $(x_K, y'_K) \in X \times Y$, either satisfies

$$g(x_K, y) \le 0 \qquad \forall\, y \in \mathcal{Y}_\epsilon(x_K)$$

or there is an y_{K+1} such that

$$y_{K+1} \in \mathcal{Y}_\epsilon(x_K) \;\wedge\; g(x_K, y_{K+1}) > 0.$$

In the former case, x_K is feasible for the ϵ-approximation (3.4) and at the same time optimal for its Y_K-relaxation (4.1). Since the objective functions of both problems coincide, we can conclude that x_K is optimal for (3.4) and terminate. In the latter case, y_{K+1} serves as a certificate for x_K's infeasibility. In this case, we set $Y_{K+1} = Y_K \cup \{y_{K+1}\}$ and iterate.

Before we consider the details of our solution procedure, we want to address two issues of practical importance. Firstly, we discuss how the disjunctive constraint (4.1b) can be expressed explicitly. Secondly, we consider a 'tightened' version of the Y_K-relaxation that can lead to faster convergence in practice.

For a given $y_k \in Y_K$, constraint (4.1b) is a disjunctive constraint of type

$$\big[\alpha(x) \le 0\big] \vee \big[\beta(x) \le 0\big]. \tag{4.2}$$

Disjunctions of this type are equivalent to the existence of a $\lambda \in [0,1]$ such that

$$\lambda\alpha(x) + (1-\lambda)\beta(x) \le 0.$$

Similarly, (4.2) is equivalent to the existence of a $\lambda \in \{0,1\}$ such that

$$\alpha(x) - M\lambda \le 0 \;\wedge\; \beta(x) - M(1-\lambda) \le 0,$$

where M denotes a sufficiently large number. Which reformulation leads to 'easier' problems depends on the remaining constraints and the available optimisation software. In our numerical tests, we employ the first variant since it does not require any additional constants.

We now consider the issue of tightening relaxation (4.1). In this problem, both x and y' are decision variables, but only x enters the objective function. Indeed, y' plays the role of an auxiliary variable that decides which constraints '$g(x,y_k) \le 0$', $y_k \in Y_K$, are to be enforced.[5] This implies that the choice $y' \in \arg\max_{y\in Y} h(x,y)$ is always optimal since it leads to the least restrictions on x. In the extreme case, $y' \in Y$ satisfies $h(x,y') \ge h(x,y_k) + \epsilon$ for all $y_k \in Y_K$, which reduces (4.1) to the essentially unconstrained problem $\min_{x\in X} f(x)$. The effect just described can result in a slow convergence until large parts of the Y-space are included in Y_K.

[5]Here, our setting deviates from [21], where the auxiliary variable naturally occurs in the objective function. As a result, the issue to be addressed in the following does not arise there.

Hence, it is desirable to impose further constraints on y'. In the following, we consider problem (4.1) with the additional constraint

$$g(x,y') \leq 0. \tag{4.1c}$$

We have already seen that feasibility of $x \in X$ for (3.4) implies existence of $y' \in Y$ such that (x, y') satisfies

$$\big[g(x,y) \leq 0\big] \vee \big[h(x,y) \leq h(x,y') - \epsilon\big] \qquad \forall\, y \in Y.$$

Since $Y_K \subseteq Y$, this directly implies feasibility of (x, y') for (4.1) without (4.1c). Moreover, since $y' \in Y$, the condition is satisfied for $y = y'$ as well, leading to

$$\big[g(x,y') \leq 0\big] \vee \big[h(x,y') \leq h(x,y') - \epsilon\big].$$

Since the second expression cannot be satisfied for $\epsilon > 0$, this implies that $g(x,y') \leq 0$. Thus, constraint (4.1c) is indeed implied by the ϵ-approximation (3.4). Although (4.1c) is not needed to guarantee convergence of our semi-infinite solution procedure, it can lead to significant performance improvements in practice. We will come back to this point in Section 5.

We are now ready to present the solution procedure for the ϵ-approximation (3.4) of the independent pessimistic bi-level problem (1.1).

Algorithm. Semi-infinite solution procedure for problem (3.4).

1. **Initialisation.** Set $Y_0 = \emptyset$ and $K = 0$ (iteration counter).
2. **Master Problem.** Solve problem (4.1):

$$\min_{(x,y') \in X \times Y} \quad f(x)$$

subject to

$$\big[g(x,y_k) \leq 0\big] \vee \big[h(x,y_k) \leq h(x,y') - \epsilon\big] \qquad \forall\, y_k \in Y_K$$
$$g(x,y') \leq 0.$$

Let (x_K, y'_K) denote the/an optimal solution.

3. **Subproblem.** Calculate $h_K = \max_{y \in Y} h(x_K, y)$. Afterwards, solve the following optimisation problem.

$$\max_{y \in Y} \; \min \big\{g(x_K,y),\, h(x_K,y) - h_K + \epsilon\big\}.$$

Let y_{K+1} denote the/an optimal solution and ϑ its objective value.

4. **Termination Criterion.** If $\vartheta \leq 0$, terminate: x_K is optimal for (3.4). Otherwise, set $Y_{K+1} = Y_K \cup \{y_{K+1}\}$ and go to Step 2.

The algorithm proceeds by iterating between Steps 2 and 3. In Step 2, the current Y_K-relaxation is solved. Step 3 checks the existence of y_{K+1} with

$$y_{K+1} \in \mathcal{Y}_\epsilon(x_K) \;\wedge\; g(x_K, y_{K+1}) > 0.$$

If such a y_{K+1} exists, it is added to Y_K in Step 4 and the algorithm iterates. Otherwise, $g(x_K, y) \leq 0$ is satisfied for all $y \in \mathcal{Y}_\epsilon(x_K)$ and the algorithm terminates with x_K as optimal solution for (3.4).

It is commonly suggested to determine the 'worst' constraint violation when solving the subproblem in Step 3, that is, to solve

$$\max_{y \in Y} \; \big\{g(x_K,y) \,:\, h(x_K,y) > h_K - \epsilon\big\},$$

instead. Note that the feasible region of this problem is not closed in general. For most practical applications, however, it suffices to solve the approximation

$$\max_{y \in Y} \left\{ g(x_K, y) \,:\, h(x_K, y) \geq h_K - \epsilon \right\}, \tag{4.3}$$

which has a closed feasible region, instead. Although this approach typically leads to a superior performance in practice, convergence cannot be guaranteed. Indeed, if a maximiser y^* of (4.3) satisfies $g(x_K, y^*) > 0$ and $h(x_K, y^*) = h_K - \epsilon$, the algorithm would enter an infinite loop. These issues notwithstanding, we will employ (4.3) in our numerical tests in Section 5.

Let us analyse the convergence properties of the algorithm. If the procedure terminates after $K < \infty$ iterations, our previous discussion implies that the final solution x_K is optimal for (3.4). Assume that the algorithm does not terminate. Since X is compact, the resulting sequence $\{x_K\}_K$ contains accumulation points. We can directly apply the argumentation developed by Blankenship and Falk in [6] to show that every accumulation point thus generated is optimal for (3.4).

5 Numerical results

In this section, we employ the semi-infinite solution procedure from Section 4 to solve the ϵ-approximation (3.4) of several independent pessimistic bi-level problems (1.1). The results will allow us to both to verify the performance of the overall algorithm and examine the use of tightening the Y_K-relaxation (4.1) by constraint (4.1c). All master and subproblems generated by our solution procedure were solved with BARON 6.0 [27].[6] We set the parameter ϵ to 10^{-3}.

Problem 1. *Convex first-stage and non-convex second-stage problem; the second-stage solution does not depend on the first-stage variables.*

$$\min_{-1 \leq x_1, x_2 \leq 1} \; \max_{-1 \leq y_1, y_2 \leq 1} \left\{ \|x - y\|_2 \,:\, y \in \operatorname*{arg\,max}_{-1 \leq y_1', y_2' \leq 1} \|y'\|_2 \right\}.$$

The second-stage problem has four global optima $(y_1^*, y_2^*) \in \{-1, 1\}^2$ that do not depend on the value of (x_1, x_2), the first-stage variables. The first-stage problem aims to determine (x_1, x_2) whose maximal distance to any of these optima is minimised. Hence, the optimal solution is $(x_1^*, x_2^*) = (0, 0)$. The progress of our algorithm is documented in Table 1. For this example, the progress is the same for both the original and tightened Y_K-relaxation. In either case, the algorithm quickly determines the optimal solution.

Table 1 Progress of our solution technique with and without tightened Y_K-relaxation (4.1c) when applied to Problem 1. The columns describe (in order) the iteration number, objective value, first-stage and second-stage solution and the degree of constraint violation.

K	objective	x_K	y_K	$g(x_K, y_K)$
1	0.000	$(0.000, 0.000)$	$(-1.000, -1.000)$	1.414
2	0.000	$(-1.000, -1.000)$	$(1.000, 1.000)$	2.828
3	1.414	$(0.000, 0.000)$	$(0.000, 0.000)$	0.000

[6]BARON Optimiser: `http://www.andrew.cmu.edu/user/ns1b/baron/baron.html`.

Problem 2. *Convex first-stage and second-stage problem.*

$$\max_{-1 \leq x \leq 1} \left\{ x \; : \; x^2 + y^2 \leq 1 \;\; \forall y \in \arg\min_{-1 \leq y' \leq 1} \|x - y'\|_2 \right\}.$$

Contrary to Problem 1, the second-stage problem is convex. For a given first-stage decision x, the optimal second-stage solution is $y^* = x$. Hence, the optimal second-stage solution depends on the first-stage variables. The first-stage objective function maximises x, but the constraint limits x to $x^* = 1/\sqrt{2} \approx 0.707$, which is the optimal solution. The progress of our algorithm with tightened Y_K-relaxation is documented in Table 2. We see that the algorithm quickly determines the optimal solution. Table 3 shows part of the progress of our algorithm with the original Y_K-relaxation. In this case, large parts of the Y-space need to be examined, which leads to a very slow progress. Here, the value of the tightened relaxation becomes clearly visible.

Table 2 Progress of our solution technique with tightened Y_K-relaxation (4.1c) when applied to Problem 2. See Table 1 for further details.

K	objective	x_K	y_K	$g(x_K, y_K)$
1	1.000	1.000	1.000	1.000
2	0.800	0.800	0.800	0.279
3	0.737	0.737	0.737	0.087
4	0.717	0.717	0.717	0.028
5	0.710	0.710	0.710	0.008
6	0.708	0.708	0.708	0.002
7	0.707	0.707	0.707	0.000

Table 3 Progress of our solution technique with original Y_K-relaxation, that is, without constraint (4.1c), when applied to Problem 2. See Table 1 for further details.

K	objective	x_K	y_K	$g(x_K, y_K)$
1	1.000	1.000	1.000	1.000
2	0.999	0.999	0.999	0.996
3	0.998	0.998	0.998	0.992
4	0.997	0.997	0.997	0.988
5	0.996	0.996	0.996	0.984
6	0.995	0.995	0.995	0.980
7	0.994	0.994	0.994	0.976
8	0.993	0.993	0.993	0.972
9	0.992	0.992	0.992	0.968
10	0.991	0.991	0.991	0.964
11	0.990	0.990	0.990	0.961
12	0.988	0.988	0.988	0.954
13	0.987	0.987	0.987	0.947
14	0.986	0.986	0.986	0.943
15	0.985	0.985	0.985	0.939

Problem 3. *Production planning without investment option (see Section 1).*

$$\max_{x \in [0,200]} \left\{ \min_{y \in \mathcal{Y}(x)} \left\{ (50 - 0.3[x + y])\, x \right\} \; : \; \frac{x_A}{x + y} \geq 0.7 \;\; \forall y \in \mathcal{Y}(x) \right\},$$

where

$$\mathcal{Y}(x) = \arg\max_{y\in[0,100]} \{(50 - 0.3[x+y])\, y\}.$$

The problem is equivalent to the first production planning problem from Section 1. Both the first-stage and the second-stage problem are convex. The progress of our algorithm with tightened Y_K-relaxation is reported in Table 4. The algorithm terminates at $(x^*, y^*) = (94.707, 36.559)$ with a leader profit of 1005.86 and a market share of 72.15%. In Section 1 we have seen that the optimal solution is $(x^*, y^*) = (89.75, 38.45)$ with a leader profit of 1035.49 and a market share of 70.01%. Here, we can see the conservative nature of the ϵ-approximation: the determined solution constitutes a feasible but suboptimal solution for the original problem. More accurate results can be obtained for smaller values of ϵ.

Table 4 Progress of our solution technique with tightened Y_K-relaxation (4.1c) when applied to Problem 3. See Table 1 for further details.

K	objective	x_K	y_K	$g(x_K, y_K)$
1	10000.00	0.000	82.755	10000.00
2	2083.04	84.315	41.753	1056.12
3	1544.56	125.712	21.057	794.16
4	1270.12	105.467	31.177	320.21
5	1170.25	62.457	52.682	204.78
6	1122.01	96.888	35.467	124.68
7	1066.08	81.091	43.366	39.21
8	1036.11	94.749	36.537	30.40
9	1026.87	81.091	43.367	6.03
10	1006.45	94.707	36.557	0.59
11	1005.86	94.707	36.559	0.00

Problem 4. *Production planning with investment option (see Section 1).*

$$\max_{x\in[0,200]} \left\{ \min_{y\in\mathcal{Y}(x)} \{(50 - 0.3[x+y])\, x\} \;:\; \frac{x}{x+y} \geq 0.7 \;\; \forall y \in \mathcal{Y}(x) \right\},$$

where

$$\mathcal{Y}(x) = \arg\max_{y\in[0,100]} \max \big\{(50 - 0.3[x+y] - 50)\, y,$$
$$(50 - 0.3[x+y] - 38)\, y - 500\big\}.$$

The problem is equivalent to the second production planning problem from Section 1. The second-stage problem is non-convex, which causes the feasible region of the first-stage problem to be non-closed. The progress of our algorithm with tightened Y_K-relaxation is reported in Table 5. The algorithm terminates at $(x^*, y^*) = (104.864, 31.481)$ with a leader profit of 950.03 and a market share of 76.91%. In section 1 we have seen that the optimal solution is $(x^*, y^*) = (103.34, 31.67)$ with a leader profit of 981.69 and a market share of 76.54%. Again, we can see the conservative nature of the ϵ-approximation.

Problem 5. *Principal-agent problem (see Section 1).*

$$\max_{x\in[0,1]} \min_{y\in\mathcal{Y}(x)} \{5(1-x)y^2\},$$

where

$$\mathcal{Y}(x) = \arg\max_{y\in[0,1]} \{10 + 5xy^2 - y - (5[y - 1/2]^+)^2\}.$$

Table 5 Progress of our solution technique with tightened Y_K-relaxation (4.1c) when applied to Problem 4. See Table 1 for further details.

K	objective	x_K	y_K	$g(x_K, y_K)$
1	10000.00	0.000	99.950	10000.00
2	2068.46	90.375	58.723	1592.13
3	1677.15	120.129	23.849	859.48
4	1199.00	104.614	31.604	243.38
5	996.54	102.726	52.548	645.43
6	971.34	144.215	11.806	510.74
7	950.03	104.864	31.481	0.00

The problem is equivalent to the principal-agent problem from Section 1. The second-stage problem is non-convex, which causes the objective function of the first-stage problem to be discontinuous. The progress of our algorithm with tightened Y_K-relaxation is reported in Table 6. The algorithm terminates at $(x^*, y^*) = (0.393, 0.521)$ with a principal profit of 0.822. We determined the same solution graphically in Section 1.

Table 6 Progress of our solution technique with tightened Y_K-relaxation (4.1c) when applied to Problem 5. See Table 1 for further details.

K	objective	x_K	y_K	$g(x_K, y_K)$
1	5.000	0.000	0.000	5.000
2	0.949	0.473	0.529	0.210
3	0.849	0.394	0.521	0.029
4	0.822	0.393	0.521	0.000
5	0.822	0.393	0.521	0.000

Summing up, all test instances were solved within moderate numbers of iterations. Moreover, the potential value of employing the tightened Y_K-relaxation has been demonstrated in one of the examples. Although we did not provide results for the application of the original Y_K-relaxation to Problems 3–5, we have encountered similar effects as in Problem 2 for these instances as well.

6 Conclusion

In this paper, we studied the independent pessimistic bi-level problem. We analysed several properties of this problem class and related it to existing models from the literature. We discussed a convergent ϵ-approximation of the problem and presented a semi-infinite solution approach. Although the ϵ-approximation was primarily employed to guarantee tractability, it is also valuable in view of imperfect information and bounded rationality, two key issues in real-world decision making. To the best of our knowledge, this represents the first solution technique proposed for this problem class. We demonstrated the usability of our solution procedure with several examples.

We identified several promising directions for future research. First and foremost, the development of efficient local optimisation techniques for pessimistic bi-level problems would greatly enhance the practical relevance of this problem class. One possible approach is to employ the semi-infinite framework suggested in Section 4, but to solve the master problem (Step 2) merely to local optimality. In this case, significantly larger problems could be solved since the subproblems are of constant size and can be solved in parallel. The size of master problem, on the other hand, grows linearly with the number of iterations and becomes increasingly non-convex due to the inclusion of disjunctive constraints. The key challenge is to analyse the convergence properties of the overall procedure.

Apart from the development of local optimisation procedures, several other directions seem interesting. Firstly, our investigation should be extended to dependent pessimistic bi-level and general multi-level problems. Secondly, a more thorough complexity analysis is needed for a better understanding of independent bi-level problems. Indeed, to the best of our knowledge, the complexity of the nonlinear independent optimistic bi-level problem is unknown. Similarly, there might be further relevant subclasses of the bi-level problem that can be solved in polynomial time. Finally, little has been reported about the constraint semantics of minimax and bi-level problems (see Section 2). A better understanding of the similarities and differences between these models could lead to novel problem formulations.

7 Acknowledgements

Support from EPSRC grant EP/C513584/1 is gratefully acknowledged.

References

[1] T. Basar and G. J. Olsder. *Dynamic Noncooperative Game Theory.* SIAM, 1999.

[2] O. Ben-Ayed and C. E. Blair. Computational difficulties of bilevel linear programming. *Operations Research*, 38(3):556–560, 1990.

[3] C. Berge and E. M. Patterson. *Topological Spaces.* Courier Dover Publications, 1997.

[4] D. P. Bertsekas. Necessary and sufficient conditions for a penalty method to be exact. *Mathematical Programming*, 9(1):87–99, 1975.

[5] B. Bhattacharjee, W. H. Green P. Lemonidis, and P. I. Barton. Global solution of semi-infinite programs. *Mathematical Programming*, 103(2):283–307, 2005.

[6] J. W. Blankenship and J. E. Falk. Infinitely constrained optimization problems. *Journal of Optimization Theory and Applications*, 19(2):261–281, 1976.

[7] K. C. Border. Correspondences. Technical report, California Institute of Technology, `http://www.hss.caltech.edu/~kcb/notes/correspondences.pdf`, February 2001.

[8] B. Colson, P. Marcotte, and G. Savard. An overview of bilevel optimization. *Annals of Operations Research*, 153:235–256, 2007.

[9] S. Dempe. *Foundations of Bilevel Programming.* Kluwer Academic Publishers, 2002.

[10] X. Deng. Complexity issues in bilevel linear programming. In A. Migdalas, P. M. Pardalos, and P. Värbrand, editors, *Multilevel Optimization: Algorithms and Applications*, pages 149–164. Kluwer Academic Publishers, 1998.

[11] D.-Z. Du and P. M. Pardalos, editors. *Minimax and Applications.* Kluwer Academic Publishers, 1995.

[12] V. Dua, K. P. Papalexandri, and E. N. Pistikopoulos. Global optimization issues in multiparametric continuous and mixed-integer optimization problems. *Journal of Global Optimization*, 30(1):59–89, 2004.

[13] N. P. Faísca, V. Dua, B. Rustem, P. M. Saraiva, and E. N. Pistikopoulos. Parametric global optimisation for bilevel programming. *Journal of Global Optimization*, 38(4):609–623, 2007.

[14] M. R. Garey and D. S. Johnson. *Computers and Intractability: A Guide to the Theory of NP-Completeness.* W. H. Freeman & Co, 1979.

[15] R. Hettich and K. O. Kortanek. Semi-infinite programming: Theory, methods, and applications. *SIAM Review*, 35(3):380–429, 1993.

[16] B. Holmstrom. Moral hazard and observability. *Bell Journal of Economics*, 10(1):74–91, 1979.

[17] R. Horst and H. Tuy. *Global Optimization: Deterministic Approaches.* Springer, 1996.

[18] R. G. Jeroslow. The polynomial hierarchy and a simple model for competitive analysis. *Mathematical Programming*, 32(2):146–164, 1985.

[19] K. C. Kiwiel. A direct method of linearization for continuous minimax problems. *Journal of Optimization Theory and Applications*, 55(2):1573–2878, 1987.

[20] Z.-Q. Luo, J.-S. Pang, and D. Ralph. *Mathematical Programs with Equilibrium Constraints*. Cambridge University Press, 1996.

[21] A. Mitsos, P. Lemonidis, and P. I. Barton. Global solution of bilevel programs with a nonconvex inner program. *Journal of Global Optimization*, 42(4):475–513, 2008.

[22] G. E. Monahan. Finding saddle points on polyhedra: Solving certain continuous minimax problems. *Naval Research Logistics*, 43(6):821–837, 1996.

[23] P. Parpas and B. Rustem. An algorithm for the global optimization of a class of continuous minimax problems. *Journal of Optimization Theory and Applications (forthcoming)*.

[24] J. O. Royset, E. Polak, and A. Der Kiureghian. Adaptive approximations and exact penalization for the solution of generalized semi-infinite min-max problems. *SIAM Journal on Optimization*, 14(1):1–34, 2003.

[25] B. Rustem and M. Howe. *Algorithms for Worst-Case Design and Applications to Risk Management*. Princeton University Press, 2002.

[26] B. Rustem, S. Zakovic, and P. Parpas. Convergence of an interior point algorithm for continuous minimax. *Journal of Optimization Theory and Applications*, 136(1):87–103, 2008.

[27] N. V. Sahinidis. Baron: A general purpose global optimization software package. *Journal of Global Optimization*, 8(2):201–205, 1996.

[28] H. Sasai. An interior penalty method for minimax problems with constraints. *SIAM Journal on Control and Optimization*, 12(4):643–649, 1974.

[29] O. Stein and G. Still. On generalized semi-infinite optimization and bilevel optimization. *European Journal of Operational Research*, 142(3):444–462, 2002.

[30] A. Tsoukalas, B. Rustem, and E. N. Pistikopoulos. A global optimization algorithm for generalized semi-infinite, continuous minimax with coupled constraints and bi-level problems. *Journal of Global Optimization*, 2008 (available via Online First).

[31] A. Tsoukalas, W. Wiesemann, and B. Rustem. Global optimisation of pessimistic bi-level problems. Working Paper, Imperial College of Science, Technology and Medicine.

[32] H. Tuy. Convex programs with an additional reverse convex constraint. *Journal of Optimization Theory and Applications*, 52(3):463–486, 1987.

[33] L. N. Vicente and P. H. Calamai. Bilevel and multilevel programming: A bibliography review. *Journal of Global Optimization*, 5(3):291–306, 1994.

[34] S. Žaković and B. Rustem. Semi-infinite programming and applications to minimax problems. *Annals of Operations Research*, 124:81–110, 2003.

Titles in This Series

55 **Panos M. Pardalos and Thomas F. Coleman, Editors,** Lectures on global optimization, 2009

54 **Noriko Yui, Helena Verrill, and Charles F. Doran, Editors,** Modular forms and string duality, 2008

53 **Mikhail Lyubich and Michael Yampolsky, Editors,** Holomorphic dynamics and renormalization: A volume in honour of John Milnor's 75th birthday, 2008

52 **Luigi Rodino, Bert-Wolfgang Schulze, and M. W. Wong, Editors,** Pseudo-differential operators: Partial differential equations and time-frequency analysis, 2007

51 **Giovanni Forni, Mikhail Lyubich, Charles Pugh, and Michael Shub, Editors,** Partial hyperbolic dynamics, laminations, and Teichmüller flow, 2007

50 **Ilia Binder and Dirk Kreimer, Editors,** Universality and renormalization, 2007

49 **Wayne Nagata and N. Sri Namachchivaya, Editors,** Bifurcation theory and spatio-temporal pattern formation, 2006

48 **Hermann Brunner, Xiao-Qiang Zhao, and Xingfu Zou, Editors,** Nonlinear dynamics and evolution equations, 2006

47 **Hans U. Boden, Ian Hambleton, Andrew J. Nicas, and B. Doug Park, Editors,** Geometry and topology of manifolds, 2005

46 **Thierry Passot, Catherine Sulem, and Pierre-Louis Sulem, Editors,** Topics in kinetic theory, 2005

45 **Ragnar-Olaf Buchweitz and Helmut Lenzing, Editors,** Representations of algebras and related topics, 2005

44 **Lajos Horváth and Barbara Szyszkowicz, Editors,** Asymptotic methods in stochastics, 2004

43 **George Janelidze, Bodo Pareigis, and Walter Tholen, Editors,** Galois theory, Hopf algebras, and semiabelian categories, 2004

42 **Saber Elaydi, Gerry Ladas, Jianhong Wu, and Xingfu Zou, Editors,** Difference and differential equations, 2004

41 **Alf van der Poorten and Andreas Stein, Editors,** High primes and misdemeanours: Lectures in honour of the 60th birthday of Hugh Cowie Williams, 2004

40 **Vlastimil Dlab and Claus Michael Ringel, Editors,** Representations of finite dimensional algebras and related topics in Lie theory and geometry, 2004

39 **Stephen Berman, Yuly Billig, Yi-Zhi Huang, and James Lepowsky, Editors,** Vector operator algebras in mathematics and physics, 2003

38 **Noriko Yui and James D. Lewis, Editors,** Calabi-Yau varieties and mirror symmetry, 2003

37 **Panos Pardalos and Henry Wolkowicz, Editors,** Novel approaches to hard discrete optimization, 2003

36 **Shigui Ruan, Gail S. K. Wolkowicz, and Jianhong Wu, Editors,** Dynamical systems and their applications in biology, 2003

35 **Yakov Eliashberg, Boris Khesin, and François Lalonde, Editors,** Symplectic and contact topology: Interactions and perspectives, 2003

34 **T. J. Lyons and T. S. Salisbury, Editors,** Numerical methods and stochastics, 2002

33 **Franz-Viktor Kuhlmann, Salma Kuhlmann, and Murray Marshall, Editors,** Valuation theory and its applications, Volume II, 2003

32 **Franz-Viktor Kuhlmann, Salma Kuhlmann, and Murray Marshall, Editors,** Valuation theory and its applications, Volume I, 2002

31 **A. Galves, J. K. Hale, and C. Rocha, Editors,** Differential equations and dynamical systems, 2002

30 **Roberto Longo, Editor,** Mathematical physics in mathematics and physics: Quantum and operator algebraic aspects, 2001

For a complete list of titles in this series, visit the
AMS Bookstore at **www.ams.org/bookstore/**.